Carlos Artemio Coello Coello, Satchidananda Dehuri, and Susmita Ghosh (Eds.)

Swarm Intelligence for Multi-objective Problems in Data Mining

Studies in Computational Intelligence, Volume 242

Editor-in-Chief
Prof. Janusz Kacprzyk
Systems Research Institute
Polish Academy of Sciences
ul. Newelska 6
01-447 Warsaw
Poland
E-mail: kacprzyk@ibspan.waw.pl

Further volumes of this series can be found on our homepage: springer.com

Vol. 221. Tassilo Pellegrini, Sören Auer, Klaus Tochtermann, and Sebastian Schaffert (Eds.)
Networked Knowledge - Networked Media, 2009
ISBN 978-3-642-02183-1

Vol. 222. Elisabeth Rakus-Andersson, Ronald R. Yager, Nikhil Ichalkaranje, and Lakhmi C. Jain (Eds.)
Recent Advances in Decision Making, 2009
ISBN 978-3-642-02186-2

Vol. 223. Zbigniew W. Ras and Agnieszka Dardzinska (Eds.)
Advances in Data Management, 2009
ISBN 978-3-642-02189-3

Vol. 224. Amandeep S. Sidhu and Tharam S. Dillon (Eds.)
Biomedical Data and Applications, 2009
ISBN 978-3-642-02192-3

Vol. 225. Danuta Zakrzewska, Ernestina Menasalvas, and Liliana Byczkowska-Lipinska (Eds.)
Methods and Supporting Technologies for Data Analysis, 2009
ISBN 978-3-642-02195-4

Vol. 226. Ernesto Damiani, Jechang Jeong, Robert J. Howlett, and Lakhmi C. Jain (Eds.)
New Directions in Intelligent Interactive Multimedia Systems and Services - 2, 2009
ISBN 978-3-642-02936-3

Vol. 227. Jeng-Shyang Pan, Hsiang-Cheh Huang, and Lakhmi C. Jain (Eds.)
Information Hiding and Applications, 2009
ISBN 978-3-642-02334-7

Vol. 228. Lidia Ogiela and Marek R. Ogiela
Cognitive Techniques in Visual Data Interpretation, 2009
ISBN 978-3-642-02692-8

Vol. 229. Giovanna Castellano, Lakhmi C. Jain, and Anna Maria Fanelli (Eds.)
Web Personalization in Intelligent Environments, 2009
ISBN 978-3-642-02793-2

Vol. 230. Uday K. Chakraborty (Ed.)
Computational Intelligence in Flow Shop and Job Shop Scheduling, 2009
ISBN 978-3-642-02835-9

Vol. 231. Mislav Grgic, Kresimir Delac, and Mohammed Ghanbari (Eds.)
Recent Advances in Multimedia Signal Processing and Communications, 2009
ISBN 978-3-642-02899-1

Vol. 232. Feng-Hsing Wang, Jeng-Shyang Pan, and Lakhmi C. Jain
Innovations in Digital Watermarking Techniques, 2009
ISBN 978-3-642-03186-1

Vol. 233. Takayuki Ito, Minjie Zhang, Valentin Robu, Shaheen Fatima, and Tokuro Matsuo (Eds.)
Advances in Agent-Based Complex Automated Negotiations, 2009
ISBN 978-3-642-03189-2

Vol. 234. Aruna Chakraborty and Amit Konar
Emotional Intelligence, 2009
ISBN 978-3-540-68606-4

Vol. 235. Reiner Onken and Axel Schulte
System-Ergonomic Design of Cognitive Automation, 2009
ISBN 978-3-642-03134-2

Vol. 236. Natalio Krasnogor, Belén Melián-Batista, José A. Moreno-Pérez, J. Marcos Moreno-Vega, and David Pelta (Eds.)
Nature Inspired Cooperative Strategies for Optimization (NICSO 2008), 2009
ISBN 978-3-642-03210-3

Vol. 237. George A. Papadopoulos and Costin Badica (Eds.)
Intelligent Distributed Computing III, 2009
ISBN 978-3-642-03213-4

Vol. 238. Li Niu, Jie Lu, and Guangquan Zhang
Cognition-Driven Decision Support for Business Intelligence, 2009
ISBN 978-3-642-03207-3

Vol. 239. Zong Woo Geem (Ed.)
Harmony Search Algorithms for Structural Design Optimization, 2009
ISBN 978-3-642-03449-7

Vol. 240. Dimitri Plemenos and Georgios Miaoulis (Eds.)
Intelligent Computer Graphics 2009, 2009
ISBN 978-3-642-03451-0

Vol. 241. János Fodor and Janusz Kacprzyk (Eds.)
Aspects of Soft Computing, Intelligent Robotics and Control, 2009
ISBN 978-3-642-03632-3

Vol. 242. Carlos Artemio Coello Coello, Satchidananda Dehuri, and Susmita Ghosh (Eds.)
Swarm Intelligence for Multi-objective Problems in Data Mining, 2009
ISBN 978-3-642-03624-8

Carlos Artemio Coello Coello, Satchidananda Dehuri,
and Susmita Ghosh (Eds.)

Swarm Intelligence for Multi-objective Problems in Data Mining

Springer

Dr. Carlos Artemio Coello Coello
Departamento de Computacion
CINVESTAV-IPN
Av. IPN No. 2508
Col. San Pedro Zacatenco
Mexico, D.F. 07360
Mexico
E-mail: ccoello@cs.cinvestav.mx

Dr. Susmita Ghosh
Reader
Department of Computer Science and Engineering
Jadavpur University
Raja S.C. Mallik Road
Kolkata, India 700 032
E-mail: susmitaghoshju@gmail.com

Dr. Satchidananda Dehuri
Reader & Coordinator
Department of information and Communication Technology
Fakir Mohan University
Vyasa Vihar
Balasore 756 019 Orissa
India
E-mail: satchi.lapa@gmail.com

ISBN 978-3-642-26053-7 e-ISBN 978-3-642-03625-5

DOI 10.1007/978-3-642-03625-5

Studies in Computational Intelligence ISSN 1860-949X

Softcover reprint of the hardcover 1st edition 2010

Typeset & Cover Design: Scientific Publishing Services Pvt. Ltd., Chennai, India.

Printed in acid-free paper

9 8 7 6 5 4 3 2 1

springer.com

To Lupita, Carlos Felipe and Víctor Eugenio
Carlos A. Coello Coello
To My Daughter Rishna, and My Wife Lopamudra
Satchidananda Dehuri
To Shinjini and Ashish
Susmita Ghosh

Preface

Multi-objective optimization deals with the simultaneous optimization of two or more objectives which are normally in conflict with each other. Since multi-objective optimization problems are relatively common in real-world applications, this area has become a very popular research topic since the 1970s. However, the use of bio-inspired metaheuristics for solving multi-objective optimization problems started in the mid-1980s and became popular until the mid-1990s. Nevertheless, the effectiveness of multi-objective evolutionary algorithms has made them very popular in a variety of domains.

Swarm intelligence refers to certain population-based metaheuristics that are inspired on the behavior of groups of entities (i.e., living beings) interacting locally with each other and with their environment. Such interactions produce an emergent behavior that is modelled in a computer in order to solve problems. The two most popular metaheuristics within swarm intelligence are particle swarm optimization (which simulates a flock of birds seeking food) and ant colony optimization (which simulates the behavior of colonies of real ants that leave their nest looking for food). These two metaheuristics have become very popular in the last few years, and have been widely used in a variety of optimization tasks, including some related to data mining and knowledge discovery in databases. However, such work has been mainly focused on single-objective optimization models. The use of multi-objective extensions of swarm intelligence techniques in data mining has been relatively scarce, in spite of their great potential, which constituted the main motivation to produce this book.

The purpose of this book is to collect contributions that are at the intersection of multi-objective optimization, swarm intelligence (specifically, particle swarm optimization and ant colony optimization) and data mining. Such a collection intends to illustrate the potential of multi-objective swarm intelligence techniques in data mining, with the aim of motivating more researchers in evolutionary computation and machine learning to do research in this field.

This volume consists of eleven chapters including an introduction (Chapter 1) that provides the basic concepts of swarm intelligence techniques and a discussion of their use in data mining. Some of the research challenges that must be faced when using swarm intelligence techniques in data mining are also addressed. The rest of the chapters were contributed by leading researchers, and were organized according to the steps normally followed in Knowledge Discovery in Databases (KDD) (i.e., data preprocessing, data mining, and post processing). Next, we provide a brief description of each of them.

Vieira, Sousa and Runkler, present in Chapter 2 an ant colony optimization approach for feature selection using fuzzy classifiers. Feature selection aims to find the minimum set of features that provide the best classification accuracy. Thus, the aim is to minimize two objectives: the number of features and the the classification error. A pheromone matrix and a specific heuristic is adopted for each of these objectives. Two pheromone matrices and two different heuristics are used for each of these objectives. A comprehensive comparison with respect to other feature selection methods is also presented, showing that the proposed approach is a competitive choice.

In Chapter 3, Torácio presents an approach for classification-rule learning based on a multi-objective particle swarm optimizer. The best rules found are stored in an unordered classifier, and Pareto dominance is applied to them, aiming that the user can keep only those rules satisfying representing the best possible trade-offs among his/her desirable objectives. Different rules' representation schemes, particle initialization procedures and stopping criteria are also discussed. The authors found that their proposed approach is competitive with respect to other (state-of-the-art) classification techniques regarding the area under the receiver output curve and the number of rules in the classifier. An interesting aspect of this work is that the rules produced by the proposed approach are good even when considered in isolation.

Zahiri and Seyedin present in Chapter 4 the use of multi-objective particle swarm optimizers for two types of tasks: designing a classifier and optimizing the performance of a conventional classifier. An interesting aspect of this work is that the authors also present a theoretical study on the relationship between a multi-objective particle swarm classifier and a Bayesian classifier. Additionally, a review of the use of single-objective particle swarm optimization algorithms for data classification is also provided.

In Chapter 5, Fieldsend presents a good introduction to decision trees, with a particular emphasis on those which only consider a rule that relates to a single feature at a node. He also discusses the use of particle swarm optimization (both in a single- and in a multi-objective formulation) to train near optimal decision trees. His results indicate that particle swarm optimization is an effective optimizer of decision trees, with the multi-objective formulation having improved search capabilities over its single-objective counterpart.

In Chapter 6, Dehuri, Coello Coello, Cho and Ghosh apply a hybridization of a polynomial neural network and a discrete particle swarm optimizer for classification tasks. The polynomial neural network is used for approximating the decision boundaries of the data, and particle swarm optimization is adopted for the simultaneous optimization of the architectural complexity and the predictive accuracy of the polynomial neural network. The proposed approach is validated using real life data sets having nonlinear class boundaries, with encouraging results.

Witt presents in Chapter 7 an overview of the existing theoretical runtime analyses of swarm intelligence techniques (namely, ant colony optimization and particle swarm optimization). This overview also points out some of the promising research directions within this area, and envisions the possibility of having a unified theory of randomized search heuristics.

Carvalho and Pozo propose the use of a multi-objective particle swarm optimizer for rule learning in Chapter 8. The proposed approach searches for rules having certain (specific) properties, with the idea of using such rules as an unordered classifier. This provides rules that are more understandable, intuitive and that have independence from each other. The main aim of the chapter is, however, to introduce a parallel version of this approach, which is intended for mining rules from large datasets.

In Chapter 9, Hetland provides a tutorial of similarity search methods based on metric indexing. The discussion includes approaches for creating lower bounds as well as some promising research areas within this field, such as parallelization and approximate methods. The relevance of such methods in multi-objective optimization and data mining is also briefly addressed.

The use of particle swarm optimization to solve the vehicle routing problem with time windows is undertaken by Muñoz-Zavala, Hernández-Aguirre and Villa-Diharce in Chapter 10. Three objectives are minimized in this case: total distance, total waiting time and number of vehicles used. A data mining strategy called space partitioning is adopted to generate a sectors model, which assigns customers to routes according to partitions performed in a polar coordinate system. Additionally, special operators are devised to combine these sectors with the particle swarm optimization algorithm. The proposed approach is found to be competitive with respect to other algorithms previously reported in the specialized literature.

The fusion of correlated measurements from multiple classifiers is addressed by Osadciw, Srinivas and Veeramachaneni in Chapter 11. The measurements from the multiple classifiers are correlated using weights that need to be carefully defined. For that sake, a multi-objective optimization function is adopted by the authors and particle swarm optimization is used to generate the weights. The proposed approach is validated with different datasets that correspond to various correlation levels, including an application in biometrics. Results indicate that the proposed approach is competitive with respect to classical fusion methods.

We hope that these chapters will constitute a valuable reference for those wishing to do research on the use of multi-objective swarm intelligence techniques in data mining and knowledge discovery in databases, since that has been the main goal of this book. Finally, we wish to thank all the authors for their high-quality contributions and to Prof. Janusz Kacprzyk for accepting to include this volume in the *Studies in Computational Intelligence* series from Springer. We also thank Dr. Thomas Ditzinger, from Springer-Verlag in Germany, who always provided prompt responses to all our queries during the preparation of this volume. Carlos A. Coello Coello also thanks Gregorio Flores for his valuable help, to CINVESTAV-IPN for providing all the facilities to prepare the final version of this book, and to his family for their support. Satchidananda Dehuri sincerely appreciates the award of BOYSCAST, DST, Govt. of India and thanks Professor Sung-Bae Cho, Soft Computing Laboratory, Department of Computer Science, Yonsei University, South Korea, for the accommodation during his visit there, especially for the office and computer facilities provided.

Mexico City, Mexico
Balasore, India
Kolkata, India
June 2009

Carlos A. Coello Coello
Satchidananda Dehuri
Susmita Ghosh
Editors

Contents

List of Contributors

André B. de Carvalho
Computer Sciences Department
Federal University of Paraná
Curitiba PR CP: 19081, Brazil
andrebc@inf.ufpr.br

Sung-Bae Cho
Soft Computing Laboratory
Department of Computer Science
Yonsei University
134 Shinchon-dong Seodaemun-gu
Seoul 120-749, South Korea
sbcho@sclab.yonsei.ac.kr

Carlos A. Coello Coello
CINVESTAV-IPN
Depto. de Computación
Evolutionary Computation Group
Av. IPN No. 2508
Col. San Pedro Zacatenco
México, D.F. 07300, México
ccoello@cs.cinvestav.mx

Satchidananda Dehuri
Department of Information
and Communication Technology
Fakir Mohan University
Vyasa Vihar
Balasore-756019, India
satchi.lapa@gmail.com

Jonathan E. Fieldsend
School of Engineering, Computing
and Mathematics
University of Exeter
Harrison Building
North Park Road
Exeter, EX4 4QF, UK
J.E.Fieldsend@exeter.ac.uk

Ashish Ghosh
Machine Intelligence Unit
Indian Statistical Institute
203 B.T. Road
Kolkata-700 108, India
ash@isical.ac.in

Susmita Ghosh
Department of Computer Science
and Engineering
Jadavpur University
Kolkata 700 032, India
susmitaghoshju@gmail.com

Arturo Hernández-Aguirre
Centro de Investigación en
Matemáticas (CIMAT)
Jalisco s/n
Mineral de Valenciana 36240
Guanajuato, Guanajuato, México
artha@cimat.mx

Magnus Lie Hetland
Norwegian University of
Science and Technology
Department of Computer
and Information Science
Sem Sælands vei 7-9
NO-7491 Trondheim, Norway
mlh@idi.ntnu.no

Angel Muñoz-Zavala
Centro de Investigación en
Matemáticas (CIMAT)
Jalisco s/n
Mineral de Valenciana 36240
Guanajuato, Guanajuato, México
aemz@cimat.mx

Lisa Osadciw
Department of Electrical
Engineering and Computer Science
Syracuse University
Syracuse, New York 13244, USA
laosadci@syr.edu

Aurora Pozo
Computer Sciences Department
Federal University of Paraná
Curitiba PR CP: 19081, Brazil
aurora@inf.ufpr.br

Thomas A. Runkler
Siemens AG, Corporate Technology
Learning Systems CT IC 4
81730 Munich, Germany
Thomas.Runkler@siemens.com

Seyed-Alireza Seyedin
Department of Electrical Engineering
Ferdowsi University of Mashhad
Mashhad, Iran
seyedin@um.ac.ir

João M. C. Sousa
Center of Intelligent Systems-IDMEC
Instituto Superior Técnico
Technical University of Lisbon
Av. Rovisco Pais 1
1049-001 Lisbon, Portugal
j.sousa@dem.ist.utl.pt

Nisha Srinivas
Department of Electrical Engineering
and Computer Science
Syracuse University
Syracuse, New York 13244, USA
nsriniva@syr.edu

Augusto de Almeida Prado G. Torácio
Department of Computer Sciences
Federal University of Parana
Curitiba, Brazil
toracio@inf.ufpr.br

Kalyan Veeramachaneni
Department of Electrical Engineering
and Computer Science
Syracuse University
Syracuse, New York 13244, USA
kveerama@syr.edu

Susana M. Vieira
Center of Intelligent Systems-IDMEC
Instituto Superior Técnico
Technical University of Lisbon
Av. Rovisco Pais 1
1049-001 Lisbon, Portugal
susana@dem.ist.utl.pt

Enrique Villa-Diharce
Centro de Investigación en
Matemáticas (CIMAT)
Jalisco s/n
Mineral de Valenciana 36240
Guanajuato, Guanajuato, México
villadi@cimat.mx

Carsten Witt
DTU Informatics
Technical University of Denmark
2800 Kgs. Lyngby, Denmark
cfw@imm.dtu.dk

Seyed-Hamid Zahiri
Department of Electrical Engineering
Birjand University, Birjand, Iran
hzahiri@birjand.ac.ir

1

An Introduction to Swarm Intelligence for Multi-objective Problems in Data Mining

Satchidananda Dehuri[1], Susmita Ghosh[2], and Carlos A. Coello Coello[3]

[1] Department of Information and Communication Technology, Fakir Mohan University, Vyasa Vihar, Balasore-756019, India
satchi.lapa@gmail.com

[2] Department of Computer Science and Engineering, Jadavpur University, Kolkata- 700 032, India
susmitaghoshju@gmail.com

[3] CINVESTAV-IPN Depto. de Computación Av. IPN No. 2508 Col. San Pedro Zacatenco, México, D.F. 07300, Mexico
ccoello@cs.cinvestav.mx

Summary. This chapter deals with the fundamentals of swarm intelligence, focusing mainly on particle swarm optimization and ant colony optimization for solving multi-objective problems in data mining and knowledge discovery in databases. Besides providing a general overview of the field, we identify some of the multiple criteria which are tightly/loosely coupled in various steps of knowledge discovery and data mining, aiming to direct the attention of researchers towards them in the near future.

1.1 Knowledge Discovery and Data Mining

During the last few decades, the developments in digital technology have popularized the use of digitized information, which has become increasingly inexpensive to store. With the advancements of computer hardware and software and the rapid propagation of computer technology in activities related to health, business, agriculture and education, among many others, large amounts of data have been collected and stored in databases. The rate at which such data is stored is growing at a phenomenal rate. As a result, traditional *ad hoc* mixtures of statistical techniques and data management tools are no longer adequate for analyzing this vast collection of data. Raw data is rarely of direct benefit. Its true value relies on our ability to extract from such raw data information useful for decision support or exploration, and on understanding the phenomenon governing the data source. In most domains, data analysis has been traditionally a manual process. Traditionally, one or more analysts would become intimately familiar with the data and, with the help of statistical techniques, would provide summaries and generate reports. Indeed, the analyst acted as a sophisticated query processor. However, such an approach rapidly breaks down as the size of the data grows and the number of dimensions increases. Databases containing a large amount of data of high dimensionality are becoming increasingly common.

C.A. Coello Coello et al. (Eds.): Swarm Intel. for Multi-objective Prob., SCI 242, pp. 1–17.
springerlink.com

When the scale of data manipulation, exploration and inferencing goes beyond human capacities, it is necessary to look for computing technologies that can automate the process. All of these have prompted the need for intelligent data analysis methodologies, which can discover useful knowledge from data in an automatic manner.

The term KDD refers to the overall process of knowledge discovery in databases and is concerned with extracting useful information from databases [15]. Data mining is a particular step in this process and comprises a set of techniques used in an automated approach to exhaustively explore and bring to the surface complex relationships embedded in very large datasets. The additional steps in the KDD process, such as data preparation, data selection, data cleaning, incorporation of appropriate prior knowledge, and proper interpretation of the results of mining, ensures that useful and, in some sense, interesting knowledge is derived from the data.

The subject of KDD has evolved, and continues to evolve, from the intersection of research from various fields such as databases, machine learning, pattern recognition, statistics, artificial intelligence, reasoning with uncertainties, knowledge acquisition for expert systems, data visualization, machine discovery, high-performance computing, evolutionary computation, multi-objective optimization and, more recently, from swarm intelligence. Fig. 1.1 shows in graphical form the relationship of these fields with data mining and KDD. KDD systems incorporate theories, algorithms, and methods from all of these fields. Many successful applications of KDD systems have been reported from diverse sectors such as marketing, biology and health, finance, banking, manufacturing, agriculture, and telecommunications. A good overview of KDD can be found in [16, 20].

Next, we briefly discuss the main concepts related to KDD, as well as the process involved within.

Definition. *KDD is the nontrivial process of identifying valid, novel, potentially useful, and ultimately understandable patterns in data* [6, 52].

The overall **KDD process** is interactive and iterative, and involves, more or less, the following steps [4]:

Domain specific knowledge: includes relevant prior knowledge and goals of the application.

Extracting/Selecting the target data set: includes extracting/selecting a data set or focusing on a subset of data instances.

Data cleansing: includes basic operations, such as noise removal and handling of missing data. Data from real-world sources are often erroneous, incomplete, and inconsistent, perhaps due to operation error or system implementation flaws. Such low quality data needs to be cleaned prior to mining.

Data integration: includes integrating multiple, heterogeneous data sources.

Data reduction and projection: includes finding useful features to represent the data (depending on the goal of the task) and using dimensionality reduction or transformation methods.

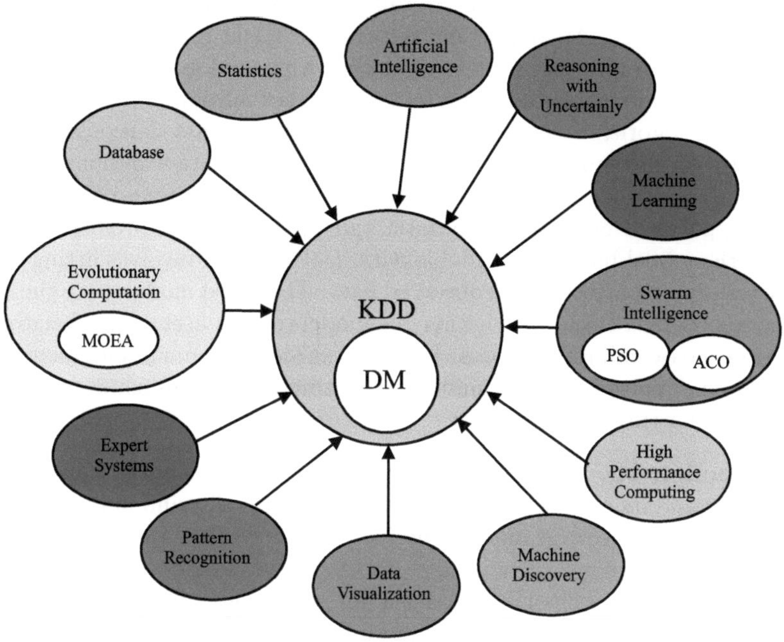

Fig. 1.1. Relationship diagram of knowledge discovery in databases (KDD) with other disciplines

Choosing the data mining function: includes deciding the purpose of the model derived by the data mining algorithm (e.g., summarization, classification, regression, clustering, link analysis, image segmentation/classification, functional dependencies, rule extraction (classification and association rules), or a combination of these).

Choosing the data mining algorithm(s): includes selecting method(s) to be used for searching patterns in data, such as deciding on which model and parameters may be appropriate.

Data mining: includes searching for patterns of interest in a particular representational form or a set of such representations.

Interpretation: includes interpreting the discovered patterns, as well as the possible visualization of the extracted patterns. One can analyze the patterns automatically or semiautomatically to identify the truly interesting/useful patterns for the user.

Using discovered knowledge: includes incorporating this knowledge into the performance of the system, taking actions based on knowledge.

Considering these steps at the level of data mining, they can be summarized into four steps: 1) data integration and acquiring domain knowledge; ii) data preprocessing; iii) data mining; and iv) post processing.

Data mining: KDD refers to the overall process of turning low-level data into high-level knowledge. An important step in the KDD process is data mining. Data mining is an interdisciplinary field with a general goal of predicting outcomes and uncovering relationships in data. It uses automated tools employing sophisticated algorithms to discover hidden patterns, associations, anomalies and/or structure from large amounts of data stored in data warehouses or other information repositories. Data mining tasks can be descriptive, i.e., discovering interesting patterns describing the data, and predictive, i.e., predicting the behavior of the model based on available data. Data mining involves fitting models to or determining patterns from observed data. The fitted models play the role of inferred knowledge. Deciding whether the model reflects useful knowledge or not is a part of the overall KDD process for which subjective human judgment is usually required. Typically, a data mining algorithm constitutes some combination of the following three components:

- **The model:** The function of the model (e.g., classification, clustering, regression) and its representational form (e.g., linear discriminants, neural networks, decision trees). A model contains parameters that are to be determined from the data.
- **The preference criterion:** A basis for preference of one model or set of parameters over another, depending on the given data. The criterion is usually some form of goodness-of-fit function of the model to the data, perhaps tempered by a smoothing term to avoid overfitting, or generating a model with too many degrees of freedom to be constrained by the given data.
- **The search algorithm:** The specification of an algorithm for finding particular models and parameters, given the data, model(s), and a preference criterion. A particular data mining algorithm is usually an instantiation of the model/preference/search components.

The most common model functions adopted in the current data mining practice include the following: 1) **Classification:** classifies a data item into one of several predefined categorical classes; 2) **Regression:** maps a data item to a real valued prediction variable; 3) **Clustering:** maps a data item into one of several clusters, where clusters are natural groupings of data items based on similarity metrics or probability density models; 4) **Rule generation:** extracts classification rules from the data; 5) **Discovering association rules:** describes association relationship among different attributes; 6) **Summarization:** provides a compact description for a subset of data; 7) **Dependency modeling:** describes significant dependencies among variables; 8) **Sequence analysis:** models sequential patterns, such as time-series analysis. The goal is to model the states of the process generating the sequence or to extract and report deviations and trends over time.

Development of new generation algorithms is expected to encompass more diverse sources and types of data that will support mixed-initiative data mining, where human experts collaborate with the computer to form hypotheses and test them.

1.2 Swarm Intelligence

Swarm intelligence (SI) methodologies model the behavior of groups of simple living beings interacting locally with each other and with their environment. Such interaction causes coherent functional global patterns as an emergent behavior [19]. The roots of swarm intelligence are deeply embedded in the biological study of self-organized behaviors in social insects. However, if social insects remain the original source of inspiration for artificial swarm intelligent systems it is important to notice that other biological systems share similar collective properties such as colonies of bacteria or amoeba [1, 2], fish schools [22, 40], bird flocks [42], sheep herds [23] or even crowds of human beings [26]. Among them, the motions of fish schools and bird flocks have, for instance, partly inspired the concept of particle swarm optimization [31]. Similarly, the foraging behavior of real ants, which are used to solve discrete optimization problems are the main inspiring point of ant colony optimization [12].

This book deals with the application of swarm intelligence techniques such as particle swarm optimization (PSO) and ant colony optimization (ACO) to the solution of multi-objective problems in data mining. Addressing the various criteria of data mining and knowledge discovery using swarm intelligence approaches is the novelty of this edited volume.

1.2.1 Particle Swarm Optimization

Next, we will discuss the theoretical foundations of particle swarm optimization and some of its potential usage in data mining and KDD from a single objective optimization perspective.

Basic Concepts of PSO

PSO is a stochastic algorithm that is used to search for the best solution by simulating the movement and flocking of birds. The algorithm works by initializing a flock of birds randomly over the search space, where every bird is called a particle. These particles fly with a certain velocity and find the global best position after performing a certain number of iterations. At each iteration k, the ith particle is represented by a vector x_i^k in multidimensional space to characterize its position. The velocity v_i^k is used to characterize its velocity. Thus PSO maintains a set of positions:

$$S = \{x_1^k, x_2^k, ..., x_N^k\}$$

and a set of corresponding velocities

$$V = \{v_1^k, v_2^k, ..., v_N^k\}.$$

Initially, the iteration counter $k = 0$, and the positions x_i^0 and their corresponding velocities v_i^0 $(i = 1, 2, \ldots, N)$, are generated randomly from the search space Ω. Each particle changes its position x_i^k, at each iteration. The new position x_i^{k+1} of the ith particle $(i = 1, 2, \ldots, N)$ is biased towards its best position

p_i^k. The best function value found by the particle so far is referred to as *personal best* or *pbest*, and the very best position found by all the particles (p_g^k) is referred to as the global best or *gbest*. The *gbest* is the best position in the population $P = \{p_1^k, p_2^k, \ldots, p_N^k\}$, where $p_i^0 = x_i^0, \forall i$.

We can say a particle in S is good or bad depending on its personal best being a good or bad point in P. Consequently, we call the ith particle ($j^{th}particle$) in S the worst (the best) if $p_i^k(p_j^k)$ is the least (best) fitted, with respect to the function value in P. We denote the *pbest* of the worst particle and the best particle in S as p_h^k and p_g^k, respectively. Hence

$$p_g^k = argmin_{i \in 1,2,..,N} f(p_i^k) \text{ and } p_h^k = argmax_{i \in 1,2,..,N} f(p_i^k).$$

At each iteration k, the position x_i^k of the ith particle is updated by a velocity v_i^{k+1} which depends on three components: its current velocity v_i^k, the cognition term (i.e., the weighted difference vectors ($p_i^k - x_i^k$)) and the social term (i.e., the weighted difference vector ($p_g^k - x_i^k$)).

Specifically, the set S is updated for the next iteration using $x_i^{k+1} = x_i^k + v_i^{k+1}$, where $v_i^{k+1} = v_i^k + r_1 \times c_1 \times (p_i^k - x_i^k) + r_2 \times c_2 \times (p_g^k - x_i^k)$.

The parameters r_1 and r_2 are uniformly distributed random numbers within $[0, 1]$ and c_1 and c_2, known as the cognitive and social parameters, respectively, are normally set to $c_1 = c_2 = 2.0$ [32]. Thus, the values $r_1 \times c_1$ and $r_2 \times c_2$ introduce some stochastic weighting in the difference vectors $(p_i^k - x_i^k)$ and $(p_g^k - x_i^k)$, respectively. The set P is updated as the new positions x_i^{k+1} are created using the following rules with a minimization of the cost function: $p_i^{k+1} = x_i^{k+1}$ if $f(x_i^{k+1}) < f(p_i^k)$, otherwise $p_i^{k+1} = p_i^k$.

This process of updating the velocities v_i^k, positions x_i^k, *pbest* p_i^k and the *gbest* p_g^k is repeated until a user-defined stopping condition is met.

We now briefly present a number of improved versions of PSO and then we show where our modified PSO can stand.

Shi and Eberhart [46] introduced the first modifications to the PSO algorithm, which consisted in adopting a constant inertia w, which controls how much a particle tends to follow its current directions compared to the memorized *pbest* p_i^k and the *gbest* p_g^k. Hence the velocity update is now given by

$$v_i^{k+1} = w.v_i^k + r_1.c_1.(p_i^k - x_i^k) + r_2.c_2.(p_g^k - x_i^k), \tag{1.1}$$

where the values of r_1 and r_2 are realized component-wise.

Again Shi and Eberhart [47] proposed a linearly varying inertia weight during the search. In this case, the inertia weight is linearly reduced during the search. This entails a more global search during the initial stages and a more local search during the final stages. They also proposed a limitation of each particle's velocity to a specified maximum velocity v^{max}. The maximum velocity was calculated as a fraction τ $(0 < \tau \leq 1)$ of the distance between the bounds of the search space, i.e., $v^{max} = \tau \times (x^u - x^l)$.

Fourie and Groenwold [18] suggested a dynamic inertia weight and a maximum velocity reduction scheme. In this modification, an inertia weight and a maximum (pre-defined) velocity are reduced by fractions α and β, respectively, if no improvement in p_g^k occur after a pre-specified number of iterations h, i.e.,

$$\text{if } f(p_g^k) = f(p_g^{k-1}) \text{ then } w_{k+1} = \alpha w_k \text{ and } v_k^{max} = \beta v_k^{max},$$

where α and β are such that $0 < \alpha$, and $\beta < 1$.

Clerc and Kennedy [7] introduced another interesting modification to PSO in the form of a constriction coefficient χ, which controls all three components in the velocity update rule. This has an effect of reducing the velocity as the search progresses. In this modification, the velocity update is given by

$$v_i^{k+1} = \chi(v_i^k + r_1 c_1(p_i^k) + r_2 c_2(p_g^k - x_i^k)),$$

where $\chi = \frac{2}{|2-\phi-\sqrt{\phi^2-4\phi}|}$, $\phi = c_1 + c_2 > 4$.

Da and Xiurun [9], also modified PSO by introducing a temperature like control parameter as in the simulated annealing algorithm. Zhang, et al. [55] modified PSO by introducing a new inertia weight during the velocity update. Generally at the beginning stages of their algorithm, the inertia weight w, should be reduced rapidly, when approaching the optimum, the inertia weight w should be reduced slowly. They adopted the following rule:

$$w = w_0 - (\tfrac{w_1}{MAXITER1}) \times k, \text{ if } 1 \le k \le MAXITER1, \text{ and}$$
$$w = (w_0 - w_1) \times exp((MAXITER1 - k)/\nu), \text{ if}$$
$$MAXITER1 < k \le MAXITER,$$

where w_0 is the initial inertia weight, w_1 is the final inertia weight considering a linear reduction scheme, $MAXITER$ is the total number of iterations; $MAXITER1$ is the number of iterations during which the inertia weight will be linearly reduced, k is a variable whose range is $[1, MAXITER]$. By adjusting k, different end values of the inertia weight are attained.

In addition, there are some methods [41] which also use the adaptive cognitive acceleration coefficient (c_1) and the social acceleration coefficients (c_2). c_1 is allowed to decrease from its initial value of c_{1i} to c_{1f} while c_2 is increased from c_{2i} to c_{2f} using the following equations:

$$c_1^k = (c_{1f} - c_{1i})\frac{k}{MAXITER} + c_{1i}, \tag{1.2}$$

and

$$c_2^k = (c_{2f} - c_{2i})\frac{k}{MAXITER} + c_{2i}. \tag{1.3}$$

Single objective PSO in data mining

Apparently, data mining and knowledge discovery do not have much in common with particle swarm optimization. However, we believe that PSO can be suitable for solving different knowledge discovery tasks, either as standalone tool or as part of a more elaborate method that involves other techniques as well. Table 1.1 summarizes a set of representative publications that use single-objective PSO for knowledge discovery in databases. Since the focus of the chapter is on multi-objective approaches rather than on single-objective ones, we will not provide any further details on these works, but the interested reader may go to their original sources for more information.

1.2.2 Ant Colony Optimization

Next, we will provide a short introduction to ant colony optimization and some of its potential use in knowledge discovery and data mining, from a single-objective optimization perspective.

Basic Concepts of ACO

In the early 1990s, ant colony optimization (ACO) was introduced by Marco Dorigo and co-authors as a novel bio-inspired metaheuristic for the solution of NP-hard problems [12]. ACO belongs to the class of metaheuristics which are approximate algorithms used to obtain good enough solutions to very hard combinatorial problems in a reasonable amount of computational time. The basic principles of ACO are inspired from the foraging behavior of real ants. When searching for food, ants initially explore the area surrounding their nest in a random manner. As soon as an ant finds a food source, it evaluates the quantity and the quality of the food and carries some of it back to the nest. During

Table 1.1. A sample of applications of single-objective PSO in data mining and knowledge discovery in databases

Reference	Methods(Hybrid/Non-hybrid)	DM and KDD	Year
[48]	Particle Swarm Based Data Mining Algorithms	Classification	2004
[38]	DCPSO	Classification & Clustering	2005
[52]	PSO based Algorithm for Classification	Classification	2006
[14]	Facing Classification Problem with PSO	Classification	2007
[51]	Feature Selection based on Rough Sets and PSO	Feature Selection	2007
[10]	An Empirical Study of PSO for Cluster Analysis	Clustering	2007
[29]	Combinatorial PSO	Clustering	2007
[34]	PSO plus Support Vector Machines	Feature Selection and Classification	2008
[8]	PSO for Attribute Selection	Feature Selection and Classification	2008
[54]	Discrete PSO for data mining	Classification	2009

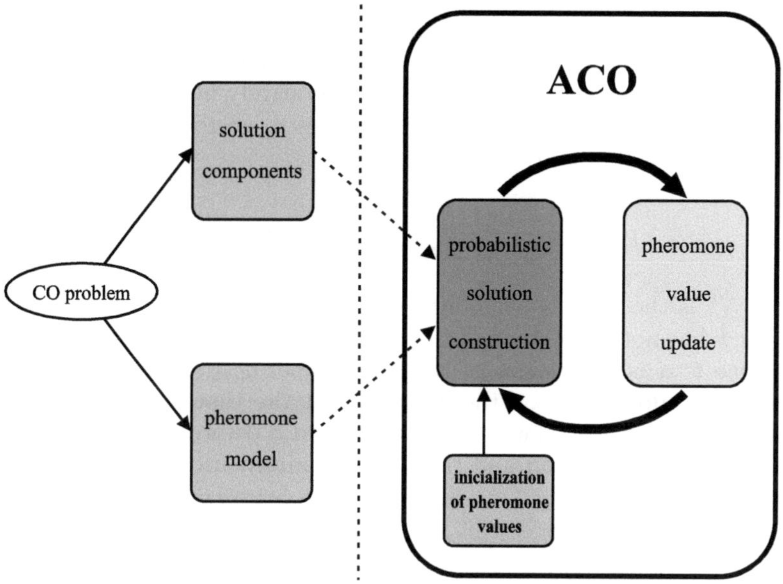

Fig. 1.2. Basic workflow of ACO [3]

the return trip, the ant deposits a chemical pheromone trail on the ground. The quantity of pheromone deposited, which may depend on the quantity and quality of the food, will guide other ants to the food source. The indirect communication between the ants via pheromone trails enables them to find shortest paths between their nest and the food sources. This characteristic of real ant colonies is exploited when modeling them in a computer in order to find approximate solutions to NP-hard problems in reasonable amounts of time.

ACO is implemented as a group of sophisticated intelligent agents which simulate the ants behavior, walking around a graph representing the problem to solve using mechanisms of cooperation and adaptation. The ACO algorithm requires the definition of the following [11]:

1. The problem needs to be represented appropriately, such that the ants are allowed to incrementally update the solutions through the use of probabilistic transition rules, based on the amount of pheromone in the trail and other problem specific knowledge. It is also important to enforce a strategy to construct only valid solutions corresponding to the problem definition.
2. A problem-dependent heuristic function η that measures the quality of the components that can be added to the current partial solution.
3. A rule set for pheromone updating, which specifies how to modify the pheromone value τ.
4. A probabilistic transition rule based on the value of the heuristic function η and the pheromone value τ that is used to iteratively construct a solution.

ACO was first introduced as a method to solve the popular NP-complete Traveling Salesman Problem (TSP). In this case, the procedure was defined as follows. Starting from its start node, an ant iteratively moves from one node to another. When being at a node, an ant chooses to go to an unvisited node at time t with a probability given by:

$$p_{i,j}^{k}(t) = \frac{(\tau_{i,j}(t))^{\alpha}(\eta_{i,j}(t))^{\beta}}{\sum_{j \in N_i^k}(\tau_{i,j}(t))^{\alpha}(\eta_{i,j}(t))^{\beta}}, j \in N_i^k \tag{1.4}$$

where N_i^k is the feasible neighborhood of the ant k, that is, the set of cities which ant k has not yet visited; $\tau_{i,j}(t)$ is the pheromone value on the edge (i,j) at the time t, α is the weight of the pheromone; $\eta_{i,j}(t)$ is the *a priori* available heuristic information on the edge (i,j) at the time t, β is the weight of the heuristic information. Two parameters α and β determine the relative influence of the pheromone trail and the heuristic information, respectively. $\tau_{i,j}(t)$ is determined by

$$\tau_{i,j}(t+1) = \rho \times \tau_{i,j}(t) + \sum_{k=1}^{n}(\Delta\tau_{i,j}^{k}(t+1)), \forall(i,j) \tag{1.5}$$

Table 1.2. A sample of applications of single-objective ACO in data mining and knowledge discovery in databases

Reference	Methods(Hybrid/Non-hybrid)	DM and KDD	Year
[37]	AntClass	Clustering	1999
[39]	Ant-Miner	Classification	2002
[35]	Ant-Miner3	Classification	2004
[50]	ACO-Miner	Classification	2004
[4]	ACM and A^4C	Clustering	2004
[45]	ACO with distributed agents	Clustering	2004
[44]	Generalization of ACO	Clustering	2005
[27]	Ant-Miner	Web Page Classification	2005
[30]	Ant Miner(I)	Classification	2005
[5]	Parallel Ant-Miner	Classification	2006
[6]	MuLAM	Classification	2006
[24]	ATTA	Clustering	2006
[33]	Ant Colony System	Association rule mining	2007
[25]	Two ACO-based approaches	Clustering	2007
[43]	Ant Colony Algorithm	Feature Selection	2007
[53]	Ant Colony decision rule Algorithm	Classification	2007
[28]	PSO/ACO	Classification	2007
[36]	AntMiner+	Classification	2007
[49]	TACO-Miner	Classification	2007
[21]	APC	Clustering	2008

$$\Delta\tau_{i,j}^{k}(t+1) = \begin{pmatrix} \frac{Q}{L_k(t)} & \text{if the edge } (i,j) \text{ is chosen by the ant } k \\ 0 & \text{otherwise} \end{pmatrix} \tag{1.6}$$

where ρ is the pheromone trail evaporation rate ($0 < \rho < 1$), and n is the number of ants. Fig. 1.2 shows the basic working principles of ACO in a graphical form [3].

Single-objective Ant Colony in Data Mining

The study of ant colonies behavior and their self-organizing capabilities is of interest to knowledge retrieval/management and decision support systems sciences, because it provides models of distributed adaptive organization, which are useful to solve difficult classification, clustering and feature selection problems. Table 1.2 illustrates some of the applications of single-objective ACO in data mining and knowledge discovery in databases reported in the specialized literature.

These successful applications of single-objective swarm intelligence methods in data mining are a clear indication of the potential of such techniques in data mining and of their potential to be extended to the multi-objective case.

1.3 Multi-Objective (MO) Problems in DM and KDD

Here, we identify several multi-objective optimization problems within data mining and knowledge discovery in databases, which can be tackled using swarm intelligence techniques.

1.3.1 Feature and Instance Selection

Feature and instance selection is an important means to select attributes that help us build robust learning models. The application of feature and instance selection helps all the phases of knowledge discovery (excluding the data integration and acquisition of the domain knowledge): data pre-processing, data mining and post processing. As a matter of fact, every phase is equally important for undertaking successful knowledge discovery, or data mining in particular. The first phase ensures that the data fed to the second phase is of good quality and of proper size; the second phase ensures that data mining can be performed effectively and efficiently; and the third phase sorts out the mined results into reliable and comprehensible knowledge being relevant to what a user needs. The three closely connected phases constitute a coherent process of knowledge discovery.

Definition: *Feature (instance) selection is the process of selection of an optimal subset of features (instances) based on certain criteria.*

The specific criteria that are selected provide the details of the measuring feature (instance) subsets. Let us discuss some of the multiple criteria normally used to evaluate the feature/instance subsets.

Seen as a multi-objective optimization problem, feature selection considers two competing objectives that we aim to minimize: the number of used features and the error produced by the classifier. The instances and features are simultaneously selected aiming to find a compact data set. The selected instances and features are used as a reference set in a specific classifier. Therefore, the goal is to improve the performance (i.e., its generalization ability) of the classifier by searching for an appropriate reference set.

In many problem domains such as medical or engineering diagnosis performance, feature selection can be appropriately assessed by *receiver operating characteristics* (ROC) analysis, in terms of classifier *specificity* and *sensitivity* [13]. Thus, in this case the multiple objectives involved in the feature selection problem are the following:

$$\text{Minimization of}\{|S|,\ \text{sensitivity and specificity}\}, \tag{1.7}$$

where $|S|$ is the cardinality of the features subset.

1.3.2 Classification

Classification in data mining is probably the most studied task. In the classification task each data instance belongs to a particular class out of the predefined sets of classes, which is indicated by the value of a goal attribute. This attribute can take on a small number of discrete values, each of them corresponding to a class. Each instance consists of two parts, namely a set of predictor attribute values and a goal attribute value. The former are used used to predict the value of the latter. Note that the predictor attributes should be relevant for predicting the class of a data instance.

The classification problem can be visualized as a multi-objective problems based on the application domains and fields of interest. Here, we will discuss some of the criteria of interest under classification rule mining.

In the context of classification rule mining there are different objectives to be optimized such as: predictive accuracy, simplicity (or comprehensibility) and interestingness. Additional criteria include sensitivity and specificity, among others.

Supervised classification can be viewed as the problem of generating appropriate class boundaries which can successfully distinguish the various classes in the feature space. In real life problems, the boundaries between different classes are usually nonlinear. It is known that any nonlinear surface can be approximated by using a number of hyperplanes. Hence, the classification problem can be viewed as searching for a number of linear surfaces that can appropriately model the class boundaries while providing the minimum number of misclassified data points. That means the aim is to minimize the number of misclassified points (*Miss*) and the number of hyperplanes (H). It may be noted that as a result of the combination of the two objectives, $Miss$ and H, into a single objective function, a situation may result, where a large number of hyperplanes may be utilized for reducing $Miss$ by a small fraction. Consequently, although such a classifier would provide good abstraction of the training data, its generalization capability is likely to be poor. Therefore the importance of treating this as a truly multi-objective optimization problem, such that the different trade-offs between the two objectives of interest can be properly analyzed.

1.3.3 Association Rule Mining

The association rule mining problem can be considered as multi-objective rather than as a single-objective one. Measures such as support count, comprehensibility and interestingness, which are used for evaluating a rule can be thought of as different objectives for the association rule mining problem. *Support count* is the number of records, which satisfies all the conditions present in the rule. This objective gives the accuracy of the rules extracted from the database. *Comprehensibility* is measured by the number of attributes involved in the rule and tries to quantify the understandability of the rule. Interestingness measures how interesting the rule is.

1.3.4 Clustering

In practice, the problem of choosing an appropriate clustering objective (viz. algorithm) can be alleviated through the application and comparison of multiple clustering methods, or through the *a posteriori* combination of different clustering results by means of ensemble methods. However, a more interesting approach may be the consideration of clustering as a multi-objective optimization problem as suggested in [17]. In a multi-objective clustering problem, $(\Omega, f_1, f_2, \ldots, f_m)$ [17], we aim to determine the clustering C^* for which

$$f_t(C^*) = min_{C \in \Omega} f_t(C), t = 1, 2, \ldots, m, \tag{1.8}$$

where Ω is the set of feasible clusterings, C is a clustering of a given set of data E, and f_t, $t = 1, 2, \ldots, m$ is a set of different (single) objective functions. Usually, no single best solution for this optimization task exists, but instead, the framework of Pareto optimality is embraced.

From a pragmatic point of view, for clustering, two complementary objectives are considered: one based on compactness, and the other one based on the connectedness of the clusters. We refrain from using a third objective based on spatial separation, as the concept of spatial separation is intrinsic (opposite) to that of the connectedness of the clusters. In order to express cluster compactness, we compute the overall deviation of a partitioning. This is simply computed as the overall summed distances between data items and their corresponding cluster center:

$$dev(C) = \Sigma_{c_k \in C} \Sigma_{i \in C_k} (\delta(i, \mu_k)), \tag{1.9}$$

where C is the set of all clusters, μ_k is the centroid of cluster C_k, and $\delta(.,.)$ is the chosen distance function. As an objective, overall deviation should be minimized. This criterion is similar to the popular criterion of intracluster variance, which squares the distance value and is more strongly biased towards spherically shaped clusters.

As an objective reflecting cluster connectedness, we can measure connectivity, which evaluates the degree to which neighboring data points have been placed in the same cluster. It is computed as:

$$Con(C) = \Sigma_{i=1}^{N}(\Sigma_{j=1}^{L} x_{i,nn_{ij}}), \tag{1.10}$$

where $x_{r,s} = 1/j$, if $\{C_k : r \in C_k \wedge s \in C_k\}$, otherwise $x_{r,s} = 0$, nn_{ij} is the jth nearest neighbor of datum i, N is the size of the clustered data set, and L is a parameter determining the number of neighbors that contribute to the connectivity measure. As an objective, connectivity should be minimized.

1.4 Conclusions

This chapter introduced the basic concepts related to swarm intelligence techniques (i.e., PSO and ACO), emphasizing their application in data mining and knowledge discovery in databases, from a single-objective perspective. Then, some potential applications of these techniques in multi-objective problems in data mining and knowledge discovery in databases were also provided. In this last case, several possible criteria to be optimized in data mining and knowledge discovery problems were briefly analyzed.

References

[1] Ben-Jacob, E., Schochet, O., Teneboum, A., Cohen, I., Czirok, A., Vicsek, T.: Generic modeling of cooperative growth patterns in bacterial colonies. Nature 368(6466), 46–49 (1994)
[2] Ben-Jacob, E., Cohen, I., Levine, H.: Cooperative self organization of microorganisms. Advances in Physics 49, 395–554 (2000)
[3] Blum, C.: Ant colony optimization: introduction and recent trends. Physics of Life Reviews 2, 353–373 (2005)
[4] Chen, L., Xu, X.H., Chen, Y.X.: An adaptive ant colony clustering algorithm. In: Proceedings of the 3rd International Conference on Machine Learning and Cybernetics, Shanghai, pp. 1387–1392 (2004)
[5] Chen, L., Tu, L.: Parallel mining for classification rules with ant colony algorithm. In: Hao, Y., Liu, J., Wang, Y.-P., Cheung, Y.-m., Yin, H., Jiao, L., Ma, J., Jiao, Y.-C. (eds.) CIS 2005. LNCS (LNAI), vol. 3801, pp. 261–266. Springer, Heidelberg (2005)
[6] Chen, A., Frietas, A.A.: A new ant colony algorithm for multi-label classification with applications in bioinformatics. In: Proceedings of 8th Annual Conference on Genetic and Evolutionary Computation, pp. 27–34. ACM Press, New York (2006)
[7] Clerc, M., Kennedy, J.: The particle swarm explosion, stability and convergence in a multidimensional complex space. IEEE Transactions on Evolutionary Computation 6(1), 58–73 (2002)

[8] Correa, E.S., Freitas, A.A., Johnson, C.G.: Particle swarm for attribute selection in Bayesian Classification: an application to protine function prediction. Journal of Artificial Evolution and Applications 8(2), 1–12 (2008)
[9] Da, Y., Ge, X.R.: An improved PSO-based ANN with simulated annealing technique. Neurocomputing Letters 63, 527–533 (2005)
[10] Dehuri, S.: An empirical study of particle swarm optimization for cluster analysis. ICFAI Journal of Information Technology 3(2), 7–24 (2007)
[11] Dorigo, M., Stützle, T.: Ant Colony Optimization. MIT Press, Cambridge (2004)
[12] Dorigo, M., Blum, C.: Ant colony optimization theory: a survey. Theoretical Computer Science 344, 243–278 (2005)
[13] Emmanouilidis, C.: Evolutionary multi-objective feature selection and ROC analysis with application to industrial fault diagnosis. In: Giannakoglou, K., et al. (eds.) Evolutioanry methods for design, optimization and control, CIMNE, Barcelona, pp. 1–6 (2002)
[14] De Falco, I., Della Cioppa, A., Tarantino, E.: Facing classification problems with particle swarm optimization. Applied Soft Computing 7(3), 652–658 (2007)
[15] Fayyad, U., Piatetsky-Shapiro, G., Smyth, P.: From data mining to knowledge discovery: An overview. In: Fayyad, U., Piatetsky-Shapiro, G., Smyth, P., Uthurusamy, R. (eds.) Advances in Knowledge Discovery and Data Mining, pp. 495–515. AAAI Press/The MIT Press (1996)
[16] Fayyad, U.M., Piatetsky-Shapiro, G., Smyth, P., Uthurusamy, R.: Advances in Knowledge Discovery and Data Mining. AAAI/MIT Press, Menlo Park (1996)
[17] Ferligoj, A., Batagelj, V.: Direct multi-criteria clustering algorithms. Journal of Classification 9, 43–61 (1992)
[18] Forie, P.C., Groenwold, A.A.: The particle swarm optimization algorithm in size and shape optimization. Structural and Multidisciplinary Optimization 23(4), 259–267 (2002)
[19] Garnier, S., Gautrais, J., Theraulaz, G.: The biological principles of swarm intelligence. Swarm Intelligence 1, 3–31 (2007)
[20] Ghosh, A., Dehuri, S., Ghosh, S. (eds.): Multi-objective Evolutionary Algorithms for Knowledge Discovery from Databases. Springer, Heidelberg (2008)
[21] Ghosh, A., Halder, A., Khotari, M., Ghosh, S.: Aggregation pheromone density based data clustering. Information Sciences 178, 2816–2831 (2008)
[22] Grunbaum, D., Viscido, S.V., Parrish, J.K.: Extracting interactive control algorithms from group dynamics of schooling fish. In: Proceedings of the cooperative control conference, pp. 103–117. Springer, Heidelberg (2005)
[23] Gautrais, J., Michelena, P., Sibbald, A., Bon, R., Deneubourg, J.L.: Allelomimetic Synchronization in Merino Sheep. Animal Behaviour 74(5), 1443–1454 (2007)
[24] Handl, J., Knowles, J., Dorigo, M.: Ant based clustering and topographic mapping. Artificial Life 12(1), 35–61 (2006)
[25] Handl, J., Meyer, B.: Ant-based and swarm-based clustering. Swarm Intelligence 1, 95–113 (2007)
[26] Helbeing, D., Molnar, P., Farkas, I.J., Bolay, K.: Self-organizing pedestrian movement. Environment and Planning B: Planning and Design 28(3), 361–383 (2001)
[27] Holden, N., Freitas, A.A.: Web page classification with an ant colony algorithm. In: Yao, X., Burke, E.K., Lozano, J.A., Smith, J., Merelo-Guervós, J.J., Bullinaria, J.A., Rowe, J.E., Tiňo, P., Kabán, A., Schwefel, H.-P. (eds.) PPSN 2004. LNCS, vol. 3242, pp. 1092–1102. Springer, Heidelberg (2004)

[28] Holden, N.P., Frietas, A.A.: A hybrid PSO/ACO algorithm for classification. In: Proceedings of the Workshop "Particle swarms the second decade" at the 2007 Genetic and Evolutionary Computation Conference (GECCO 2007), pp. 2745–2750. ACM, New York (2007)

[29] Jarboui, B., Cheikh, M., Siarry, P., Rebai, A.: Combinatorial particle swarm optimization (CPSO) for partitional clustering problem. Applied Mathematics and Computation 192(2), 337–345 (2007)

[30] Jiang, W., Xu, Y., Xu, Y.: A novel data mining method based on ant colony algorithm. In: Costabile, M.F., Paternó, F. (eds.) INTERACT 2005. LNCS, vol. 3585, pp. 284–291. Springer, Heidelberg (2005)

[31] Kennedy, J., Eberhart, R.: Particle swarm optimization. In: Proceedings of the IEEE International Conference on Neural Networks, pp. 1942–1948. Bureau of Labor Statistics, Washington (1995)

[32] Kennedy, J., Eberhart, R.C.: The particle swarm: social adaptation in information processing systems. In: Corne, D., Dorigo, M., Glover, F. (eds.) New Ideas in Optimization, pp. 379–387. McGraw-Hill, Cambridge (1999)

[33] Kuo, R.J., Shih, C.W.: Association rule mining through the ant colony system for national health insurance research database in Taiwan. Computers and Mathematics with Applications 54, 11–12 (2007)

[34] Lin, S.W., Ying, K.C., Chen, S.C., Lee, Z.J.: Particle swarm optimization for parameter determination and feature selection of SVM. Expert Systems with Applications 35(4), 1817–1824 (2008)

[35] Liu, B., Abbass, H.A., McKay, B.: Classification rule discovery with ant colony optimization. IEEE Computational Intelligence Bulletin 3(1), 31–35 (2004)

[36] Martens, D., De Backer, M., Haesen, R., Vanthienen, J., Snoeck, M., Baesens, B.: Classification with ant colony optimization. IEEE Transactions on Evolutionary Computation 11(5), 651–665 (2007)

[37] Monmarche, N.: On data clustering with artificial ants. In: Freitas, A.A. (ed.) AAAI 1999 and GECCO 1999 Workshops on Data Mining with Evolutionary Algorithms: Research Directions, Orlando, Florida, pp. 23–26 (1999)

[38] Omran, M.G.H., Salman, A.A., Engelbrecht, A.P.: Dynamic clustering using PSO with application in unsupervised image classification. In: Proceedings of the World Academy of Science, Engineering and Technology, vol. 9, pp. 199–204 (2005)

[39] Parpinelli, R.S., Lopes, H.S., Freitas, A.A.: Data mining with ant colony optimization algorithm. IEEE Transactions on Evolutionary Computation 6(4), 321–332 (2002)

[40] Parrish, J.K., Viscido, S.V., Grunbaum, D.: Self organized fish schools: an examination of emergent properties. Biological Bulletin 202(3), 296–305 (2002)

[41] Ratnaweera, A., Halgamuge, S.K., Watson, H.C.: Self-organizing hierarchical particle swarm optimizer with time varying acceleration coefficients. IEEE Transactions on Evolutionary Computation 8(3), 240–255 (2004)

[42] Reynolds, C.W.: Flocks, herds and school: a distributed behavioral model. Computer Graphics 21(4), 25–34 (1987)

[43] Robbins, K.R., et al.: The ant colony algorithm for feature selection in high-dimension gene expression data for disease classification. Mathematical Medicine and Biology 24(4), 413–426 (2007)

[44] Runkler, T.A.: Ant colony optimization of clustering models. International Journal of Intelligent Systems 20(12), 1233–1251 (2005)

[45] Shelokar, P.S., Jayaraman, V.K., Kulkarni, B.D.: An ant colony approach for clustering. Analytica Chimica Acta 509(2), 187–195 (2004)

[46] Shi, Y., Eberhart, R.C.: A modified particle swarm optimizer. In: Proceedings of the IEEE International Conference on Evolutionary Computation, pp. 69–73. IEEE Press, Pisacataway (1998)
[47] Shi, Y., Eberhart, R.C.: Parameter selection in particle swarm optimization. In: Porto, V.W., Waagen, D. (eds.) EP 1998. LNCS, vol. 1447, pp. 591–600. Springer, Heidelberg (1998)
[48] Sousa, T., Silva, A., Neves, A.: Particle swarm based data mining algorithms for classification tasks. Parallel Computing 30(5-6), 767–783 (2004)
[49] Thangavel, K., Jaganathan, P.: Rule mining algorithm with a new ant colony optimization algorithm. In: Proceedings of the International Conference on Computational Intelligence and Multimedia Applications (ICCIMA 2007), pp. 135–140 (2007)
[50] Wang, Z., Feng, B.: Classification rule mining with an improved ant colony algorithm. In: Webb, G.I., Yu, X. (eds.) AI 2004. LNCS (LNAI), vol. 3339, pp. 357–367. Springer, Heidelberg (2004)
[51] Wang, X., Yang, J., Teng, X., Xia, W., Jensen, R.: Feature selection based on rough sets and particle swarm optimization. Pattern Recognition Letters 28(4), 459–471 (2007)
[52] Wang, Z., Sun, X., Zhang, D.: Classification rule mining based on particle swarm optimization. In: Wang, G.-Y., Peters, J.F., Skowron, A., Yao, Y. (eds.) RSKT 2006. LNCS (LNAI), vol. 4062, pp. 436–441. Springer, Heidelberg (2006)
[53] Xie, L., Mei, H.B.: The application of the ant colony decision rule algorithm on distributed data mining. In: Communications of the IIMA (2007)
[54] Yeh, W.C., Chang, W.W., Chung, Y.Y.: A new hybrid approach for mining breast-cancer pattern using discrete particle swarm optimization and statistical method. Expert Systems with Applications 36(4), 8204–8211 (2009)
[55] Zhang, J.R., Jun, Z., Lok, T.M., Lyu, M.R.: A hybrid particle swarm optimization-back-propagation algorithm for feed-forward neural network training. Applied Mathematics and Computation 185, 1026–1037 (2007)

2

Multi-Criteria Ant Feature Selection Using Fuzzy Classifiers

Susana M. Vieira[1], João M.C. Sousa[1], and Thomas A. Runkler[2]

[1] Center of Intelligent Systems - IDMEC
Instituto Superior Técnico
Technical University of Lisbon
Av. Rovisco Pais 1, 1049-001 Lisbon, Portugal
{susana,j.sousa}@dem.ist.utl.pt

[2] Siemens AG, Corporate Technology
Learning Systems CT IC 4, 81730 Munich - Germany
Thomas.Runkler@siemens.com

Summary. One of the most important techniques in data preprocessing for data mining is feature selection. Real-world data analysis, data mining, classification and modeling problems usually involve a large number of candidate inputs or features. Less relevant or highly correlated features decrease, in general, the classification accuracy, and enlarge the complexity of the classifier. The goal is to find a reduced set of features that reveals the best classification accuracy for a fuzzy classifier. This chapter presents an ant colony optimization (ACO) algorithm for feature selection, which minimizes two objectives: the number of features and the error classification. Two pheromone matrices and two different heuristics are used for each objective. The performance of the method is compared to other feature selection methods, revealing similar or higher performance.

2.1 Introduction

Feature selection has been an active research area in data mining [8], pattern recognition [24] and statistics communities [23]. The main idea of feature selection is to choose a subset of available features, by eliminating features, with little or no predictive information and also redundant features that are strongly correlated. Many practical pattern classification tasks (e.g., medical diagnosis) require learning of an appropriate classification function that assigns a given input pattern (typically represented by using a vector of feature values) to one of a set of classes. The choice of features used for classification has an impact on the accuracy of the classifier and on the time required for classification. The challenge is selecting the minimum subset of features with little or no loss of classification accuracy. The feature subset selection problem consists of identifying and selecting a useful subset of features from a larger set of often mutually redundant, possibly irrelevant, features with different associated importance.

The methods found in the literature can generally be divided into two main groups: model-free methods and model-based methods. Model–free methods use

C.A. Coello Coello et al. (Eds.): Swarm Intel. for Multi-objective Prob., SCI 242, pp. 19–36.
springerlink.com

the available data only and are based on statistical tests, properties of functions, etc. These methods do not need to develop models to find significant inputs. The methods discussed in this chapter belong to the group of model-based methods. Models with different sets of features are compared, and the model that minimizes the model output error is selected. These methods include exhaustive search in which all combinations of subsets were evaluated. This method guarantees an optimal solution, but finding the optimal subset of features is NP–hard. Decision tree search methods, with the proper branch conditions, limit the search space to the best performed branches, but do not guarantee to find the global best solution [22].

For a large number of features, evaluating all states is computationally non–feasible requiring heuristic search methods [6]. Feature selection algorithms have been reviewed in [21]. More recently, nature inspired algorithms have been used to select features [2, 18, 32, 39].

Nature inspired algorithms like ant colony optimization have been successfully applied to a large number of difficult combinatorial problems like quadratic assignment, traveling salesman problems, routing in telecommunication networks, scheduling, machine learning and feature selection [10]. Ant colony optimization is particularly attractive for feature selection since no reliable heuristic is available for finding the optimal feature subset, so it is expected that the ants discover good feature combinations as they proceed through the search space.

An ACO approach for feature selection problems was presented in [2], where a term called updated selection measure (USM) is used for selecting features. A major application of the algorithm in [2] is in the field of texture classification and classification of speech segments. Another application of ACO to feature selection can be found in [17], where an entropy-based modification of the original rough set-based approach for feature selection problems was presented. A different application is presented in [28], where an ACO approach is used for labeling point features. Further, in [32] a relatively simple model of ACO is presented, where the major difference from previous works is in the calculation of the heuristic value, which is treated as a simple function of the classification accuracy.

This chapter presents an ACO approach for feature selection. In this approach, two objectives are considered: minimizing the number of features and minimizing the error classification. Two pheromone matrices and two different heuristics are used for each objective. This approach is an extension from previous work, where a single objective function was used and the subset size was not determined by the algorithm [37].

The approach described in this chapter is an extended version of [36]. The chapter is organized as follows. Next section presents a brief description of fuzzy classification. An introduction of ACO applications in feature selection problems is discussed in Sect. 2.3, where the methodology is described. In Sect. 2.4 the results are presented. Some conclusions are drawn in Sect. 2.5 and the possible future work is discussed.

2.2 Fuzzy Models for Classification

Rule-based expert systems are often applied to classification problems in fault detection, biology, medicine, etc. Fuzzy logic improves classification and decision support systems by allowing the use of overlapping class definitions and improves the interpretability of the results by providing more insight into the classifier structure and decision making process [27, 33]. The automatic determination of fuzzy classification rules from data has been approached by several different techniques: neuro–fuzzy methods, genetic–algorithm based rule selection and fuzzy clustering in combination with GA–optimization [29].

In this chapter, an approach that addresses simplicity and accuracy issues is described. Interpretable fuzzy rule-based classifiers are obtained from observation data.

2.2.1 Model Structure

Takagi-Sugeno (TS) fuzzy models [35], where the consequents are crisp functions of the antecedent variables are applied. Rule antecedents are fuzzy descriptions in the n–dimensional feature space and rule consequents are functions representing the degree of approximation to a given class label from the set $1, 2, \ldots, N_c$, where N_c is the number of classes:

$$R_i : \textbf{If } x_1 \text{ is } A_{i1} \textbf{and } \ldots \textbf{and } x_n \text{ is } A_{in} \textbf{then}$$
$$y_i = \mathbf{a}_i \mathbf{x} + b_i, \quad i = 1, \ldots, K \tag{2.1}$$

Here, $\mathbf{x} = [x_1, x_2, \ldots, x_n]^T$ is the feature vector, y_i is the output off the ith rule, $A_{i1}, \ldots, A_{in}$ are the antecedent fuzzy sets, $\mathbf{a}_i$ is a parameter vector and b_i is a scalar offset. The **and** connective is modeled by the product operator. The degree of activation of the ith rule is calculated as

$$\beta_i(\mathbf{x}) = \prod_{j=1}^{n} \mu_{A_{ij}}(x_j) \tag{2.2}$$

where $\mu_{A_{ij}} \in [0, 1]$ is the membership function of the fuzzy set A_{ij} in the antecedent of R_i.

The model output, y, is computed by aggregating the individual rules contribution:

$$y = \frac{\sum_{i=1}^{K} \beta_i y_i}{\sum_{i=1}^{K} \beta_i}, \tag{2.3}$$

The output of the classifier is given by the following classification rule, as proposed in [29]:

$$\hat{y}_i = \begin{cases} 1, & \text{if } \; y_i < 1.5 \\ 2, & \text{if } \; 1.5 \geq y_k < 2.5 \\ \ldots & \\ N_c, & \text{if } \; y_k \geq N_c + 0.5 \end{cases} \tag{2.4}$$

2.2.2 Parameter Estimation

Given N available input–output data pairs $(\mathbf{x}_k, y_k)$, the n–dimensional pattern matrix $\mathbf{X} = [\mathbf{x}_1, \ldots, \mathbf{x}_N]^T$, and the corresponding class vector $\mathbf{y} = [y_1, \ldots, y_N]^T$ are constructed .

The number of rules K, the antecedent fuzzy sets A_{ij}, and the consequent parameters $\mathbf{a}_i$ and b_i are determined by means of fuzzy clustering in the product space of the input and output variables [26]. Hence, the data set $\mathbf{Z}$ to be clustered is composed from $\mathbf{X}$ and $\mathbf{y}$:

$$\mathbf{Z} = [\mathbf{X}, \mathbf{y}]^T . \tag{2.5}$$

Given the data $\mathbf{Z}$ and the number of clusters K, several fuzzy clustering algorithms can be used. In this chapter, the fuzzy c-means (FCM) [5] clustering algorithm is used to compute the fuzzy partition matrix $\mathbf{U}$. The matrix $\mathbf{Z}$ provides a description of the system in terms of its local characteristic behavior in regions of the data identified by the clustering algorithm, and each cluster defines a rule.

The fuzzy sets in the antecedent of the rules are obtained from the partition matrix $\mathbf{U}$, whose ikth element $\mu_{ik} \in [0, 1]$ is the membership degree of the data object $\mathbf{z}_k$ in cluster i. One-dimensional fuzzy sets A_{ij} are obtained from the multidimensional fuzzy sets defined point-wise in the ith row of the partition matrix by projections onto the space of the input variables x_j:

$$\mu_{A_{ij}}(x_{jk}) = \mathrm{proj}_j^{\mathbb{N}_{n+1}}(\mu_{ik}), \tag{2.6}$$

where proj is the point-wise projection operator [19]. The point-wise defined fuzzy sets A_{ij} are approximated by suitable parametric functions in order to compute $\mu_{A_{ij}}(x_j)$ for any value of x_j.

The consequent parameters for each rule are obtained as a weighted ordinary least-square estimate. Let $\theta_i^T = \left[\mathbf{a}_i^T; b_i\right]$, let $\mathbf{X}_e$ denote the matrix $[\mathbf{X}; \mathbf{1}]$ and let $\mathbf{W}_i$ denote a diagonal matrix in having the degree of activation, $\beta_i(\mathbf{x}_k)$, as its kth diagonal element. Assuming that the columns of $\mathbf{X}_e$ are linearly independent and $\beta_i(\mathbf{x}_k) > 0$ for $1 \le k \le N$, the weighted least-squares solution of $\mathbf{y} = \mathbf{X}_e\theta + \varepsilon$ becomes

$$\theta_i = \left[\mathbf{X}_e^T \mathbf{W}_i \mathbf{X}_e\right]^{-1} \mathbf{X}_e^T \mathbf{W}_i \mathbf{y} . \tag{2.7}$$

Rule bases constructed from clusters can be redundant due to the fact that the rules defined in the multidimensional premise are overlapping in one or more dimensions. A possible approach to solve this problem is to reduce the number of features n of the model, as addressed in this chapter.. This chapter describes the use of an ant colony optimization approach to perform feature selection, as explained in the following section.

The number of fuzzy rules (or clusters) that best suits the data must be determined prior to the construction of the fuzzy classifiers. For this purpose the following criterion, as proposed in [34], is used to determine the number of clusters:

$$S(K) = \sum_{k=1}^{N} \sum_{i=1}^{K} (\mu_{ik})^m (\| \mathbf{x}_k - v_i \|^2 - \| v_i - \bar{\mathbf{x}} \|^2), \tag{2.8}$$

where N is the number of data points to be clustered, K is the number of clusters ($K \geq 2$), $\mathbf{x}_k$ is the kth data point (usually vector), $\bar{\mathbf{x}}$ is the mean value for the inputs, v_i the center of the ith cluster, μ_{ik} is the grade of the kth data point belonging to i^{th} cluster and m is an adjustable weight. The parameter m has a great importance in this criterion. The bigger the m the bigger the optimum number of clusters. Usually, this value is around 2.

The number of clusters K is increased from two up to the number that gives the minimum value for $S(K)$. Note that this minimum can be local. However, this procedure diminishes the number of rules and consequently the complexity of the fuzzy model. At each iteration, the number of clusters are determined using the fuzzy c-means [5] algorithm and the process stops when $S(K)$ increases from one iteration to the next one. The first term of the right-hand side of (2.8) is the variance of the data in a cluster and the second term is the variance of the clusters themselves. The optimal clustering achieved is the one that minimizes the variance in each cluster and maximizes the variance between clusters.

The performance criterion used to evaluate the fuzzy model is the classification accuracy C_a, given by the percentage of correct classifications:

$$C_a = \frac{(N_n - MIS)}{N_n} \times 100\%, \tag{2.9}$$

where N_n is the number of used samples and MIS is the number of misclassifications (number of classification errors in test samples).

2.3 Ant Feature Selection

Ant algorithms were first proposed by Dorigo [9] as a multi-agent approach to difficult combinatorial optimization problems, such as traveling salesman problem, quadratic assignment problem or supply chain management [30, 31]. The ACO methodology is an optimization method suited to find minimum cost paths in optimization problems described by graphs [11]. This chapter presents an implementation of ACO applied to feature selection (AFS), where the best number of features is determined automatically. In this approach, two objectives are considered: minimizing the number of features and minimizing the error classification. Two ant colonies are considered. The first determines the number (cardinality) of features and the second selects the features based on the cardinality given by the first colony. Thus, two pheromone matrices and two different heuristics are used. A different approach for computing a heuristic value is presented to determine the cardinality of features. Heuristic value calculations are application specific and help the algorithm reach the optimal solution quickly by reducing the search domain. The best number of features is called *features cardinality* N_f. The determination of the *features cardinality* is addressed in the first colony sharing the same minimization cost function, which in this case aggregates both the maximization of the classification accuracy and the minimization of the features cardinality. The first colony determines the size of the subsets of the ants in the second colony. The second colony selects the features that will be part of the subsets.

2.3.1 AFS Algorithm

The algorithm described in this chapter deals with the feature selection problem as a multi–criteria problem with a single objective function. Therefore, a pheromone matrix is computed for each criterion, and different heuristics are used. The two main criteria of a feature selection problem are the size of the subset of features which are the features that are selected to build the fuzzy classifier. The objective function of this optimization algorithm aggregate both criteria, the minimization of the classification error rate and the minimization of the features cardinality:

$$J^k = \frac{MIS^k}{N_n} + \frac{N_f}{n} \tag{2.10}$$

where N_n is the number of used data samples and n is the total number of features. To evaluate the classification error, a fuzzy classifier is built for each solution following the procedure described in Sect. 2.2.

Let $\mathbf{x} = [x_1, x_2, \ldots, x_n]^T$ be the set of given n features, and $\mathbf{w} = [w_1, w_2, \ldots, w_{N_f}]^T$, be a subset of features where ($\mathbf{w} \subseteq \mathbf{x}$). It is desirable that $N_f << n$.

Probabilistic rule

Consider a problem with N_f nodes and two colonies of g ants. Initially, g ants of the first colony randomly select the number of nodes N_f to be used by the g ants of the second colony. The probability that an ant k chooses the features cardinality $N_f(k)$ is given by

$$p_i^k(t) = \frac{[\tau_{n_i}]^{\alpha_n} \cdot [\eta_{n_i}]^{\beta_n}}{\sum_{l \in \Gamma_n^k} [\tau_{n_l}]^{\alpha_n} \cdot [\eta_{n_l}]^{\beta_n}} \tag{2.11}$$

where τ_{n_i} is the pheromone concentration matrix and η_{n_i} is the heuristic function matrix, for path (i). The values of the pheromone matrix are limited to $[\tau_{n_{min}}, \tau_{n_{max}}]$, with $\tau_{n_{min}} = 0$ and $\tau_{n_{max}} = 1$. Γ_n^k is the feasible neighborhood of ant k (available number of features to be selected), which acts as the memory of the ants and contains all the trails that the ants have not passed and can be chosen. The parameters α_n and β_n measure the relative importance of trail pheromone and heuristic knowledge, respectively.

After all the g ants, from the first colony have chosen the features cardinality $N_f(k)$, each ant k from the second colony select $N_f(k)$ features (nodes). The probability that an ant k chooses feature j as the next feature to visit is given by

$$p_j^k(t) = \frac{[\tau_{f_j}(t)]^{\alpha_f} \cdot [\eta_{f_j}]^{\beta_f}}{\sum_{l \in \Gamma_f^k} [\tau_{f_l}(t)]^{\alpha_f} \cdot [\eta_{f_l}]^{\beta_f}} \tag{2.12}$$

where τ_{f_j} is the pheromone concentration matrix and η_{f_j} is the heuristic function matrix for the path (j). Again, the pheromone matrix values are limited to $[\tau_{f_{min}}, \tau_{f_{max}}]$, with $\tau_{f_{min}} = 0$ and $\tau_{f_{max}} = 1$. Γ_f is the feasible neighborhood of ant k (available features), which contains all the features that the ants have not selected and can be chosen. Again, the parameters α_f and β_f measure the relative importance of trail pheromone and heuristic knowledge, respectively.

Updating rule

After a complete tour, when all the g ants have visited all the $N_f(k)$ nodes, both pheromone concentration in the trails are updated by

$$\tau_{n_i}(t+1) = \tau_{n_i}(t) \times (1-\rho_n) + \Delta\tau_{n_i}(t) \tag{2.13}$$

$$\tau_{f_j}(t+1) = \tau_{f_j}(t) \times (1-\rho_f) + \Delta\tau_{f_j}(t) \tag{2.14}$$

where $\rho_n \in [0,1]$ is the pheromone evaporation of the features cardinality , $\rho_f \in [0,1]$ is the pheromone evaporation of the features and $\Delta\tau_{n_i}$ and $\Delta\tau_{f_j}$ are the pheromone deposited on the trails (i) and (j), respectively, by the ant q that found the best solution J^q for this tour:

$$\Delta\tau_{n_i}^q = \begin{cases} \mathcal{Q}_i & \text{if node } (i) \text{ is used by the ant } q \\ 0 & \text{otherwise} \end{cases} \tag{2.15}$$

$$\Delta\tau_{f_j}^q = \begin{cases} \mathcal{Q}_j & \text{if node } (j) \text{ is used by the ant } q \\ 0 & \text{otherwise} \end{cases} \tag{2.16}$$

The number of nodes $N_f(k)$ that each ant k has to visit on each tour t is only updated every I_n tours (iterations), in order to allow the search for the best features for each features cardinality N_f. The algorithm runs I times. Both colonies share the same cost function given in (2.10).

Heuristics

The heuristic value used for each feature (ants visibility) for the second colony, is computed as

$$\eta_{f_j} = 1/MIS_j \tag{2.17}$$

with $j = 1, \ldots, n$.

For the features cardinality (first colony), the heuristic value is computed using the Fisher discriminant criterion for feature selection [12]. Considering a classification problem with two possible classes, class 1 and class 2, the Fisher discriminant criterion is described as

$$F(i) = \frac{|\mu_1(i) - \mu_2(i)|^2}{\sigma_1^2 + \sigma_2^2} \tag{2.18}$$

where $\mu_1(i)$ and $\mu_2(i)$ are the mean values of feature i for the samples in class 1 and class 2, and σ_1^2 and σ_2^2 are the variances of feature i for the samples in class 1 and 2. The score aims to maximize the between-class difference and minimize the within-class spread. Other currently proposed rank-based criteria generally come from similar considerations and show similar performance [12]. Since our

Algorithm 1. Ant Feature Selection

```
/*Initialization*/
set the parameters ρ_f, ρ_n, α_f, α_n, β_f, β_n, I, I_n, g.
for t = 1 to I do
  for k = 1 to g do
    Choose the subset size N_f(k) of each ant k using (2.11)
  end for
  for l = 1 to I_n do
    for k = 1 to g do
      Build feature set L_f^k(t) by choosing N_f(k) features using (2.12)
      Compute the fuzzy model using the L_f^k(t) path selected by ant k
      Compute the cost function J^k(t)
      Update J^q
    end for
    Update pheromone trails τ_{n_i}(t + 1) and τ_{f_j}(t + 1), as defined in (2.13) and
    (2.14).
  end for
end for
```

goal is to work with several classification problems, which contain two or more possible classes, a one versus-all strategy is used to rank features. In other words, for a C-class prediction problem, a particular class is compared with the other $C-1$ classes that are considered together. The features are weighted according to the total score summed over all C comparisons:

$$\sum_{j=1}^{C} F_j(i) \tag{2.19}$$

where $F_j(i)$ denotes the Fisher discriminant score for the ith feature at the jth comparison.

In Algorithm 1, a brief description of the ant feature selection algorithm is presented.

2.4 Experimental Results

2.4.1 Data Sets and Parameters

The effectiveness of the presented method is applied to data sets taken from the well known benchmarks in the UCI repository [3]. The obtained results are compared with other algorithms, namely with particle swarm optimization for rough set-based feature selection algorithm (PSORSFS) [39], positive region-based attribute reduction algorithm (POSAR) [16], conditional entropy-based attribute reduction (CEAR) [38], discernibility matrix-based attribute reduction (DISMAR) [14], GA-based attribute reduction (GAAR) [4], and top–down (TD) and bottom–up (BU) decision tree search feature selection [22]. More details can be found in the cited references. For all the data sets, the results are

Table 2.1. Values of parameters used in the experiments.

Data set	α_n	β_n	ρ_n	τ_{n0}	α_f	β_f	ρ_f	τ_{f0}	I	I_n	g
Wine	0.3	0.9	0.05	0.5	0.3	0.9	0.1	0.5	50	5	20
Breast Cancer	0.3	0.9	0.05	0.5	0.3	0.9	0.1	0.5	50	5	20
Vote	0.2	0	0.06	0.5	0.3	0.9	0.09	0.5	100	5	30
M-of-N	0.4	0.9	0.05	0.5	0.3	0.9	0.1	0.5	50	5	20
Lung	0.3	0.9	0.05	0.5	0.3	0.9	0.1	0.5	150	10	20

Table 2.2. Description of the used data sets.

Data sets used	# features	# classes	# samples
Wine	13	3	178
Breast Cancer	9	2	699
Vote	16	2	300
M-of-N	13	10	1000
Lung	56	3	32

obtained using ten runs of the presented algorithm. The parameters used in the experiments for the databases are presented in Table 2.1. The classification error rates of the used classifiers are obtained by performing cross validation (10-fold cross validation (CV10)). For the sake of comparison with other published algorithms, the results presented for the wine data set, are not obtained using cross validation. In this case a test data set that is different from the one used for training is used. The experimental results are presented as the best, the worst and the mean value of correct classifications C_a.

2.4.2 Cross Validation

For the cross validation procedure, the data set with N_n samples is divided into 10 mutually exclusive sets of approximately equal size, with each subset consisting of approximately the same proportions of labels as the original data set, known as stratified cross validation [20]. The classifier is trained 10 times, with a different subset left out as the test set and the other samples used to train the classifier at each time. During the training phase, the classifier is trained on 9 out of 10 folds in which classification accuracy is used, as defined in (2.9). The prediction performance of the classifier is estimated by considering the average classification accuracy of the 10 cross-validation experiments, described as

$$E_{CV} = \left(\frac{1}{N_n} \sum_{i=1}^{10} C_i \right) \times 100\% \tag{2.20}$$

where C_i is the number of correctly classified samples:

$$C_i = N_n - MIS \tag{2.21}$$

2.4.3 Wine Data

The wine data set is a widely used classification data available online in the repository of the University of California [3], and contains the chemical analysis of 178 wines grown in the same region in Italy, derived from three different cultivars. Thirteen continuous attributes are available for classification: alcohol, malic acid, ash, alkalinity of ash, magnesium, total phenols, flavanoids, nonflavanoids phenols, proanthocyanism, color intensity, hue, OD280/OD315 of dilluted wines and proline.

The AFS algorithm is applied to select the relevant features within the wine classification data set and is compared to other approaches available in the literature. The data set characteristics are presented in Table 2.2. Ten runs with one-half training instance and one-half testing set was used to evaluate the performance of the models during the feature selection process. The final model is validated using cross validation. As can be seen in Table 2.3, the obtained results are better than those in feature selection based on interclass separability [27], and classification without feature selection, using a fuzzy classifier [15]. Further, the obtained classifier uses far less rules than the 2 first approaches (3 compared to 60) and less features. Ant feature selection uses a similar number of features as in [26], obtaining better classification accuracy. The approach in [22] has better classification accuracy using a tree search method with the top–down approach using 11 features. The bottom–up approach in [22] uses slightly less features and the classification accuracy is worse.

For the sake of better understanding of the algorithm, an example of the process of the ant feature selection searching for optimal solutions for wine data set is given in Figures 2.1 and 2.2, where it can be observed that the average classification error decreases, indicating the convergence of the algorithm. In Fig. 2.1 the average number of incorrect classifications for all the k ants at each iteration t is presented. Fig. 2.2 shows the minimum number of incorrect classifications of all ants at each iteration t. Finally, Fig. 2.3 presents the evolution of the search for the best number of features.

Table 2.3. Classification rates on the Wine data for ten independent runs.

Method	Number of features	Classification accuracy (%)		
		Max.	Mean	Min.
Fuzzy model	13 (all)	98.9	93.7	89.0
AFS approach	4-8	100	99.8	98.9
Corcoran and Sen [7]	13	100	99.5	98.3
Ishibuchi et al. [15]	13	99.4	98.5	97.8
Roubos et al. [26]	4-7	99.4	-	98.3
Mendonça et al. [22] (Top–down)	11	100	99.9	99.4
Mendonça et al. [22] (Bottom–up)	4	100	98.5	92.7

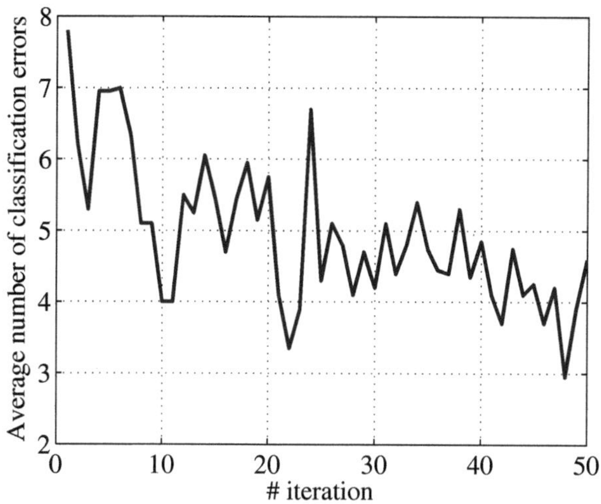

Fig. 2.1. Average classification accuracy for each iteration in the Wine data set.

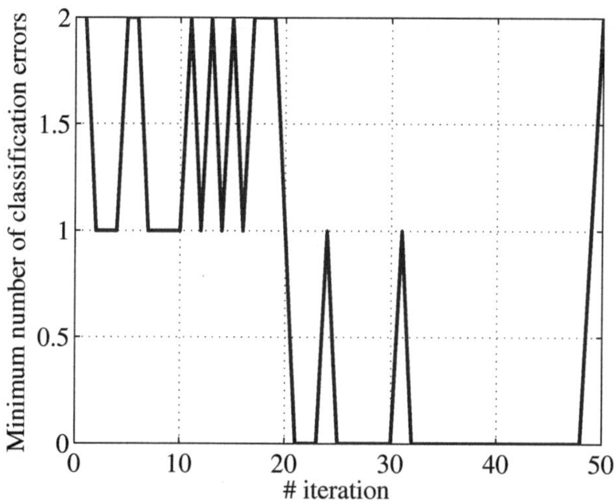

Fig. 2.2. Minimum number of errors for each iteration in the Wine data set.

2.4.4 Breast Cancer Data

The breast cancer data set is also available in the repository of the University of California [3] and it was obtained from the University of Wisconsin Hospitals, Madison from Dr. William H. Wolberg. The Wisconsin breast cancer data is widely used to test the effectiveness of classification algorithms. The aim of the classification is to distinguish between benign and malignant cancers based on the available nine measurements (attributes): clump thickness, uniformity of cell size, uniformity of cell shape, Marginal Adhesion, Single Epithelial Cell Size,

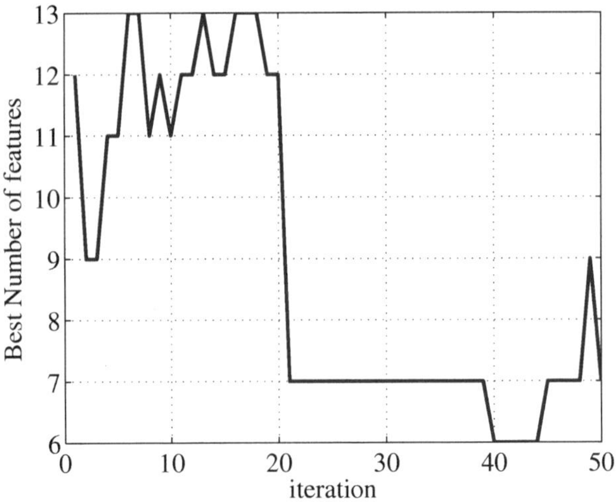

Fig. 2.3. Best number of features for each iteration in the Wine data set.

Bare Nuclei, Bland Chromatin, Normal Nucleoli and Mitoses. The attributes have integer value in the range (1,10). The original database contains 699 instances, however 16 of these are omitted because these are incomplete, which is common in data mining. The class distribution is 65.5% benign and 34.5% malignant. Ten runs with one-half training instance and one-half testing set was used to evaluate the performance of the models during the feature selection process. The final model is validated using cross validation. As can be seen in Table 2.4, the obtained results are always better than those in [39] using PSO and rought set-based feature selection, and in [1] using Gath-Geva clustering

Table 2.4. Classification rates on the Breast Cancer data for 10–fold cross validation.

Method	Number of features	Classification accuracy (%)		
		Max.	Mean	Min.
Fuzzy model	all (9)	97.0	97.0	97.0
AFS approach	3	100.0	96.0	91.4
Wang et al. [39](POSAR)	4	95.94	-	-
Wang et al. [39](CEAR)	4	94.20	-	-
Wang et al. [39](DISMAR)	5	95.94	-	-
Wang et al. [39](GAAR)	4	95.65	-	-
Wang et al. [39](PSORSFS)	4	95.80	-	-
Abonyi et al. [1] (GG: R = 4)	9	98.57	95.14	88.57
Abonyi et al. [1] (Sup: R = 4)	8-9	98.57	95.57	90.00

Table 2.5. Classification rates on the Vote data for 10–fold cross validation.

Method	Number of features	Classification accuracy (%)		
		Max.	Mean	Min.
Fuzzy model	all (16)	93.0	91.6	91.0
AFS approach	4	100	94.3	87.1
Wang et al. [39](POSAR)	9	94.3	-	-
Wang et al. [39](CEAR)	11	92.3	-	-
Wang et al. [39](DISMAR)	8	93.7	-	-
Wang et al. [39](GAAR)	9	94.0	-	-
Wang et al. [39](PSORSFS)	8	95.3	-	-

based classifier, supervised clustering, and feature selection based on interclass separability.

2.4.5 Vote Data Set

The vote data set is a widely used classification data available in the repository of the University of California [3], and is a binary class problem (democrat vs. republican), with sixteen binary features and 435 instances. In the votes data set, there is a high proportion between each class. There are some missing values in this data set. The samples with missing values were discarded, and the total number of instances used is 300. Ten runs with one-half training instance and one-half testing set was used to evaluate the performance of the models during the feature selection process. The final model is validated using cross validation.

Table 2.5 presents the results obtained with AFS and compares them to other algorithms presented in the literature. In Table 2.5, it can be observed that the obtained results are always better than those in [39], recall that these algorithms were described in section 2.4.4. Note that the number of features used by the AFS approach is lower than in the other algorithms, and that the obtained accuracy is better.

2.4.6 M-of-N Data Set

An artificial data set that is M-of-N concept [25] was also tested, which is an at least M-of-N among 13 features, and 1000 instances are used. Table 2.6 shows the results of the feature selection algorithm. Ten runs with one-half training instance and one-half testing set was used to evaluate the performance of the models during the feature selection process. The final model is validated using cross validation. As can be seen in Table 2.6, the obtained results are similar to those in [39]. In this data set the number of features used by AFS is higher than in the other algorithms, but the accuracy is maintained.

Table 2.6. Classification rates on the M-of-N data for 10–fold cross validation.

Method	Number of features	Classification accuracy (%)		
		Max.	Mean	Min.
Fuzzy model	all (13)	100	100	100
AFS approach	9	100	100	100
Wang et al. [39](POSAR)	7	100	-	-
Wang et al. [39](CEAR)	7	100	-	-
Wang et al. [39](DISMAR)	6	100	-	-
Wang et al. [39](GAAR)	6	100	-	-
Wang et al. [39](PSORSFS)	6	100	-	-

2.4.7 Lung Data Set

This data was used by Hong and Young [13] to illustrate the power of the optimal discriminant plane even in ill-posed settings. It has 32 instances and 56 attributes. There are some missing attribute values. The data is classified in three different classes. The class distribution is: 9 observations in class 1, 13 observations in class 2 and 10 observations in class 3. Table 2.7 shows the results of the feature selection algorithm. Ten runs with one-half training instance and one-half testing set was used to evaluate the performance of the models during the feature selection process. The final model is validated using cross validation. All the previous used data sets are rather small in terms of number of features. The use of the Lung data set is important once it has a bigger number of features. The successful application of the AFS algorithm to this data set reenforces the utility and validity of the presented algorithm.

In Fig. 2.4, an example of the evolution of the algorithm tours for the Lung data set is presented. In Fig. 2.5, the evolution of the search for the best feature cardinality is presented for one of the runs of the presented AFS algorithm. Note that in Fig. 2.5 the best solution found by the algorithm is not always the

Table 2.7. Classification rates on the Lung data for 10–fold cross validation.

Method	Number of features	Classification accuracy (%)		
		Max.	Mean	Min.
AFS approach	5-11	93.3	84.0	73.3
Wang et al. [39](POSAR)	4	86.7	-	-
Wang et al. [39](CEAR)	5	73.3	-	-
Wang et al. [39](DISMAR)	4	73.3	-	-
Wang et al. [39](GAAR)	6	70.0	-	-
Wang et al. [39](PSORSFS)	4	90.0	-	-

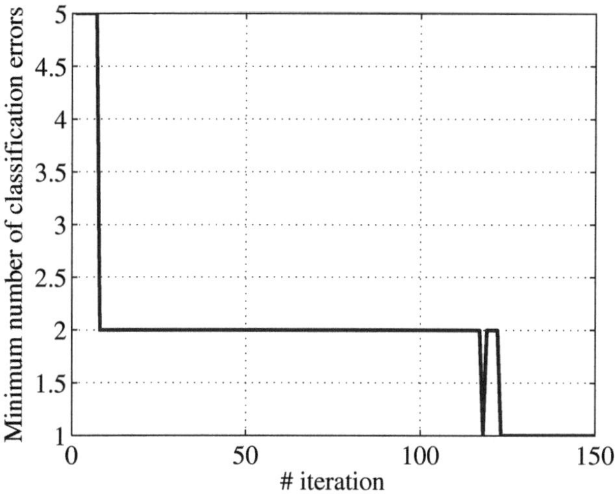

Fig. 2.4. Average classification accuracy for each iteration in Lung data set.

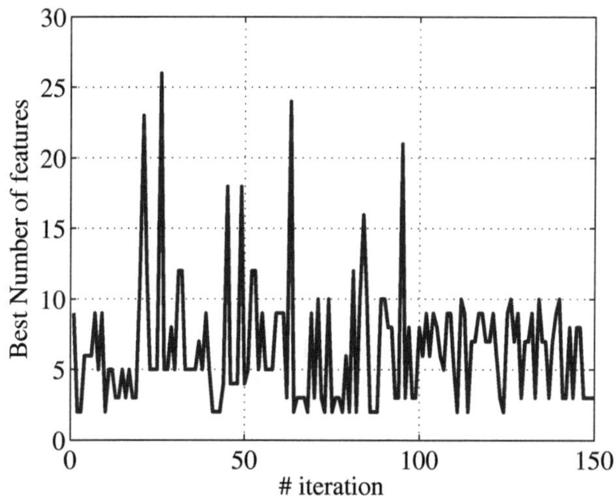

Fig. 2.5. Best number of features for each iteration in Lung data set.

solution with less number of features, due to the weight of the model performance in the objective function. In the case presented in Figs. 2.4 and 2.5, the best solution found has nine features and a classification accuracy of 93.3%.

As can be seen in Table 2.7, the obtained results are similar or better than those in [39].

2.4.8 Discussion

Four real data sets (Wine, Wisconsin Breast Cancer, Vote and Lung), and one artificial data set (M–of–N) were used to test the ant feature selection algorithm. It can be observed that real data sets can have many weakly relevant features rather than strongly relevant or totally irrelevant features. The ant feature selection algorithm discards a bigger percentage of features for the case of real data sets than for the artificial data set. However, the selected features are not always the same, once there are features that are weakly relevant and have a similar influence in the classifier.

The time of convergence of the presented algorithm can be reduced with the number of used ants. The number of ants is strongly related to the number of features in the data set. The performance of the fuzzy models with ant feature selection is equal or better than the performance of the fuzzy models using all features, as can be seen in Tables 2.3, 2.4, 2.5,2.6 and 2.7. In the M–of–N data set the performance is the same, but AFS has the advantage of selecting the more relevant features and discarding the non–relevant features.

For all data sets, the presented algorithm performs, in general, similar or better then other algorithms cited in the literature, as can be observed in Tables 2.3, 2.4, 2.5, 2.6 and 2.7.

2.5 Conclusions

A feature selection algorithm based on ant colony optimization is presented in this chapter. The problem is divided into two objectives: choosing the features cardinality and selecting the most relevant features. The feature selection algorithm uses fuzzy classifiers. The ant feature selection algorithm was applied to five classification databases that are considered benchmarks. The performance of the presented algorithm was compared to previous works. The ant based feature selection algorithm yielded similar os better classification rates.

As the presented ant feature selection algorithm is generic, it can be combined with other intelligent classifiers, such as neural networks classifiers.

Acknowledgements

This work is supported by the Portuguese Government and FEDER under the programs: Programa de financiamento Plurianual das Unidades de I&D da FCT´ (POCTI-SFA-10-46-IDMEC) and by the FCT grant SFRH/25381/2005, Fundação para a Ciência e a Tecnologia, Ministério do Ensino Superior, da Ciência e da Tecnologia, Portugal.

References

[1] Abonyi, J., Szeifert, F.: Supervised fuzzy clustering for the identification of fuzzy classifiers. Pattern Recognition Letters 24(14), 2195–2207 (2003)

[2] Al-Ani, A.: Feature subset selection using ant colony optimization. International Journal of Computational Intelligence 2(1), 53–58 (2005)

[3] Asuncion, A., Newman, D.J.: UCI machine learning repository (2007)
[4] Bazan, J., Nguyen, H.S., Nguyen, S.H., Synak, P., Wroblewski, J.: Rough set algorithms in classification problem. In: Polkowski, L., Tsumoto, S., Lin, T.Y. (eds.) Rough Set Methods and Applications, pp. 49–88. Physica-Verlag, New York (2000)
[5] Bezdek, J.C.: Pattern Recognition With Fuzzy Objective Functions. Plenum Press, New York (1981)
[6] Boz, O.: Feature subset selection using sorted feature relevance. In: Proc. of ICMLA 2002, international conference of machine learning and applications, Los Angeles, USA, pp. 147–153 (2002)
[7] Corcoran, A.L., Sen, S.: Using real-valued genetic algorithms to evolve rule sets for classification. In: Proc. 1st IEEE Conf. Evolut. Comput., pp. 120–124 (1994)
[8] Dash, M., Choi, K., Scheuermann, P., Liu, H.: Feature selection for clustering-a filter solution. In: Proc. of Second International Conference on Data Mining, ICDM 2002, pp. 115–122 (2002)
[9] Dorigo, M.: Optimization, Learning and Natural Algorithms (in Italian). PhD thesis, Dipartimento di Elettronica, Politecnico di Milano, Italy (1992)
[10] Dorigo, M., Birattari, M., Stützle, T.: Ant colony optimization. IEEE Computational Intelligence Magazine 1(4), 28–39 (2006)
[11] Dorigo, M., Stützle, T.: The Ant Colony Optimization Metaheuristic: Algorithms, Applications, and Advances. In: Glover, F., Kochenberger, G. (eds.) Handbook of Metaheuristics. International Series in Operations Research and Management Science. Kluwer, Dordrecht (2003)
[12] Duda, R.O., Hart, P.E., Stork, D.G.: Pattern Classification, 2nd edn. Wiley–Interscience Publication, Chichester (2001)
[13] Hong, Z.Q., Yang, J.Y.: Optimal discriminant plane for a small number of samples and design method of classifier on the plane. Pattern Recognition 24(4), 317–324 (1991)
[14] Hu, K., Lu, Y.C., Shi, C.Y.: Feature ranking in rough sets. AI Commun. 16(1), 41–50 (2003)
[15] Ishibuchi, H., Nakashima, T., Murata, T.: Performance evaluation of fuzzy classifier systems for multidimensional pattern classification problems. IEEE Trans. Syst. Man, Cybern., B 29, 601–618 (1999)
[16] Jensen, R., Shen, Q.: Finding rough set reducts with ant colony optimization. In: Proceedings of the 2003 UK Workshop on Computational Intelligence, Bristol, UK, September 2003, pp. 15–22 (2003)
[17] Jensen, R., Shen, Q.: Fuzzy-rough attribute reduction with application to web categorization. Fuzzy Sets and Systems 141(3), 469–485 (2004)
[18] Jensen, R., Shen, Q.: Fuzzy-rough data reduction with ant colony optimization. Fuzzy Sets and Systems 149, 5–20 (2005)
[19] Klir, G.J., Yuan, B.: Fuzzy Sets and Fuzzy Logic: theory and applications. Prentice-Hall, Upper Saddle River (1995)
[20] Kohavi, R.: A study of cross-validation and bootstrap for accuracy estimation and model selection. In: Proc. International Joint Conf. Artificial Intelligence (1995)
[21] Liu, H., Yu, L.: Toward integrating feature selection algorithms for classification and clustering. IEEE Transactions on Knowledge and Data Engineering 17(4), 491–502 (2005)
[22] Mendonça, L.F., Vieira, S.M., Sousa, J.M.C.: Decision tree search methods in fuzzy modeling and classification. International Journal of Approximate Reasoning 44(2), 106–123 (2007)

[23] Miller, A.: Subset Selection in Regression, 2nd edn. Chapman & Hall/CRC, Boca Raton (2002)
[24] Mitra, P., Murthy, C.A., Pal, S.K.: Unsupervised feature selection using feature similarity. IEEE Trans. Pattern Analysis and Machine Intelligence 24(3), 301–312 (2002)
[25] Murphy, P.M., Pazzani, M.J.: Id2-of-3: Constructive induction of M-of-N concepts for discriminators in decision trees. In: Proceedings of the Eighth International Workshop on Machine Learning, Evanston, IL, June 1991, pp. 183–187 (1991)
[26] Roubos, J.A., Setnes, M., Abonyi, J.: Learning fuzzy classification rules from data. In: John, R., Birkenhead, R. (eds.) Developments in Soft Computing, pp. 108–115. Springer, Heidelberg (2001)
[27] Roubos, J.A., Setnes, M., Abonyi, J.: Learning fuzzy classification rules from labeled data. International Journal of Information Sciences 150(1), 77–93 (2003)
[28] Schreyer, G.R., Raidl, M.: Letting ants labeling point features. In: Proceedings of the 2002 Congress on Evolutionary Computation, 2002. CEC 2002, vol. 2, pp. 1564–1569 (2002)
[29] Setnes, M., Roubos, J.A.: GA-fuzzy modeling and classification: complexity and performance. IEEE Transactions on Fuzzy Systems 8(5), 509–522 (2000)
[30] Silva, C.A., Sousa, J.M.C., Runkler, T.A.: Rescheduling and optimization of logistic processes using GA and ACO. Engineering Applications of Artificial Intelligence 21(3), 343–352 (2008)
[31] Silva, C.A., Sousa, J.M.C., Runkler, T.A., Sá da Costa, J.M.G.: Distributed optimization of a logistic system and its suppliers using ant colony optimization. International Journal of Systems Science 37(8), 503–512 (2006)
[32] Sivagaminathan, R.K., Ramakrishnan, S.: A hybrid approach for feature subset selection using neural networks and ant colony optimization. Expert Systems with Applications 33, 49–60 (2007)
[33] Sousa, J.M.C., Kaymak, U.: Fuzzy Decision Making in Modeling and Control. World Scientific and Imperial College, Singapore (2002)
[34] Sugeno, M., Yasukawa, T.: A fuzzy-logic-based approach to qualitative modeling. IEEE Transactions on Fuzzy Systems 1(1), 7–31 (1993)
[35] Takagi, T., Sugeno, M.: Fuzzy identification of systems and its applications to modelling and control. IEEE Transactions on Systems, Man and Cybernetics 15(1), 116–132 (1985)
[36] Vieira, S.M., Sousa, J.M.C., Runkler, T.A.: Fuzzy classification in ant feature selection. In: Zurada, J.M., Yen, G.G., Wang, J. (eds.) WCCI 2008. LNCS, vol. 5050, pp. 1763–1769. Springer, Heidelberg (2008)
[37] Vieira, S.M., Sousa, J.M.C., Runkler, T.A.: Ant Colony Optimization Applied to Feature Selection in Fuzzy Classifiers. In: Melin, P., et al. (eds.) IFSA 2007. LNCS (LNAI), vol. 4529, pp. 778–788. Springer, Heidelberg (2007)
[38] Wang, G.Y., Zhao, J.: Theoretical study on attribute reduction of rough set theory: comparison of algebra and information views. In: Proc. Third IEEE Int. Conf. on Cognitive Informatics (2004)
[39] Wang, X., Yang, J., Teng, X., Xia, W., Jensen, R.: Feature selection based on rough sets and particle swarm optimization. Pattern Recognition Letters 28, 459–471 (2007)

3

Multiobjective Particle Swarm Optimization in Classification-Rule Learning

Augusto de Almeida Prado G. Torácio

Department of Computer Sciences
Federal University of Parana
Curitiba, Brazil
toracio@inf.ufpr.br

The chapter presents a convenient algorithm of classification-rule learning that uses the Multiple Objective Particle Swarm Optimization technique to obtain a set of the best classification rules as an alternative way to cover algorithms, avoiding the loss of quality of the rules. The best rules are memorized in an unordered classifier, based on the concept of Pareto dominance, which permits the selection of rules with several characteristics that can be defined by the user. The rules are selected from the classifier at the same time that they are discovered and, differently from the cover algorithms, the selection does not remove examples from the dataset. We study classification-rule learning, describing some paradigms, different rule's representations and different particle initialization procedures. Then, we describe the way the particles move in the search space and the fitness functions. After that, we show some aspects of the required global repositories and describe aspects of different stopping criteria. Then, we present the proposed algorithm MOPSO-RL, showing its complexity and restrictions. Comparisons with other algorithms from the literature show that the proposed algorithm is competitive related to the area under ROC (Receiver Output Curve) and regarding the number of rules in the classifier. Besides, the selected rules have high support and weighted relative accuracy (Wracc), which denotes that the rules are important even when considered in isolation. In this way, it produces a set that is good for classification, with very good rules that can bring knowledge even if they are analyzed in isolation.

3.1 Introduction

The focus of this work is on classification-rule learning, which is part of supervised symbolic learning. In this type of learning, each example is described as a pre-determined fixed set of features called attributes. Thus, learning determines rules that form a relation between their attributes and their classes. In classification rule learning, these rules are generally in form of {attribute Operator Value} (propositional logic) or in first order logic, which have the enormous advantage of being much more comprehensible to humans than other forms of

C.A. Coello Coello et al. (Eds.): Swarm Intel. for Multi-objective Prob., SCI 242, pp. 37–64.
springerlink.com

representation. It is, therefore, widely used in data mining. Thus, many other disciplines can benefit with this type of learning.

The process of rule learning aims to enable the machine to properly classify positive and negative examples of the described concept using a set of rules. Thus, the algorithm corrects all its errors during the training process, and can guide its induction to the correct direction. The disadvantage is that, normally, one will need an extensive number of labeled examples to get a good classifier [1]. In this process, some learning algorithms seek to obtain a set of classification rules, based on the coverage of the training dataset. In general, this is done by removing, from the dataset, the already covered examples, so that the new learnt rules focus only on uncovered examples.

This can make that good rules are discarded during the selection process, because these algorithms, also called cover algorithms, select the rules with the best performance in the sub-dataset. In general, these classifiers are ordered (i.e., there is an order of the rules for classification), which harms the analysis of the rules, because a rule cannot be seen in isolation. And, even in cases of unordered classifiers, there is still a loss of quality of the rules.

Another way to undertake the learning task, avoiding the loss of quality of the rules, is by searching for all the rules from the dataset, sorting them by quality, and making a sub-set of that selection. This tends to raise the performance of the classifier and its rules. However, besides the fact that we need two steps for performing the classification (generation and sub-selection), the sub-selection is usually done by coverage of the rules (equal to the previous way), removing examples from the training dataset and falling in the same issue of having to exclude some good rules.

Another problem concerns the evaluation metric for the classification rules. In the task of learning classification rules, the goal is to create rules that together have good performance for classification. The great majority of the methods tries to optimize the performance of the classification by optimizing the precision in the training set [2]. But, recent studies have shown some problems with the precision metric for rules induction [3], especially when the classes are not balanced or there are different costs for the classification errors. The solution for these cases is to use probabilistic classifiers [2].

A frequent measure used to evaluate performance of a probabilistic classifier is the area under the ROC curve (Receiver Operating Characteristic). A ROC curve is a graph that relates the false positive rate (FP) (axis-x) and the true positive rate (TP) (axis-y) of a classifier [4]. A ROC curve can be obtained from a unique rule, a partial classifier or a complete classifier. In order to obtain a good performance, many recent classification algorithms have the goal to maximize the area under the ROC curve (AUC - Area Under Curve).

So, this chapter proposes a method of classification that tries to maximize AUC, using a multiobjective optimization algorithm such as MOPSO [5]. Besides, this method makes a different type of selection of rules, which does not remove examples from the training dataset. Furthermore, we look for unordered classifiers, making easier the interpretation of the rules so that they can be read in isolation.

This method could use genetic algorithms to do this task. But, we believe that particle swarm optimization has advantages compared to a genetic algorithm, because:

- The genetic algorithm maintains in the population the best individuals. As the generations pass, many individuals of the same species are present in the population. In our case, this is unwanted.
- With the parameters manipulation of MOPSO, we avoid this uniformity of the population, maintaining certain evolutionary intelligence, with the global and local leader utilization.

Particle swarm optimization has been previously used for the classification task in [6]. However, this work does not explore the ROC curve or any multiobjective issues. On the other hand, recently, an increasing interest has emerged in applying the concept of Pareto-optimality to machine learning inspired by the successful developments in evolutionary multiobjective optimization. These researches include multiobjective feature selection, multiobjective model selection in training multilayer perceptrons, radial-basis-function networks, support vector machines, decision trees and intelligent systems [7]. In the literature, few works deal with multiobjective evolutionary algorithms for rule learning, among them [8–10]. The first work focuses on the rule selection phase; it presents a genetic-based multiobjective rule selection algorithm to find a smaller rule subset with higher accuracy than the heuristically extracted rule sets. The algorithm has the objective to maximize the accuracy, and to minimize the number of rules. In [10], multiobjective association rules generation and selection with NSGA-II (Non-Dominated Sorting Genetic Algorithm–II) are discussed. In [11], a multiobjective optimization evolutionary algorithm with Pareto concepts is used to discover interesting classification rules for a target class. It presents an implementation of NSGA-II with positive confidence and sensitivity as its objectives. This work is extended in [9] using multiobjective metaheuristics to produce sets of interesting classification rules. A measure of dissimilarity of rules was introduced to promote diversity on the population.

This work is organized as follows. In Sect. 3.2, we describe many concepts of classification-rule learning. Section 3.3 presents particle swarm optimization and the Multiple Particle Swarm Optimization technique (MOPSO). In Sect. 3.4, we describe all the concepts related to our algorithm that use MOPSO for classification-rule learning. Sect. 3.5 presents the experiments done with our algorithms and the comparisons with other algorithms from the literature. Then, we conclude the work in Sect. 3.6.

3.2 Classification Rule Learning

Classification-rule learning is the process of finding, by means of a set of training examples, a set of rules that can be used for classification and prediction.

A classification-rule is a pair <condition => class>, where condition is the set of conditions that makes an instance to be classified as a value of the class set. A two-class problem classification rule would be something like IF <CONDITION> THEN <YES> ELSE <NO>, where CONDITION can be any boolean logic expression.

The rules are commonly used in data mining because of their simplicity, intuitive aspect, modularity and because they can be obtained directly from a dataset [2]. Therefore, rule induction has been established as a fundamental component of many data mining systems. Furthermore, it was the first machine learning technique to become part of successful commercial applications [12].

In the rule learning process, we associate learning to a training data set. This data set contains examples of predictions the system will do. Based on these examples, we look for patterns that a classifier system can learn. So, the algorithm needs as its input a training data set, with the values of attributes and classes, and returns as its output, a classification rule set for the problem.

Briefly, for example, with the following training examples:

1. [Outlook=Sun,Temperat.=hot,Humidity=high,Wind=no ->Play=NO],
2. [Outlook=Sun,Temperat.=hot,Humidity=high,Wind=yes ->Play=NO],
3. [Outlook=Rain,Temperat.=cold,Humidity=high,Wind=no ->Play=YES],

the algorithm must be able to induce situations like: If Outlook = Sun Then Play = No.

So, any new instance, even unknown, that has outlook = sun, will be classified as No. Evidently, the induction can be imperfect.

In recent years, many techniques have emerged to deal with rule learning. Generally, we can classify them in three main groups: separate-and-conquer, divide-and-conquer and based on association-rules [13]. In the first group, also called cover algorithms, the search procedure generally consists of a greedy algorithm that, at each iteration, finds the best rule and removes the covered examples from the dataset. This process is reiterated with the remaining examples of the dataset [14]. That means, it finds the best rule, separates the covered instances and conquers recursively the remaining instances until there is no instance to conquer.

To form a classifier in this strategy, we create a set of the best rules. This set can be ordered (when the order of the rules to the classification is important) or unordered (when the order is not important to the classification). In an ordered classifier, also called decision list, the classification is given by the first rule that covers the premise. In an unordered classifier, the rules that cover the premise must agree and assign a class to the entry.

A classical algorithm of this group is CN2 [15]. This algorithm induces a decision list, using entropy as its search heuristic. There is also an extended version of the algorithm that induces an unordered classifier and uses Laplace error correction as its evaluation function, as described in [12]. In the second group, divide-and-conquer, a global classifier is constructed, following a top-down approach. The result is generally represented as a decision tree that divides completely the instance space [16]. The great representative of this group is the algorithm C4.5 [17], which is almost a standard in symbolic learning algorithms comparison. It uses the information gain as its measure of quality to construct a decision tree and uses a further step of pruning, based on error reduction. In this approach, each branch of the tree can be considered a rule [16] and there are no rules overlap. The advantages of rule induction compared to tree induction are the intelligibility and little storage space requirements. But, the rules induction process tends to be slower than the tree induction processes [18].

The third group offers an alternative to other two groups. It consists of classifiers based on association rules, or associative classifiers [19]. Generally, these classifiers are obtained through the utilization of exhaustive searches in the possible rules space, collecting all rules that satisfy some conditions. The problem of this approach is generally that the number of generated rules is huge. Some works, such as ROCCER [16] and MORLEA [20], look for reducing this set of rules, by means of rules sub-selection. But, these algorithms have separate-and-conquer characteristics.

3.2.1 Rule Evaluation

Being imperfect the induction of the classifier, its rule evaluation is necessary, because the algorithm can do classification mistakes. So that we can consider a rule (or classifier) good or bad, we use certain evaluation metrics. Considering each rule as L ->R, being L the condition and R the class, we can obtain a contingency matrix of this rule, for a two-class problem, as shown in Table 3.1. For problems having more than two classes, a contingency matrix is obtained for each class. In this case, any reference to the class R is interpreted as a positive class and any reference to another class is considered as negative.

In this table, L is the set of examples which the rule classifies as positive and its complement $\overline{L}$ is the set of examples which the rule classifies as negative. R are the examples that belong to the positive class. $\overline{R}$ are the examples that belong to the negative class. So, TP (true positive) is the set of positive class examples that are classified as positive. In the same manner, FN (false negative) is the set of positive class examples that are classified as negative; FP (false positive), the negative class examples that are classified as positive; finally, TN (true negative) is the set of negative class examples classified as negative. Finishing, r is the number of examples of the set R, r=|R|; in the same way: $\overline{r}$=|$\overline{R}$|, l=|L|, $\overline{l}$=|$\overline{L}$|; n is the total number of examples.

Table 3.1. Contingency Matrix of a Rule L -> R

	L	$\overline{L}$	
R	TP	FN	r
$\overline{R}$	FP	TN	$\overline{r}$
	l	$\overline{l}$	n

Table 3.2. A dataset with three classes

Index	Attr1	Attr2	Attr3	Class
1	A	1	BA	C1
2	B	1	BC	C1
3	B	2	BC	C2
4	A	3	BA	C2
5	C	1	BC	C2
6	A	2	BA	C4
7	C	2	BC	C3
8	A	1	BA	C3

Table 3.3. Contingency Matrix of the example rule

	L	$\overline{L}$	
R	1	2	3
$\overline{R}$	3	2	5
	4	4	8

Considering a problem with more than two classes, we may transform its dataset into a two-class dataset. This is done concerning the class that we are analyzing. Table 3.2 presents a dataset with four classes. If we consider a rule {If Attr1 = A and Attr3 = BA Then C2} then the contingency matrix of the rule is as presented in Table 3.3, because we are considering C2 as positive and non-C2 as negative. TP is 1 because of example 4; FN is 2 because of examples 3 and 5; FP is 3 because of examples 1, 6 and 8; and TN is 2 because of examples 2 and 7.

Based on the contingency matrix, it is possible to describe several classifier evaluation measures. Table 3.4 presents some examples of the most used measures [18]. For our example rule, we calculate a positive confidence of 1/4 and a negative confidence of 2/4. That means the rule hits only 25% of the examples it covers and only half of those that it does not cover.

In the case of class imbalance, where some classes are represented by a large number of examples and the others are represented by a few, it is easy to realize that the performance analysis of the classifier should consider more measures than only accuracy. This is because it is easy to obtain high accuracy, by simply predicting the majority class. For example, it is straightforward to obtain a classifier with an error rate of 1% in a domain where the majority class has 99% of the instances, by simply predicting each new instance as belonging to the majority class [21]. So, specially when the misclassification costs are unknown or are likely to change, the use of the accuracy rate as a basic performance measure may lead to misleading conclusions. Other aspects of performance, such as discrimination capacity and decision tendencies, are confused when we consider accuracy as a basic performance measure [21].

Table 3.4. Rules Evaluation Measures

$\frac{TP}{l}$	positive confidence
$\frac{TP+1}{l+2}$	Laplace corrected positive confidence
$\frac{TN}{\bar{l}}$	negative confidence
$\frac{TN+1}{\bar{l}+2}$	Laplace corrected negative confidence
$\frac{l}{n}$	Coverage
$\frac{TP}{n}$	Support
$\frac{TP}{r}$	Sensitivity
$\frac{TN}{\bar{r}}$	Specificity
$\frac{TP+TN}{n}$	Total Accuracy

To properly discriminate between positive and negative examples, we can use ROC (Receiver Operating Characteristic) analysis, because it provides a tool for analysis and evaluation of classifiers in imprecise environments. The basic idea is to disjoin relative error rate (false positive) from hit rate (true positive), using them as the axes of a bi-dimensional space. So, a classifier's performance is a pair of measures instead of a single error rate. Thus, with a classifier threshold biasing the final class selection, we plot the percentage of TP and FP on the graph. Varying the threshold, we obtain many points on the graph, forming a curve of performance. With this curve, it is possible to know the decision tendencies of the classifier for several scenarios. To compare ROC curves, we calculate area under curve (AUC), which represents the probability that a randomly chosen positive example will be rated higher than a negative one. The greater the AUC value, the greater the classifier probability of low error rate. But, it is important to know that a classifier with a maximum AUC (AUC=1) may not be the classifier with the lowest error rate [21].

3.2.2 Ordered versus Unordered Classification

A rule-learning algorithm can be generally defined in a three-layers model. The first layer is the discovery algorithm. This algorithm searches for the best classification rules learnt from the dataset. It is in this step that we can use optimization algorithms such as particle swarm optimization and genetic algorithms. The second layer is the selection algorithm. It selects the rules found by the discovery algorithm to join the classifier set. It is in this step that it is defined if the classifier is ordered or unordered. For example, if a separate-and-conquer algorithm removes all the covered examples of the dataset each time it finds a rule to the classifier, then there is only sense to use an ordered classifier. On the contrary, if the algorithm removes only the correctly covered examples, i.e., only those of the same class, then it makes no sense to use an ordered classifier. In the same way, when the algorithm does not remove any example from the dataset, like ours, the classification should be unordered.

The third layer is the classification and validation step. To classify a new instance, the ordered classifier tries, in order, each rule of the set until some rule satisfies the condition of the new instance. That means the rules of the classifier do not make sense in isolation, except for the first one. That is because the rules compose just one rule in {IF X THEN Y ELSE IF W THEN Z...} form. For example, consider the following ordered classifier:
If feather = yes then class = bird
Else if legs = two then class = human
Else...
The {if legs=two then class = human} rule, when considered in isolation, is incorrect, because birds also have two legs. The problem of incomprehensibility increases with the number of rules, becoming too difficult to understand the context of a rule. This problem does not happen in an unordered classifier, because the rules can be considered in isolation. So, if the objective of the classifier is the discovery of interesting rules in a classification context, the unordered form is preferred. With this kind of classifier, all rules must be tested, and those which satisfy the condition of the instance to be classified are collected. With these rules, we apply some method of choice of the class, for example:

Random - choose randomly a rule from the collected rules to classify the instance.

First Rule - take the first collected rule that fits the condition of the instance to classify it.

Class Coverage - most usual procedure. The objective is to analyze which class is more covered by the collected rules. For example, consider a rule that covers the classes {yes,no}, respectively, {15,1}, another rule that covers {20,0}, and another rule that covers {1,10}. Notice that the most covered class is {yes}, that would be the class chosen by the classifier. This method has the disadvantage of not considering possible exception rules that cover few examples [12].

Simple Vote - method in which each rule votes in the desired class. The winner class is the chosen class [2].

Weighted Vote - It is equivalent to the simple vote method, but, in this case, each rule votes with the positive confidence measure that it has. For example, consider the classes {yes,no} and two rules. The first predicts the class of the instance as {yes}, the second as {no}. But, the positive confidence of the first rule is 0.75 and the second, 0.64. Thus, the class {yes} wins, because its score is greater [2].

Weighted Vote with Negative Vote - To overcome the problem when a classifier does not classify non-covered examples by any rule, we introduce this method. It is equivalent to the weighted vote, but, in this case, there is also the negative vote, that decreases the class score. The idea of this method is that if a rule has high negative confidence then, if the rule does not cover an example, it is very probable that it does not pertain to the class of the rule. So, we have to decrease the class score towards others. This may work better when there are overlapped rules, because there are many opinions in the vote. So, for example, a rule has 0.75 positive confidence and 0.80 negative confidence. In this case, for each instance in which the rule satisfies the conditions, it votes adding 0.75 to the class score. But, for each instance that the rule does not satisfy the conditions, it votes also, decreasing the class score by 0.80. In this method, all rules interact, not only those which satisfy the conditions.

3.3 Particle Swarm Optimization

Particle swarm optimization (PSO) is a population based stochastic optimization technique developed by Russell Eberhart and James Kennedy in 1995 [22], inspired by the social behavior of bird flocking or fish schooling. PSO shares many similarities with evolutionary computation techniques such as Genetic Algorithms (GAs). The system is initialized with a population of random solutions and searches for optima by updating generations. However, unlike GAs, PSO has no evolution operators such as crossover and mutation. In this model, there is a set of particles (called swarm) that are possible solutions for the problem. These particles move through an n-dimensional space based on their neighbors' best positions and on their own best position. For that, on each generation, the position and velocity of particles are updated, considering the best position already obtained by the particle and the best global position obtained by all particles of the swarm. The best particles are found based on the fitness function, which is the problem's objective function.

Each particle p, at some iteration t, has a position in R^n, $x(t)$, and a displacement velocity in this space, $v(t)$. It has also a small memory that contains its best position already fetched, $p(t)$, and the best position, $g(t)$, already fetched by all the particles that p knows, i.e., the best $p(t)$ of all particles attached to the neighborhood of p, $(N(p))$. It is important to notice that $x(t)$, $v(t)$, $p(t)$ and $g(t)$ are n-dimensional vectors.

The movement of the particles is based on the following equations:

$$\boldsymbol{v}(t+1) = \omega.\boldsymbol{v}(t) + \phi 1(\boldsymbol{p}(t) - \boldsymbol{x}(t)) + \phi 2(\boldsymbol{g}(t) - \boldsymbol{x}(t)),$$

$$\boldsymbol{x}(t+1) = \boldsymbol{x}(t) + \boldsymbol{v}(t+1),$$

where $\phi 1$ and $\phi 2$ are the coefficients that determine the influence of the particle's best $(\boldsymbol{p}(t))$ and the particle's global best $(\boldsymbol{g}(t))$ on the particle's velocity, respectively. The coefficient ω is the inertia of the particle, i.e., how much its previous velocity affects the current velocity.

To make possible to know if one solution is better than another, we have the fitness function. In multiobjective problems, there is more than one optimization function. So, the best particles are found based on the Pareto dominance concept

[23]. Let Π be the set of the problem's possible solutions, one solution $(x) \in \Pi$ for a multiobjective problem is a non-dominated solution if there is no solution $(y) \in \Pi$ that dominates (x). A solution (x) dominates another solution (y) if (for a minimization):

$$\forall \alpha \in \Phi, \alpha(\boldsymbol{x}) \leq \alpha(\boldsymbol{y}),$$

$$\exists \alpha \in \Phi, \alpha(\boldsymbol{x}) < \alpha(\boldsymbol{y}),$$

where Φ represents the set of all the objective functions. That means, one solution dominates another if it doesn't lose in any objective, but wins in at least one. In most cases, there is no single best solution, but a set of non-dominated solutions. These best solutions are in a region of the objective space, known as the Pareto Front, which contains all the non-dominated solutions of the problem.

A particle swarm algorithm for the solution of multiobjective problems was presented in [5]. In MOPSO (Multiple Objective Particle Swarm Optimization), in contrast with PSO, in this case there are many fitness functions. Differently from PSO, in MOPSO there is no global best, but a repository with the non-dominated solutions found.

3.4 MOPSO Rule Learning Algorithm Description

The proposed algorithm is an implementation of MOPSO which aims to overcome some of the problems from separate-and-conquer approaches. The case is that, as said before, the separate-and-conquer algorithms need the continuous update of the dataset, removing the examples already covered by other rules. This may cause a progressive decrease of the rules' quality. This happens because the rules that are learnt from that sub dataset are apparent, because they do not consider all the training examples, and, therefore, their measures of error are not valid in the problem's context. And, even evaluating the particles with the complete dataset, we waive many other good rules.

Besides, with these algorithms, the classifiers are decision lists, which, with no doubt, are good for classification, but do not allow the search of interesting rules, for example, for specialists, because the rules of the classifier can not be read in isolation. For data mining, this is very important. Still, this can be solved using unordered classifiers, but there is still the problem of the loss of quality of the rules.

So, our algorithm tries to overcome these problems in a more direct way, by not removing the examples of the dataset when a rule is selected. So, the subsequent rules that are found are learnt with the entire dataset, maintaining the rule's quality. Because of that, the selected rules are more interesting, with a greater relative accuracy.

Another data mining requisition is the intelligibility of the rules. This algorithm should provide very comprehensible classifiers, because they are unordered and their rules can be read in isolation.

The algorithm can be defined as a function $f(Ex[M, N], P, G, f_1, f_2, ..., f_k)$, being $k \geq 2$ the number of problem's objectives and f_i the i-th objective function, with $i = 1$ to k. For a problem with two objectives, the algorithm would be $f(Ex[M,N], P, G, f1, f2)$. The parameter $Ex[M,N]$ is the set of examples of the training dataset with M examples and N attributes. P represents the number of the particles for each class. G is the number of generations. The function returns a set $C[M1,N]$, which is the classifier, containing $M1$ rules.

With this configuration, the velocity equation parameters are random. But, if wanted, different values to these parameters can be defined, defining the algorithm as $f(Ex[M,N], P, G, \omega, \phi1, \phi2, f_1, f_2, ...f_k)$.

3.4.1 Rule Representation

In our algorithm, we follow the Michigan approach to represent a particle. In this approach, each particle is a rule, not a classifier. Another possibility would be the Pittsburgh approach, where each particle represents an entire classifier, that means, a rule set. The advantage of the Michigan approach is that the individual is simpler and its operators are more intuitive. But, it has the disadvantage that it does not consider interactions between the rules, which is possible with the Pittsburgh approach [24].

So, a rule is represented as the position of the particle in the search space. This space is defined by the dataset. If a dataset has N attributes, the search space is N-dimensional. Furthermore, the size of each dimension is defined as the number of possible values for each attribute. So, the first dimension would have the number of possible values for the first attribute; the second dimension, the second attribute; and so on. To represent the cases in which the attribute does not appear in the rule, there is the possibility of using the '?' value for each attribute, except for the objective attribute.

For a better visualization, an example is presented, based on Table 3.5. We can see that the dataset has fourteen examples and five attributes. 'Play' is the objective attribute of the dataset. So, the particles can move in a five dimension space, considering:

1. Attribute outlook with values: '?'(0), sunny (1), cloudy (2) and rainy (3).
2. Attribute temperature with values: '?' (0), hot (1), cool (2) and cold (3).
3. Attribute humidity with values: '?' (0), high (1) and normal (2).
4. Attribute wind with values: '?' (0), no (1) and yes (2).
5. Attribute play (class) with values: no (0) and yes (1).

The numbers in parentheses are the identifiers of each attribute. So, the rule **If Outlook=Sunny AND Wind=Yes THEN Play=Yes**, would be represented as [**Sunny, ?, ?, Yes, YES**], which would be equivalent to [**1,0,0,2,1**]. With this numerical representation, it is possible to use the same formulas of the velocity and position of Multiple Objective Particle Swarm Optimization.

3.4.2 Particle Initialization

In this step, the particles are created in the swarm and spread in the N-dimensional space. This can be done in many ways. Here, we present three of them.

Table 3.5. Training dataset

index	outlook	temperature	humidity	wind	play
1	Sunny	hot	high	No	No
2	Sunny	hot	high	Yes	No
3	cloudy	hot	high	No	Yes
4	rainy	cool	high	No	Yes
5	rainy	cold	normal	No	Yes
6	rainy	cold	normal	Yes	No
7	cloudy	cold	normal	Yes	Yes
8	Sunny	cool	high	No	No
9	Sunny	cold	normal	No	Yes
10	rainy	cool	normal	No	Yes
11	Sunny	cool	normal	Yes	Yes
12	cloudy	cool	high	Yes	Yes
13	cloudy	hot	normal	No	Yes
14	rainy	cool	high	Yes	No

Random Initialization

This is the simpler type of particle initialization. Basically, it consists of defining random values in the attributes of each particle, and then, spreading them in the search space.

It has many disadvantages, like the formation, depending on the dataset, of rules which are too specific, which makes harder both the learning and evolution processes. Furthermore, it runs the risk of producing rules that do not cover any example or that are too grouped with other rules, prejudicing the space exploration. In this case, it is desirable to use a roulette for choosing an attribute value, giving more priority to the generic attribute '?'.

More Generic Mutation Initialization

This kind of initialization tries to correct some problems of the random initialization. It consists, basically, of initializing all attributes of the particles, except for the class, with the generic '?'. After that, randomly, some attributes of the particles are changed. This permits the generation of more generic particles, controlling the specialization problem, and, besides, reducing the number of rules that do not cover examples of the dataset, because this tends to cover more examples.

The choice of the attributes is done based on a roulette, where the most frequent attributes of the dataset and the generic value have more priority.

Initialization by Covering

The initialization by covering tries to correct the problem of specific rules and the coverage of the examples of the dataset. Firstly, each particle is initialized with an example from the dataset, chosen randomly.

After that, the particles are generalized until they reach certain specified number of attributes. For example, example 3 of the dataset from Table 5.3, [**cloudy,hot,high,No,YES**], could be reduced, defining the number of attributes equal to two, to [**cloudy, ?,?,No,YES**]. In the same way, in example 5 [**rainy,cold,normal,No,YES**], by defining the number of attributes equal to one, it could be reduced to [**?,cold,?,?,YES**].

Thus, we make sure that the rule covers at least one example of the dataset. This helps the improvement of the quality of the rules already in the initial swarm.

3.4.3 Particles' Movement

The movement of the particles is done based on the velocity and position formulas, with a complete neighborhood topology, i.e., each particle is a neighbor of every other particle:

$$\boldsymbol{vi} = \omega.\boldsymbol{vi} + \phi 1(\boldsymbol{pi} - \boldsymbol{xi}) + \phi 2(\boldsymbol{gi} - \boldsymbol{xi}) \tag{3.1}$$

$$\boldsymbol{xi} = \boldsymbol{xi} + \boldsymbol{vi} \tag{3.2}$$

where $\boldsymbol{pi}$ is the best position already reached by particle i; $\boldsymbol{gi}$ is a position of a particle of the repository, chosen as a guide of i. There are many forms to make this choice as demonstrated in [25]; $\boldsymbol{xi}$ is the particle's actual position; $\boldsymbol{vi}$ is the particle's velocity. The coefficients $\phi 1$ and $\phi 2$ determine the influence of the particle's best ($\boldsymbol{pi}$) and the particle's global best ($\boldsymbol{gi}$) on the particle's velocity, respectively. The coefficient ω is the inertia of the particle, i.e., how much its previous velocity affects the current velocity.

Each attribute i of the particle is calculated in isolation from the others, using its numerical value. As an example, let's consider the dataset of Table 5.3 and the particle [**1,0,0,2,YES**]. Let's also consider its initial velocity [**0,0,0,0,0**], its best position reached [**2,0,0,2,YES**] and the guide particle [**2,1,2,2,YES**]. In this case, its velocity would be modified to:

$$\boldsymbol{v} = [0,0,0,0,0] + \phi 1([2,0,0,2,YES] - [1,0,0,2,YES]) +$$

$$+\phi 2([2,1,2,2,YES] - [1,0,0,2,YES])$$

$$\boldsymbol{v} = [0,0,0,0,0] + \phi 1[1,0,0,0,0] + \phi 2[1,1,2,0,0]$$

Considering $\phi 1 = 2$ e $\phi 2 = 2$, we have:

$$\boldsymbol{v} = [2,0,0,0,0] + [2,2,4,0,0] = [4,2,4,0,0]$$

So, the next position will be:

$$\boldsymbol{x} = [1,0,0,2,YES] + [4,2,4,0,0] = [5,2,4,2,YES]$$

Here, there is a problem. The position [**5,2,4,2,YES**] does not exist in the space. We could bypass this situation, by disqualifying this solution from the fitness function, by giving it a zero value. But, this would be an unnecessary computation, which would maintain in the swarm an individual we know is invalid. Another idea would be to limit the search space, bringing any attribute that exceeds the front to one of its boundaries. Thus, the position of the particle would be, for example, [**3,2,2,2,YES**], which is a valid particle. But, this would reduce drastically the diversity of the swarm, because as the optimization advances, several particles would be sent to one of the boundaries of the search space. To deal with this problem, the new value of the attribute that was exceeded could be randomly chosen, or we can think about circular axes in the space.

Circular axes means to say that when a particle exceeds the limit of an axis (attribute), then it is moved back to the beginning of the range, and then it moved from there until reaching the difference in excess. In the case of our example particle, we have for the first attribute: 5 - 3 (maximum value of the attribute) = 2 (exceeded quantity). So, the value of the attribute **outlook** would be equal to 1 (sunny). This is equivalent to the **mod** function, because 5 **mod** 4, the number of values for the attribute, is 1. Thus, the position of the particle would pass from [**5,2,4,2,YES**] to [**1,2,1,2,YES**].

Another way to do this movement is that the particles move to the opposite direction in the axis when the limit is reached. In the case of the first attribute, the particle would reach 3 and, after that, instead of going to four and five, it would go back to the positions two and one. So, in this case, the particle would pass from [**5,2,4,2,YES**] to [**1,2,0,2,YES**].

In the next sections, we consider only the use of circular axes.

3.4.4 Fitness Functions

The fitness functions of the particles are functions that identify the rules quality. They are related generally with their quantity of correct predictions on the dataset. For example: positive confidence (how much the rule predicts correctly the examples that it predicts as positive), negative confidence (how much the rule predicts correctly the examples that it predicts as negative), sensitivity (how much the classifier predicts correctly the positive examples) and specificity (the same with negative examples).

The positive confidence maybe obtains best results when associated with the weighted vote (WV). This is because the chosen rules are more certain when they say yes. So, when using the weighted vote scheme, the rule only votes proportionally to its confidence and if it covers the example. However, many examples may not be covered. With the sensitivity, many false positives may appear, because the rules cover many examples, including negative ones. Generally, we need to compensate the sensitivity with the specificity. However, for a vote system, specially weighted vote, the sensitivity with specificity should generate false positive votes, since the extreme points (1,0) and (0,1) belong to the front. The weighted vote with negative vote (WVNV) generally makes sense when associated with the maximization of negative confidence too, because the rules are more certain when they say no, and they vote in a way that is proportional to this certainty. Joining the negative confidence to the positive confidence, with WVNV, we avoid the lack of classification of the non-covered instances, since the rule votes even when it does not cover the example. Thus, in the case that the discovered front does not cover correctly the examples of the dataset, it is recommendable to use some form of negative vote, so that we do not have non-classified examples.

3.4.5 Global Groups

In the algorithm, there are two global groups or repositories: the group of global bests, which is composed by the non-dominated particles of the swarm, and the group of particles of the classifier, that is formed by the particles that were selected to compose the classifier. For each group, there is a threshold that determines, for each fitness function, the accepted value for the rule to pertain to the group. Generally, the threshold is zero, so the dominance is equivalent to Pareto dominance [23]. Though, this threshold could be changed to try to obtain better results.

A particle pertains to a group if it is non-dominated with respect to a threshold E. For it, a particle $\boldsymbol{x}$ dominates another particle $\boldsymbol{y}$, if:

$$\begin{aligned} &\forall\alpha \in \Phi, (\alpha(\boldsymbol{x})/(1-E)) \leq \alpha(\boldsymbol{y}) \\ &\exists\alpha \in \Phi, (\alpha(\boldsymbol{x})/(1-E)) < \alpha(\boldsymbol{y}) \end{aligned} \tag{3.3}$$

being Φ the set of fitness functions and $0 \leq E < 1$.

The group of global bests influences directly the movement of the particles in the space, since the guide of each particle is chosen from this group. The other group is the classifier which is the return of the algorithm, used also for validation.

So, the same particle may, for example, not be accepted as the global best, but can be accepted as a classification rule, and viceversa, depending on the value of the threshold. If the thresholds of both groups are equal, then the groups will be equal.

With the increase of the threshold, the number of particles of the group increases and can, in the case of the classifier, bring more accurate results.

3.4.6 The Algorithm

The algorithm of rule learning with MOPSO (MOPSO-RL) starts with the creation of the swarms and the initialization of the particles. This is done considering the dataset information. It is interesting to mention that an optimization with 10 particles, for a problem with two classes, will create two swarms with 10 particles each and will have two groups of global bests and two classifier groups. If there were three classes, there would be three swarms with 10 particles each, and so on.

After spreading the particles, it starts the iterative process for a defined number of generations or until another stop criterion is reached. So, at each generation, the fitness of each particle of the swarm is evaluated, based on the defined objectives, using the dataset.

After that, we try to insert each particle in the group of global bests and in the group of classifiers, referring to the class of the particle. If a particle is non-dominated, in terms of the definition of dominance defined in equation 3.3, and no equal particle already exists, then it is included in the group. In this process, if the particle dominates any other particle of the group, the dominated particle must be excluded from the group.

The next step is the particles' movement. In this step, the global best of the particle is chosen, by means of sigma distance vector method that can be seen in [25]. This method guarantees certain diversity. But, to make sure that the particles will not choose always the same leader, we create some sort of roulette with the sigma distances. So, the lower the distance between the sigma vectors of a particle p of the swarm and a non-dominated particle P, the greater the probability of P to be chosen as leader of p.

With this leader, with the best position of the particle and with its actual position, the velocity based on equation 3.1 is calculated. With this velocity, the position of the particle is updated as in the equation 3.2, using circular axes. The $\omega, \phi 1, \phi 2$ parameters, by default, are chosen randomly each time a particle moves. Though, fixed values may be defined.

After the particles movement, a new generation begins with the evaluation of the rules, followed by their inclusion in the global groups and their movement. At the end of generations, the rules in the classifier groups of each class are joined into one single group. This group is returned by the algorithm as the found classifier. After that, the validation of the algorithm takes place, using some statistical method. The pseudocode of the algorithm is as follows:

1. Creation of the swarms and particles initialization.
2. While the stop criterion has not been reached:
   a. evaluate the particles.
   b. for each particle i of the swarm, do:
      i. Try to insert the particle in each global group, based on the thresholds $E1$ and $E2$.
      ii. Choose the global guide of the particle, from the global bests group.
      iii. Particles' movement.
         - $\boldsymbol{vi} = \omega.\boldsymbol{vi} + \phi 1(\boldsymbol{pi} - \boldsymbol{xi}) + \phi 2(\boldsymbol{gi} - \boldsymbol{xi})$
         - $\boldsymbol{xi} = \boldsymbol{xi} + \boldsymbol{vi}$
3. Return the union of the classifier groups.

3.4.7 Stopping Criterion

The simplest stopping criterion for the algorithm, in terms of its implementation difficulty is to define a limited number of generations. So, the algorithm stops when this number is reached. The problem with this criterion is that it keeps the algorithm from improving when it is far away from converging, or, on the contrary, it wastes time in unnecessary optimizations, when the algorithm has already converged.

To avoid this, it is possible to define a criterion based on the convergence of the algorithm. In this case, the algorithm stops if it does not obtain certain quantity of new rules during a subsequent generation period. This quantity can be a percentage of the total quantity of discoverable rules in one generation, that is the number of particles times the number of classes. So, two parameters need to be defined. The first one is the parameter that expresses the percentage of the number of possible rules. As said before, this value relates to the number of particles in the swarm. The second represents the number of subsequent generations in which a number of new rules lower than the first parameter should be obtained, so that the algorithm stops. The disadvantage of this approach is that the execution times in large datasets could be very high and, sometimes, the improvement achieved is not significant.

3.4.8 Result Validation and Classification

Once the optimization of the dataset has finished, we have, in the group of classifiers, the rules that were chosen to classify new instances. In our unordered classifier, we must choose a classification method, such as weighted vote, for example, to make possible the choice of a class in case of conflicting prediction. But, to have a measure of precision of the classifier for possible new entries that are not on the dataset, we have to validate it by means of a statistical method.

The optimization happens with the entire training dataset. Then, the generated classifier is validated with the test dataset. This shows the capacity of the classifier to classify totally new instances that were not present on the training dataset.

The resubstitution method, which consists of testing the classifier in the training set itself, is extremely optimistic and gives a totally apparent statistical measure [18]. To estimate a more real measure, we may use other statistical methods, such as [18]:

Simple Validation - the dataset is divided in a percentage p of examples for the training set and $(1 - p)$ for the test set, being $p > 1/2$.

Random sampling - L hypothesis are inducted, being $L << N$, from each training set with t random examples. After L iterations, the error is calculated as the mean of the hypothesis errors.

Cross validation - it is a midpoint between the simple validation and leave-one-out. It consists in a division of the dataset in r equal partitions mutually exclusive of n/r examples as the size. The test is done in the r folds, using, for each test, the other $(r - 1)$ folds as the training set. The statistical measure is calculated as the arithmetic mean of the r executions.

Stratified Cross Validation - it is equivalent to the cross validation, though, in this case, the proportion of the classes of the original dataset is maintained to all folds.

Leave-one-out - it is a special case of cross validation. It is equivalent to the n-fold cross validation, when the n is the number of examples of the original dataset. So, the classifier is inducted with $(n - 1)$ examples and tested with the remaining example. This process is repeated for all the dataset examples, that means, n times. Therefore, it is a costly method, used generally for small datasets.

Bootstrap - the initial set is of the same size of the original dataset, where the examples are chosen randomly. There can be duplicated examples, so the examples that do not appear in this training set make the test set. The process is repeated a number of times, estimating the statistical values as the mean of the iteration statistical values.

3.4.9 Algorithm Complexity

The training algorithm is an iterative process that consumes the number of generations times the number of particles. So, we can consider, a priori, the number of steps as $P \times G$, where P is the number of particles and G, the number of generations. However, each particle is evaluated with the examples of the training set. So, being N the number of examples of the training set, we can consider $P \times G \times N$ as the number of steps. There is still a comparison of each particle with the particles of the global repositories. In this way, and simplifying, being A the mean number of particles in the repositories, the complexity can be $P \times G \times N + P \times G \times A$. Still, this is done for each attribute of the domain. Being M the number of attributes, thus, the number of steps is $P \times G \times M \times N + P \times G \times M \times A$.

3.4.10 Algorithm Restriction

It is important to notice that the proposed algorithm deals only with categorical attributes. So, the discretization of the continuous attributes of the dataset is needed before the learning process. This simplifies learning, though there is some information lost. So, if there is a continuous attribute such as **temperature**, this attribute must be discretized, for example: **temperature** = [-20 to 20], [20 to 40] and [40 to 60]. This creates three discrete values of **temperature**. The discretization process is generally a pre-processing of the data. It depends strictly on the dataset, and, many times, it needs the aid of a specialist of the domain.

Another disadvantage of the algorithm, a priori, is the quantity of parameters to set. Although we define some default parameters, there are still many PSO parameters to set.

However, as already said, we believe that particle swarm optimization has advantages compared with genetic algorithms, because its goal is that a particle is never equal from one generation to another and that it does not conflict with any other particle. Even with elitism, that is not so easily reached with a GA, precisely because the population tends to evolve into a single species.

3.5 Experiments

To verify the algorithm's performance, we performed some experiments with the datasets of Table 3.6. These datasets were obtained from the UCI repository

Table 3.6. UCI Training datasets

index	dataset	# attr.	# instances	Maj. Class (%)	# possible rules
1	breast	10	683	65.00	1,800,000,000
2	bupa	7	345	57.98	77,760
3	ecoli	8	336	89.58	1,458,000
4	german	21	1.000	70.00	$3.99x10^{13}$
5	glass	10	214	92.07	10,206,000
6	haberman	4	306	73.53	1,000
7	heart	14	270	55.55	192,893,400
8	ionosphere	34	351	64.10	$2.60x10^{32}$
9	new-thyroid	6	215	96.06	26,400
10	pima	9	768	65.10	6,667,920
11	flag	29	194	91.24	386,614,654,382,880,000
12	nursery	9	12960	97.45	230,400
13	kr-vs-kp	37	3196	52.22	400,252,360,791,997,656
14	letter-a	17	20000	96.06	42,443,058,438,000,000
15	satimage	37	6435	90.27	$2.27x10^{36}$
16	vehicle	19	846	76.48	59,019,018,854,400,000

[26]. Their continuous attributes were discretized (because the algorithm only deals with discrete values) using Weka's Discretization function, with default parameters. Besides, the datasets that had more than two classes were reduced to two-class problems, taking the less frequent class as positive, and joining the others as negative. Notice that Table 3.6 shows the number of possible rules for each dataset. With this, it is possible to notice that the experiments were done with very small datasets, just like **haberman**, with one thousand possible rules, and huge datasets also, just like **ionosphere**, with 200 nonillion possible rules.

All particles were initialized with initialization by covering with two attributes in the condition. For the stopping criterion, we used a generation limit. For the algorithm's validation and AUC calculation, we used 10-fold stratified cross-validation. So, the dataset is divided in 10 folds. At each run, a fold is used for testing and the rest is the training set. So, we make ten runs for each class of the dataset, since the AUC values are calculated in isolation. The final result of these values is the arithmetic mean of the ten runs.

In this domain, each attribute of the problem receives a number in the discrete search space. But, this does not mean that a solution with minimum distance from other good solution is a good solution, too. This depends strictly on the used dataset and the problem. Therefore, we have the idea that each dataset has a different parameter configuration to obtain better results. So, the velocity parameters of the MOPSO do not influence directly the classifier, since the best performance stands in rules that were chosen to integrate the classifier. However, these parameters influence directly the space exploration and, consequently, the time to obtain the front.

An experiment was performed with our algorithm, with the first ten datasets from Table 3.6. We used 100 particles in 50 generations, with the parameters $\phi 1$

Table 3.7. Front Areas for dataset classes. Influence of $\phi 1$ and $\phi 2$.

dataset	(0.5,0.5)	(1.5,1.5)	(1,1)	(2,2)	(0..2,0..2)
breast C1	77.76 (3.55)	76.51 (4.17)	75.41 (4.59)	77.99 (3.41)	75.94 (4.44)
breast C2	78.48 (2.51)	79.19 (3.01)	83.98 (3.34)	85.40 (1.84)	80.81 (3.87)
bupa C1	45.23 (0.60)	44.44 (1.05)	44.11 (1.02)	44.07 (0.97)	44.51 (1.13)
bupa C2	55.91 (2.23)	58.08 (1.49)	58.26 (1.52)	55.35 (0.86)	58.31 (1.68)
ecoli C1	31.06 (0.00)	31.05 (0.00)	30.06 (3.78)	31.05 (0.00)	31.05 (0.00)
ecoli C2	75.66 (2.12)	74.33 (2.63)	75.45 (2.72)	71.79 (2.59)	74.33 (2.05)
german C1	39.55 (0.60)	38.84 (1.46)	39.01 (1.17)	39.02 (1.62)	39.15 (0.98)
german C2	69.42 (1.92)	67.74 (2.13)	68.44 (1.90)	66.38 (2.58)	67.29 (2.75)
glass C1	10.40 (0.19)	10.40 (0.21)	10.46 (0.16)	10.46 (0.16)	10.47 (0.19)
glass C2	70.01 (2.35)	65.55 (6.20)	67.41 (3.95)	63.43 (5.34)	64.24 (6.29)
haberman C1	60.19 (0.00)	60.19 (0.01)	60.19 (0.01)	59.94 (0.95)	60.19 (0.00)
haberman C2	43.56 (0.00)	43.56 (0.00)	43.56 (0.02)	43.56 (0.00)	43.56 (0.00)
heart C1	72.49 (0.93)	72.39 (1.17)	73.87 (0.48)	73.65 (0.58)	73.80 (0.50)
heart C2	68.43 (0.39)	67.82 (0.68)	68.49 (0.67)	68.52 (0.40)	68.50 (0.57)
ionosphere C1	69.77 (1.82)	69.58 (1.74)	69.92 (1.91)	69.68 (2.10)	68.57 (1.95)
ionosphere C2	77.52 (12.44)	69.43 (13.95)	66.29 (16.72)	67.91 (14.96)	74.41 (12.94)
new-thyroid C1	29.69 (0.00)	28.82 (0.89)	29.63 (0.32)	29.45 (0.00)	29.69 (0.00)
new-thyroid C2	89.95 (0.00)	89.95 (0.00)	89.69 (1.41)	89.95 (0.00)	89.95 (0.00)
pima C1	66.08 (2.53)	64.75 (3.37)	64.27 (3.68)	64.54 (3.44)	65.78 (2.78)
pima C2	43.05 (0.88)	42.47 (0.90)	42.95 (1.06)	42.89 (0.90)	42.58 (1.06)
Average	58.71 (20.58)	57.76 (20.12)	58.07 (20.43)	57.75 (20.38)	58.16 (20.30)

and $\phi 2$ varying in $\{0.5, 1, 1.5, 2\}$, with $\phi 1 = \phi 2$; or $\phi 1$ and $\phi 2 = random[0..2]$. The inertia $\omega = 1$. In the random [0..2] configuration for the $\phi 1$ and $\phi 2$ parameters, the values are chosen separately for each particle to each new position update. The optimizations with all datasets were performed 30 times, taking the mean. The objectives of the algorithm were the positive confidence and the negative confidence (with Laplace err correction).

Table 3.7 presents the zero threshold E front area of the best rules obtained by the algorithm for each class of the datasets. This performance measure was calculated with the metric S [27]. This metric calculates the performance of the front based on the area formed above the curve of the non-dominated points. It is a non-cardinal metric, because it is not directly influenced by the number of non-dominated solutions.

Analyzing Table 3.7, it is possible to realize that there was no statistical difference between the parameters in the obtainment of the front with the parameters of 100 particles in 50 generation. The same can be said related to the number of rules obtained in the front, presented in Table 3.8. Probably, there is a difference between the time each parameter configuration obtains the area.

Our experiment with the inertia parameter ω showed, also, that there was no statistical difference between the values of the front area considering $\omega = \{0.5, 1, 1.5, 2, [0..1]\}$, 30 runs, using 100 particles in 50 generations, $\phi 1$ and $\phi 2 =$ random $[0..2]$, positive and negative confidence.

Table 3.8. Number of rules in the front. Influence of $\phi 1$ and $\phi 2$.

dataset	(0.5,0.5)	(1.5,1.5)	(1,1)	(2,2)	(0..2,0..2)
Breast C1	1.83 (0.53)	1.90 (0.61)	2.03 (0.76)	1.90 (0.61)	1.97 (0.67)
breast C2	4.40 (0.67)	5.33 (0.88)	6.80 (0.89)	6.50 (0.90)	5.20 (1.03)
bupa C1	5.43 (0.68)	5.93 (1.93)	6.77 (2.31)	5.73 (1.60)	5.60 (1.77)
bupa C2	8.03 (2.98)	7.20 (1.49)	6.83 (1.39)	8.67 (2.71)	7.40 (1.45)
ecoli C1	10.87 (0.90)	11.47 (0.82)	11.57 (0.77)	11.90 (0.76)	10.80 (1.03)
ecoli C2	8.37 (2.57)	6.37 (2.74)	7.50 (3.01)	8.50 (4.48)	6.67 (2.73)
german C1	6.63 (1.69)	7.33 (2.12)	7.53 (2.06)	5.73 (1.66)	6.77 (1.72)
german C2	14.67 (3.29)	13.57 (2.80)	15.17 (3.67)	13.63 (3.38)	13.40 (2.67)
glass C1	1.63 (0.49)	1.63 (0.49)	1.77 (0.43)	1.77 (0.43)	1.83 (0.38)
glass C2	8.50 (6.55)	8.10 (2.80)	9.90 (3.53)	8.07 (2.07)	9.20 (3.21)
haberman C1	5.97 (0.18)	6.03 (0.18)	6.00 (0.26)	5.70 (1.02)	6.00 (0.00)
haberman C2	2.00 (0.00)	2.00 (0.00)	2.03 (0.18)	2.00 (0.00)	2.00 (0.00)
heart C1	9.97 (1.99)	10.63 (2.24)	15.03 (1.85)	14.10 (1.75)	14.23 (1.91)
heart C2	14.30 (1.64)	11.40 (1.81)	12.23 (1.72)	13.00 (1.62)	12.77 (1.45)
ionosphere C1	2.93 (0.69)	3.10 (1.03)	2.47 (1.04)	2.80 (0.85)	2.87 (1.17)
ionosphere C2	5.63 (1.61)	5.23 (1.25)	5.27 (1.17)	5.30 (1.53)	4.97 (1.25)
new-thyroid C1	2.80 (0.41)	2.97 (0.18)	2.93 (0.25)	3.00 (0.00)	3.00 (0.00)
new-thyroid C2	4.87 (0.43)	4.97 (0.18)	4.90 (0.31)	4.47 (0.57)	4.93 (0.25)
pima C1	12.97 (2.99)	11.00 (2.59)	12.50 (2.99)	12.10 (2.60)	11.43 (2.10)
pima C2	6.17 (0.83)	6.43 (0.86)	6.47 (0.86)	6.20 (1.00)	5.83 (1.12)
average	6.90 (4.03)	6.63 (3.51)	7.29 (4.23)	7.05 (4.06)	6.85 (3.92)

Other parameters that possibly influence the measure of the front area are the number of particles and the number of optimizations. Table 3.9 shows the results of the front area obtained for 10 particles in 10, 50 and 100 generations, and with 50 and 100 particles in 10 generations. The results show the behavior of the algorithm with the parameters $\phi 1$ and $\phi 2 = \{0.5, 1, 1.5, 2, random[0..2]\}$ for the ten first datasets of Table 3.6. The algorithm's objectives were the positive confidence and the negative confidence (Laplace). The experiments were repeated 30 times and the ω parameter was fixed to 1.

Analyzing the table, it is possible to notice that the results of the front were very close to each other, concerning the different parameter configurations. However, it is interesting to notice the relation of the front with the increase in the number of particles and generations. We notice that a larger number of particles generated a larger front area (and a larger number of rules in the front) using the same number of generations. For instance, using 50 particles in 10 generations brought better front area than using 10 particles in 100 generations. Besides, the computational time in the first case was much lower than the second. Comparatively, and based on empirical evidence, although the complexity had been equal in both cases, the time of 100 particles in 10 generations was lower than the time of 10 particles in 100 generations.

So, increasing the number of particles is better than increasing the number of generations. This is valid until the obtainment of the Pareto Front. From

Table 3.9. Front Area (metric S)

	(0.5,0.5)	(1.5,1.5)	(1,1)	(2,2)	(0..2,0..2)	average
10 p. 10 g.	48.12 (17.70)	47.71 (18.62)	48.03 (17.58)	47.56 (18.00)	48.51 (18.25)	47.99
10 p. 50 g.	50.75 (18.96)	51.24 (18.88)	51.76 (19.23)	51.48 (18.65)	51.90 (18.89)	51.43
50 p. 10 g.	55.50 (19.34)	54.23 (19.61)	54.93 (19.77)	54.10 (19.25)	54.81 (19.75)	54.71
10 p. 100 g.	53.27 (19.24)	52.31 (18.92)	52.74 (19.37)	52.11 (18.78)	53.14 (18.83)	52.71
100 p. 10 g.	57.18 (19.95)	56.52 (20.38)	56.42 (20.31)	55.97 (20.03)	56.79 (19.83)	56.58
Average	52.96	52.4	52.78	52.24	53.03	

thereafter, no improvement is possible. The more the algorithm approaches the Pareto Front, the lesser the difference between increasing the number of particles or generations.

To compare our technique with other rule learning algorithms from the literature, we performed an experiment that consisted of analyzing AUC and the classifier size (number of rules). To do it, we used the weighted vote with negative vote method for the classification. With this method, we obtained a numerical rank for each test instance, concerning the class we were analyzing. This rank can be used with a threshold to produce a binary classifier. If the instance rank goes beyond the threshold, the classifier produces a yes, otherwise, it produces a no. Each threshold value generates a different point in the ROC plane. So, varying the threshold from $-\infty$ to $+\infty$ produces a curve on the ROC plane and we are able to calculate the area under curve (AUC) [2].

With this measure, we compared the performance of other algorithms obtained from the work of Prati and Flash [16]. For that sake, we used the same datasets of their work that were sent by the authors to us. The comparisons between the algorithms were done at a 95% confidence level. We used 450 particles during 100 generations for the optimization. The executions were repeated 100 times, taking the mean and standard deviations. But, we performed the test with kr-vs-kp, letter-a and satimage datasets running 30 times, using 450 particles during 50 generations, because of the huge computational time involved. We verified the normality of the AUC results with the shapiro.test of the statistical R software [28].

The velocity parameters $\omega, \phi 1$ and $\phi 2$ were taken randomly at each moment that the particle moved itself. ω was taken from 0 to 1, $\phi 1$ and $\phi 2$ from 0 to 2. The E threshold of the global groups was defined as zero, which is equivalent to the Pareto dominance definition. We used two objectives to define the problem: Laplace err correction positive confidence and Laplace err correction negative confidence. This means that we selected in our classifier the best rules related to these measures. The algorithms used in the comparison were:

- Roccer - it is described in [16]. It uses the ROC curve to select rules from an Apriori generated set. With this, it makes sure that only rules that will increase the true positive rate or that will decrease the false positive rate will be inserted.
- CN2 - It is a typical separate-and-conquer algorithm. It is described in [15]. It induces a decision list using entropy as its heuristic function. In [12], it was modified to use the Laplace err correction as its heuristic.
- CN2-Unordered - unordered version of CN2, described in [12].
- C4.5 - it is a typical divide-and-conquer algorithm, proposed in [17]. It uses the information gain to produce a decision tree and a pruning step based in err reduction.

Table 3.10 shows the comparison of AUC values among the algorithms. Values in **boldface** indicate situations in which the proposed technique overcomes (with certain statistical significance) another algorithm. Values underlined indicate situations in which another algorithm overcomes (with certain statistical confidence) ours. Analyzing the table, it is possible to notice that the proposed algorithm is competitive with respect to all the others presented. For an unordered classifier, the algorithm presented very good results in the classification. However, it presents worst results with the largest datasets: nursery, Kr-vs-kp, letter-a, satimage and vehicle, which have larger numbers of examples. That is due possibly to the lack of coverage of the front rules in these datasets, because the discovered front is not sufficient to cover many of the test examples. If another stop criterion related to the coverage of the front was defined, maybe the results would be better. Besides, a larger number of particles or generations could, also, improve the results. However, we believe that we can obtain better results by increasing the classifier threshold E, which increases the number of selected rules. This would make the classifier more accurate, but much larger, harming, in some cases, the classifier intelligibility.

Table 3.11 presents the comparison of the MOPSO-RL with others, related to the number of rules, or size, of the classifier. The MOPSO-RL, with the objectives of positive confidence and negative confidence (with Laplace err correction) generated good size classifiers, lesser than the values obtained by other algorithms. Still, the lowest mean is the **haberman** dataset with 7.60, which denotes that the algorithm does not generate very small classifiers, having just one rule, for example, which, depending on the goal of the classification, could be undesirable. Even for an unordered classifier, the results of the proposed technique related to the size of the classifier were very good. Compared to the other unordered classifier, CN2 NO, we notice that the size of MOPSO-RL with those two objectives was much better. Analyzing Table 3.11 together with Table 3.10, we notice that with the datasets that the algorithm presented lower AUC related to others, the size of the classifier was generally much lower, justifying the results.

Table 3.10. Experimental Results - comparison of AUC

dataset	MOPSO-RL	Roccer	C4.5	CN2	CN2 NO
breast	98.56 (1.07)	98.63 (1.88)	97.76 (1.51)	99.13 (0.12)	99.26 (0.81)
bupa	71.09 (7.16)	65.30 (7.93)	**62.14 (9.91)**	**62.21 (8.11)**	**62.74 (8.85)**
ecoli	83.60 (9.69)	90.31 (11.56)	**50.00 (0.00)**	85.15 (11.38)	90.17 (6.90)
german	74.59 (5.56)	72.08 (6.02)	71.63 (5.89)	70.90 (4.70)	75.25 (5.38)
glass	75.11 (13.67)	79.45 (12.98)	**50.00 (0.00)**	79.64 (13.24)	73.74 (15.40)
haberman	66.28 (6.51)	66.41 (11.54)	**55.84 (6.14)**	59.28 (10.13)	59.83 (9.87)
heart	86.21 (6.76)	85.78 (8.43)	84.81 (6.57)	82.25 (6.59)	83.61 (6.89)
ionosphere	90.38 (6.67)	94.18 (4.49)	86.09 (9.97)	92.18 (7.54)	96.23 (2.97)
new-thyroid	90.14 (9.78)	98.40 (1.70)	87.85 (10.43)	98.43 (2.58)	99.14 (1.19)
pima	70.26 (5.24)	70.68 (5.09)	72.07 (4.42)	71.97 (5.44)	70.96 (4.62)
flag	71.65 (21.41)	61.83 (24.14)	**50.00 (0.00)**	**42.78 (24.43)**	53.22 (24.12)
nursery	91.57 (2.04)	97.85 (0.44)	99.42 (0.14)	99.99 (0.01)	100.00 (0.00)
kr-vs-kp	95.61 (2.01)	99.35 (0.36)	99.85 (0.20)	99.91 (0.17)	99.85 (0.16)
letter-a	88.78 (2.93)	96.08 (0.52)	95.49 (1.96)	99.44 (0.63)	99.34 (0.28)
satimage	84.52 (3.40)	89.39 (2.38)	90.15 (1.70)	91.48 (0.90)	91.48 (1.45)
vehicle	89.16 (4.73)	96.42 (1.47)	94.76 (3.00)	96.49 (2.41)	97.38 (2.05)
Average	82.97	85.13	77.98	84.51	83.20

Table 3.11. Experimental Results - number of rules in the classifier

dataset	MOPSO-RL	Roccer	C4.5	CN2	CN2 NO
breast	9.43 (0.95)	**48.40 (2.32)**	**37.80 (12.62)**	**24.30 (2.71)**	**32.80 (2.74)**
bupa	18.48 (1.73)	3.90 (0.99)	15.00 (10.53)	**83.60 (5.87)**	**100.70 (3.77)**
ecoli	24.50 (4.80)	6.70 (0.82)	0.00 (0.00)	**33.60 (3.13)**	**35.30 (2.83)**
german	44.30 (4.64)	23.70 (6.75)	**78.20 (18.50)**	**106.80 (4.37)**	**143.90 (6.98)**
glass	27.72 (12.89)	2.40 (0.52)	0.00 (0.00)	21.90 (3.18)	21.60 (1.26)
haberman	7.60 (2.01)	0.80 (0.42)	5.60 (5.72)	**57.00 (5.98)**	**75.90 (3.90)**
heart	32.23 (3.22)	**68.20 (4.42)**	13.20 (4.49)	**36.10 (1.79)**	**42.80 (2.49)**
ionosphere	12.15 (2.47)	**67.10 (5.38)**	**23.40 (3.98)**	**22.90 (1.29)**	**36.90 (2.28)**
new-thyroid	8.00 (0.00)	**8.50 (0.53)**	4.00 (0.00)	**15.30 (0.82)**	**18.70 (0.67)**
pima	23.68 (5.74)	4.00 (0.82)	**49.40 (20.27)**	**168.40 (7.90)**	**169.70 (10.08)**
flag	47.17 (18.22)	1.70 (2.26)	0.00 (0.00)	13.40 (2.12)	20.20 (1.62)
nursery	13.99 (1.47)	**18.90 (1.66)**	**114.60 (3.10)**	**17.60 (0.52)**	**112.40 (2.27)**
kr-vs-kp	60.25 (24.88)	43.60 (7.97)	29.20 (2.15)	28.10 (1.73)	30.10 (1.85)
letter-a	12.15 (1.93)	**78.40 (3.06)**	**183.80 (15.22)**	**120.20 (2.70)**	**126.30 (3.56)**
satimage	18.15 (2.70)	**143.90 (51.28)**	**531.10 (84.63)**	**199.80 (6.99)**	**158.80 (6.09)**
vehicle	10.81 (1.72)	**48.40 (4.48)**	**89.50 (12.12)**	**41.20 (2.82)**	**49.80 (3.29)**
Average	23.16	35.54	73.43	73.49	61.89

So, to demonstrate that our algorithm can obtain AUC results for the datasets kr-vs-kp, letter-a, nursery, vehicle and satimage comparable to the others, we present the AUC results with a classifier threshold $E = 0.01$. The algorithm presented the mean of 97.45 (0.91) for kr-vs-kp, 99.14 (0.37) for nursery, 96.24 (0.99) for letter-a, 90.26 (1.99) for satimage and 92.96 (2.96) for vehicle. This increased the total mean of AUC to 84.62, reaching the others. However, the number of rules in the classifier considerably increases as well: 667.67 (76.79) for kr-vs-kp, 1758.03 (246.98) for nursery, 516.56 (159.28) for letter-a, 160.00 (20.11) for satimage and 93.46 (9.75) for vehicle. This raised the total mean of number of rules to 215.69, basically because of the mean of nursery.

We now must show that our algorithm generates a good classification set that has, at the same time, important rules. The Table 3.12 shows the total support mean and weighted relative accuracy (WRAcc) of all obtained rules from all executions of the algorithm with classifier threshold $E = 0$. The weighted relative accuracy measures the importance of the rule as the difference between the number of expected true positives and the observed ones, varying from 0 to 0.25 [16]. The support varies from 0 to 1 and measures the relative coverage of each rule.

The Table 3.12 shows that MOPSO-RL, with the analyzed approach, presented the greatest support and weighted relative accuracy of all the analyzed algorithms. Possibly, the support would be better with the utilization of objectives that preserve the rules coverage. In this way, our algorithm has the great advantage of generating unordered classifiers with very important and significant rules with a good classification performance, generating, still, small sets of rules. So the discovered classifiers are interesting and easily comprehensible.

Another aspect to analyze is the computational time of the optimizations. The MOPSO-RL execution time is influenced, in a simplified manner, by the dataset size in number of attributes and instances (just as all the other algorithms) and also by the number of particles and generations. With this, the dataset that needed more time to compute (on an Intel Celeron 1.3 GHz, 512 RAM), with 450 particles in 100 generations, was the **nursery** dataset with a mean of 570 seconds per fold, followed by **german** with 100 seconds, which is a good value. For comparison purposes, Roccer, with the **german** dataset, takes about 10 minutes per fold [14]. Concerning the optimizations with 450 particles in 50 generations, the most lingering dataset was letter-a with a mean of 800 seconds

Table 3.12. Experimental Results - Support and WRAcc

Algorithm	Support (%)	WRAcc
MOPSO-RL	20.80 (8.75)	0.0487 (0.038)
Roccer	13.67 (13.89)	0.0355 (0.018)
C4.5	3.73 (6.01)	0.0094 (0.013)
CN2NO	3.90 (2.52)	0.0110 (0.009)
CN2	3.10 (2.18)	0.0085 (0.007)

per fold, followed by satimage with 550 seconds. These values were almost the same with the classifier threshold $E = 0.01$.

3.6 Conclusion

In this chapter, we have seen many steps and aspects of the rule learning process, using a multiple objective particle swarm technique. It is possible to notice that the proposed approach is very simple and easy to implement. In the experiments, we noticed that this method generates good size unordered classifiers with high support and WRAcc, which is a good advantage over other algorithms, because its classifiers are much more intelligible. Besides, the experiment shows that this algorithm is competitive related to AUC values and can be configured to be more precise if wanted, increasing the threshold E.

A natural extension of this work would be the description of a better stopping criterion, which may avoid the low coverage in the datasets with a high number of examples. Another extension could be work with another type of attributes: non-categorical. Still, it is interesting to undertake a more in-depth study of the behavior of the particle swarm parameters and the utilization of other objectives in the optimization, such as novelty and other measures.

Another aspect is the study of parallelization methods, so that we can reduce the computational time and define a high number of particles and generations to approach the true Pareto front.

3.7 Summary

In this chapter, we proposed an algorithm that learns an unordered classifier, selecting rules through a Multiple Objective Particle Swarm Optimization technique. The rules are selected in the classifier without removing examples from the training dataset, so that the quality of the rules is maintained. We introduced rule learning concepts and showed the difference between ordered and unordered classification. Ordered classifiers generally are precise, but they are difficult to understand, because the rules must follow an order and cannot be considered in isolation. This prejudices data mining. Unordered classifiers need a method of classification. We described many methods and proposed the Weighted Vote with Negative Vote as the best choice with positive and negative confidence as objectives.

We proposed our algorithm based on MOPSO, describing the particle representation as a numeric vector, where the values of the attributes have an index number, following the Michigan approach. We described particle initialization by coverage, generic mutation and random. After particle initialization, we showed the particle's movement and how circular axes work, when a particle goes out of the space. The particle's movement uses the velocity and position formulas of MOPSO. We defined the global groups of the swarm which permits the separation of the global repository, where the leaders of the particles are, and classifier group, where the rules, selected to the classifier, are. With a classifier threshold E, it is possible to select more rules to the classifier, not modifying the optimization process.

The experiments described showed that a greater number of particles is better than a greater number of generations. Other experiment showed that MOPSO-RL is competitive with respect to other algorithms from the literature. The datasets in which the algorithm had a low AUC with respect to the other approaches were those that had a very small number of rules in the classifier. So, we performed the experiments with these datasets, increasing the classifier threshold $E = 0.01$. This brought us a good AUC value toward other algorithms, but the number of rules in the classifier overgrew. Finally, we analyzed the importance of the rules in the classifier, by means of support and weighted relative accuracy. The values of these two metrics were very good with respect to the values from the other algorithms, showing that our algorithm produces small unordered classifiers with good significance, which is very important in data mining.

References

[1] Russel, S., Norvig, P.: Artificial Intelligence: A modern approach, 946 p. Prentice-Hall, New Jersey (2003)

[2] Fawcett, T.: Using Rule Sets to Maximize ROC Performance. In: IEEE International Conference on Data Mining (ICDM 2001), pp. 131–138 (2001)

[3] Provost, F., Fawcett, T.: The case against accuracy estimation for comparing induction algorithms. In: Proceedings of the 15th International Conference on Machine Learning, pp. 445–453. Morgan Kaufmann, San Francisco (1998)

[4] Westin, L.K.: Receiver operating characteristics (ROC) analysis: Evaluating discriminance effects among decision support systems. Umea University, Umea (2001)

[5] Coello, C.A., Lechuga, M.S.: MOPSO: A proposal for Multiple Objective Particle Swarm Optimization. In: IEEE World Congress on Computational Intelligence, 2002. Proceedings of the 2002 Congress on Evolutionary Computation, pp. 1051–1056. IEEE Press, Hawaii (2002)

[6] Sousa, T.F., Silva, A.P., Silva, A.F.: Particle Swarm Based Data Mining Algorithms for Classification Tasks. Parallel Computing 30, 767–783 (2004)

[7] Jin, Y. (ed.): Multi-Objective Machine Learning. Springer, Berlin (2006)

[8] Ishibuchi, H., Nojima, Y.: Accuracy-Complexity Tradeoff Analysis by Multiobjective Rule Selection. In: Proc. of ICDM 2005 Workshop on Computational Intelligence in Data Mining, pp. 39–48 (2005)

[9] de la Iglesia, B., Reynolds, A., Rayward-Smith, V.J.: Developments on a Multi-Objective Metaheuristic (MOMH) Algorithm for Finding Interesting Sets of Classification Rules. In: Coello Coello, C.A., Hernández Aguirre, A., Zitzler, E. (eds.) EMO 2005. LNCS, vol. 3410, pp. 826–840. Springer, Heidelberg (2005)

[10] Ishibuchi, H., Kuwajima, I., Nojima, Y.: Multiobjective association rule mining. In: Proc. of PPSN Workshop on Multiobjective Problem Solving from Nature, Reykjavik, Iceland, September 9, 2006, 12 pages (2006)

[11] de la Iglesia, B., Philpott, M.S., Bagnall, A.J., Rayward-Smith, V.J.: Data Mining Rules using Multi-Objective Evolutionary Algorithms. In: Proc. of 2003 Congress on Evolutionary Computation, pp. 1552–1559 (2003)

[12] Clark, P., Boswell, R.: Rule Induction with CN2: Some Recent Improvements. In: Machine Learning - Proceedings of the European Conference, pp. 151–163. Springer, Berlin (1991)
[13] Pham, T.H., Clemente, J.C., Satou, K., Ho, T.B.: Computational Discovery of transcriptional regulatory rules. Bioinformatics 21(1), 101–107 (2005)
[14] Prati, C.R., Flash, P.: Roccer: A ROC convex hull rule learning algorithm. In: ECML/PKDD 2004 Workshop on Advances in Inductive Rule Learning, Pisa (Italy), pp. 144–153 (2004)
[15] Niblett, T., Clark, P.: The CN2 induction algorithm. Machine Learning 3(4), 261–283 (1989)
[16] Prati, C.R., Flash, P.A.: ROCCER: An Algorithm for Rule Learning Based on ROC Analysis. In: Proceedings of the 19th International Joint Conference on Artificial Intelligence (IJCAI 2005), August 2005, pp. 823–828 (2005)
[17] Quinlan, J.R.: C4.5 programs for Machine Learning. Morgan Kaufmann, San Francisco (1993)
[18] Monard, M.C., Baranauskas, J.A.: Inducão de Regras e Árvores de Decisão. In: Rezende, S. (ed.) Sistemas Inteligentes: Fundamentos e Aplicacões, 525 p. Editora Manole, Barueri (2003) (in Portuguese)
[19] Liu, B., Hsu, W., Ma, Y.: Integrating classification and association rule mining. In: Proceedings of the 4th Int. Conf. on Knowledge Discovery and Data Mining, New York, pp. 80–86 (1998)
[20] Batista, G.E.A., Prati, R.C., Monard, M.C., Giusti, R., Milaré, C.R.: Classificacão Associativa Utilizando Selecão e Construcão de Regras: um Estudo Comparativo. In: Encontro Nacional de Inteligência Artificial (ENIA). Rio de Janeiro. Anais do Congresso da Sociedade Brasileira de Computacão, pp. 1321–1330 (2007) (in Portuguese)
[21] Prati, R.C., Batista, G.E.A.P.A., Monard, M.C.: A Study with Class Imbalance and Random Sampling for a Decision Tree Learning System. In: Bramer, M. (ed.) Artificial Intelligence in Theory and Practice II, IFIP 20th World Computer Congress, TC 12: IFIP AI 2008, Stream, Milano, Italy, September 7-10, 2008, pp. 131–140 (2008)
[22] Kennedy, J., Eberhart, R.C.: Particle Swarm Optimization. In: Proceedings of the 1995 IEEE International Conference on Neural Networks, pp. 1492–1948. IEEE Press, Los Alamitos (1995)
[23] Pareto, V.: Manuel D'Économie Politique. Marcel Giard, Paris (1927)
[24] Freitas, A.A.: Data Mining and Knowledge Discovery with Evolutionary Algorithms, 264 p. Springer, Berlin (2002)
[25] Fieldsend, J.E.: Multi-Objective particle Swarm optimization methods (March 2004)
[26] Asuncion, A., Newman, D.J.: UCI Machine Learning Repository. Irvine, University of California (2007), http://www.ics.uci.edu/mlearn/MLRepository.html
[27] Zitzler, E.: Evolutionary algorithms for multiobjective optimization: methods and applications. PhD Thesis Swiss Federal Institute of Technology. Zürich, Switzerland
[28] The R Project for Statistical Computing (Accessed: February 2008), http://www.r-project.org

4

Using Multi-Objective Particle Swarm Optimization for Designing Novel Classifiers

Seyed-Hamid Zahiri[1] and Seyed-Alireza Seyedin[2]

[1] Department of Electrical Engineering
Birjand University, Birjand, Iran
hzahiri@birjand.ac.ir
[2] Department of Electrical Engineering
Ferdowsi University of Mashhad, Mashhad, Iran
seyedin@um.ac.ir

Summary. The effectiveness and powerfulness of the multi-objective swarm intelligence algorithms are motivations for utilizing them to estimate the optimum decision functions/decision rules in such a way that various performance aspects of classifiers (e.g., score of recognition and reliability) are *simultaneously* optimized.

This chapter explains the applications of multi-objective swarm intelligence techniques (especially particle swarm optimization) on designing novel classifiers (named direct applications) and optimizing the performance aspects of conventional classifiers (named indirect applications). Also, a review of some of the past and ongoing related research is presented.

4.1 Basic Concepts

In this section, the basic concepts of *data classification* are described and then the *conventional classification techniques* are explained.

4.1.1 Data Classification

Data classification and prediction algorithms are means, in general computational for *reducing* the amount of information. The input is an information-rich sequence of data, and the output may be a single number or, in the simplest case, a *yes* or *no*, representing a choice between two categories [1].

One of the most useful ways to represent pattern classifiers is in terms of a set of *discriminant functions* $g_i(x)$, $i = 1, \ldots, M$, where M is the number of classes. The classifier is said to assign a feature vector x to class C_i if $g_i(x) > g_j(x)$ for all $j \neq i$ [16].

If the feature space (with four dimentions or more) can be partitioned linearly, the discriminant functions are called *hyperplanes*. Otherwise, the decision functions are *hypercurves*.

A general hyperplane is in the form:

$$d(x) = W.X^T = w_1x_1 + w_2x_2 + \ldots + w_Nx_N + w_{N+1} \tag{4.1}$$

C.A. Coello Coello et al. (Eds.): Swarm Intel. for Multi-objective Prob., SCI 242, pp. 65–92.
springerlink.com

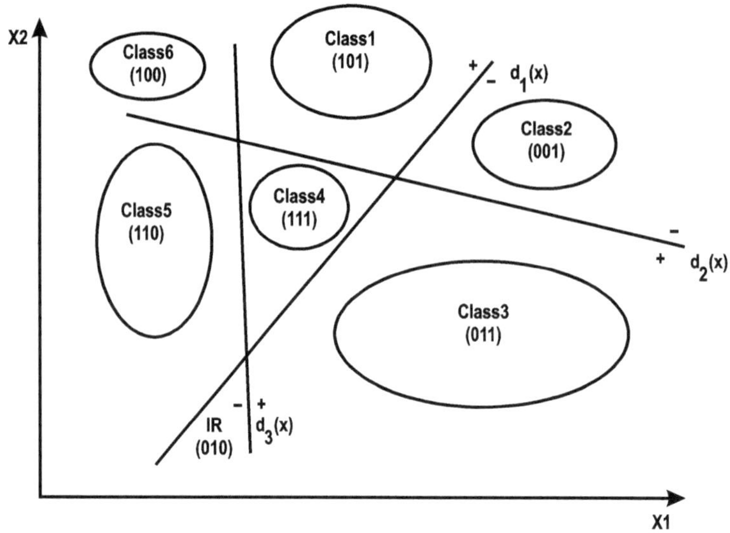

Fig. 4.1. Feature space X is partitioned into seven disjoint regions by the given decision lines ($d_1(x), d_2(x)$, and $d_3(x)$).

where $X = [x_1, x_2, \ldots, x_N, 1]$ and $W = [w_1, w_2, \ldots, w_N, w_{N+1}]$ are called the *augmented feature vector* and the *weight vector*, respectively. N is the feature space dimension.

In a general case, there are a number of hyperplanes that separate the feature space into different regions. Each region distinguishes an individual class as shown in Fig. 4.1. In Fig. 4.1, each class belongs to a region, which is encoded by the sign of three hyperplanes. In this figure, IR denotes the indeterminate region. Note that since each hyperplane provides two regions, H hyperplanes provide a maximum of 2^H regions. Hence for a M class problem, H needs to be greater than or equal to $\log_2 M$.

There are two main divisions of classifications: *supervised classification* (or discrimination) and *unsupervised classification* (or clustering). In supervised classification, we have a set of data samples specified by their types using corresponding labels. These data samples are used as exemplars in the classifier design. In unsupervised classification, the data are not labeled and we seek to find groups in the data and the features that distinguish one group from another.

Designing a classifier consists of three major phases of training, testing, and performance evaluation. In the training phase, the labeled feature vectors are used to estimate the decision functions/decision rules and in the testing stage, the unknown input feature vectors are used to evaluate the performance of the classifier. The performance of a classifier can be estimated by some aspects. The most important performance aspects are *generalization, overfitting, scalability, interpretability*, and *reliability* (or *robustness*). *Generalization* indicates how well the classifier classifies unseen data and is expressed by the recognition score (or error rate). Overfitting is related to the number of decision hyperplanes

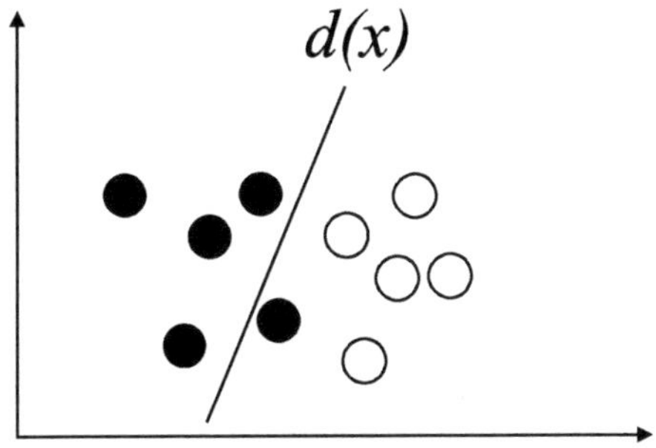

Fig. 4.2. An example in which the recognition score of "white circles" is 100%, but its reliability is not 100% and the recognition score of "black circles" is not 100% but its reliability is 100%.

(or decision rules) of a classifier. *Overfitting* occurs when an overly complex model is obtained by a large number of decision hyperplanes or decision rules and such model allows perfect classification of the training samples, but it is unlikely to provide a good classification of unknown patterns. *Scalability* refers to the ability to construct the model efficiently when given a large amount of data. *Interpretability* denotes the level of understanding and insight that is provided by the classifier. Finally, *reliability* is a measure of how well the designed classifier estimates the posterior probabilities of class membership. In fact, this measure shows how accurate are the decisions of a classifier. There are some cases in which a classifier is able to classify many training points successfully (i.e., its recognition score is high), but its decision has not the desired validity (i.e., its reliability is low). The major reason for this conflict is the overlapping of the classes.

For more descriptions, see Fig. 4.2, where two classes of "white" and "black" circles are separated by a decision function $d(x)$ in a two-dimensionsal feature space. It is shown that all "white" circles are correctly classified by $d(x)$ and that the recognition score of "white" circles is 100% by this decision line. But the reliability of this class is not 100%, because a member of the "black" circles has entered into the region of "white" circles. The recognition score of "black" circles in Fig. 4.2 is not 100%, but its reliability is 100%. This means that when the classifier recognizes an input training pattern as a "black" circle, this decision is true "almost everywhere".

The above mentioned performance metrics and other similar aspects should really be a part of classifier design and not an aspect to be considered separately, as it often is. Considering more performance aspects while designing a classifier (i.e., during training) yields more valuable results (during testing).

4.1.2 Conventional Classifiers

There are various typess of *heuristic*, *statistical*, and *synthetic* approaches for data classification which are known as *conventional classifiers*. Among them,

Bayesian classifiers, k-nearest neighbor (k-NN), multi-layer perceptron (MLP), and rule-based classifiers (e.g., fuzzy classifiers) are well-known classifiers with wide range of applications in pattern recognition problems. A brief review of these methods is presented below:

Bayesian classifier

The Bayesian classifier is a statistical approach which allows to design optimal classifiers (if a full statistical model is known) by predicting the probability that a given sample belongs to a particular class. The decision making process of the Bayesian classifier is as follows:

$$x \in C_i \quad \text{if } P(C_i|x) \geq P(C_j|x) \quad \text{for } \forall j = 1, 2, \ldots, M, \quad j \neq i \tag{4.2}$$

where x is an unknown input pattern, C_i is the ith reference class, M is the number of reference classes, and $P(C_i|x)$ (called *a posteriori*) is the conditional probability which represents the probability that x belongs to the respective class C_i. The a *posteriori* is obtained using Bayes' formula:

$$P\left(C_i|x\right) = \frac{p\left(x|C_i\right) P\left(C_i\right)}{p\left(x\right)} \tag{4.3}$$

We assume that the *a priori* probabilities $P(C_i) \quad i = 1, 2, \ldots, M$ are known. The other statistical quantities assumed to be known are class-conditional probability density functions $p(x|C_i) \quad i = 1, 2, \ldots, M$ describing the distribution of the feature vectors in each of the classes. $p(x)$ is the probability density function of x and is calculated by the following relation:

$$p\left(x\right) = \sum_{i=1}^{M} p\left(x|C_i\right) P\left(C_i\right) \tag{4.4}$$

There are some parametric and non-parametric techniques which can be utilized for estimating density functions [39] (e.g., maximum likelihood, Gaussian mixture models, histograms, and Parzen windows).

Using the above relations, an equivalent statement of the Bayesian classification rule is derived:

$$x \in C_i \quad \text{if } p\left(x|C_i\right) P\left(C_i\right) \geq p\left(x|C_j\right) P\left(C_j\right) \quad \forall j = 1, 2, \ldots, M \quad j \neq i \tag{4.5}$$

The Bayesian classifier minimizes the cost of class membership of an unknown pattern. However, it needs prior knowledge about the distribution of the patterns in the feature space for optimal performance. We know that, in general, such prior knowledge is not available.

k-nearest-neighbor (k-NN)

The k-nearest-neighbor (k-NN) rule starts at the test point and grows until it encloses k training samples, and it labels the test point by a majority vote of

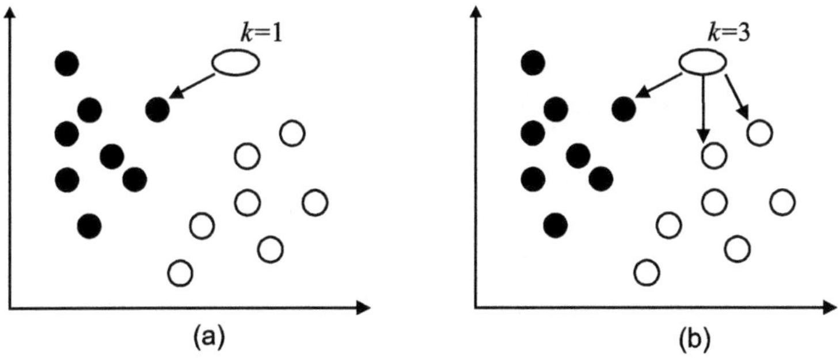

Fig. 4.3. (a) For $k = 1$, the input pattern is recognized as black, (b) for $k = 3$, the input pattern is recognized as white.

these samples. If $k = 1$, the classifier is called the *nearest neighbor classifier*. It is known that with an unlimited number of training samples, the error rate of the nearest neighbor classifier is never worse than twice the Bayesian classifier error [2].

The k-NN classifier has a simple structure, but it is not suitable for large databases because its delay and computational costs are high. Another weakness of k-NN is that its performance strongly depends on the value of k. Fig. 4.3 illustrates this matter with a simple example. In Fig. 4.3(a) the nearest neighbor classifier ($k = 1$) assigns the unknown pattern to the class of black circles and in Fig. 4.3(b), a 3-NN classifier recognizes the unknown pattern as a white circle.

Multi-layer perceptron (MLP)

Several neural networks' structures have been used for pattern recognition problems. When using neural networks, the classifier is represented as a network of cells modelling neurons of the human brain. A multi-layer perceptron (MLP) is a type of network suitable for this purpose. Each MLP consists of an input layer, hidden layers and an output layer. Fig. 4.4 shows a simple MLP that has four inputs in the input layer, six nodes in the hidden layer and produces one output in the output layer.

The nodes of each layer are linked to the other nodes of the forward layer with a weight value. During the training stage, data samples are considered as input vectors of the MLP and the network learns through a training algorithm (e.g., using the backpropagation method) which is used to estimate the proper set of weight vectors.

Although it is known that complex decision curves can be estimated using neural network classifiers, the performance of such classifiers strongly depends on the structure of the network (number of hidden layers, number of neurons, training algorithm, and so on).

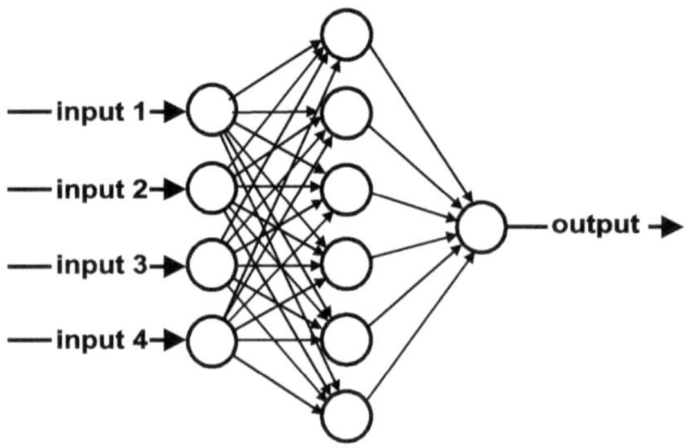

Fig. 4.4. An example of a MLP structure that has one hidden layer.

Rule-based classifiers

Rule-based classifiers can be roughly divided into two groups:

a) **Decision trees**: A decision tree is a class discriminator that recursively partitions the training set until each partition consists entirely or dominantly of examples from one class. Indeed, decision tree cuts the feature space with straight lines (hyperplanes) parallel to the axes and an impurity measure is used to select the best cut.
Upon the arrival of a feature vector, the search of the region to such feature vector will be assigned is achieved via a sequence of decisions along a path of nodes of an appropriately constructed tree. The algorithm consists of two phases:
   - Build an initial tree from the training data such that each leaf node is pure.
   - Prune this tree to increase its accuracy on test data.

   The sequence of decisions is applied to individual features, and the questions to be answered are in the form of "is feature $x_i \leq \alpha$?", where α is a threshold value. Such trees are known as *ordinary classification trees*. Other types of trees are also possible that split the space into convex polyhedral cells or into pieces of spheres. Fig. 4.5 shows an example of a decision tree. There are several algorithms available in the literature for extracting effective decision trees, such as CN2, C4.5, CART, SPRINT, and ID3 [2].
b) **Fuzzy classifiers:** Fuzzy logic provides a general concept for description and measurement. Most fuzzy logic systems encode human reasoning into a program in order to make decisions in data classification. Fuzzy logic comprises fuzzy sets, which are a way of representing non-statistical uncertainty and approximate reasoning, which includes the operations used to make inferences in fuzzy logic. Unlike traditional crisp logic, in fuzzy logic, fuzzy sets membership occurs by a degree within the range [0,1], which is represented

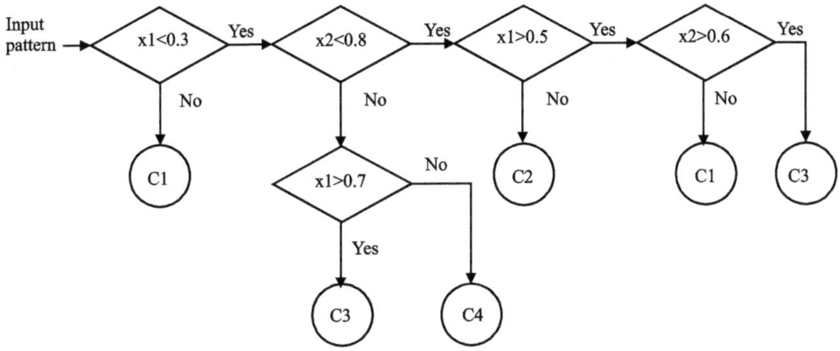

Fig. 4.5. An example of a decision tree.

by a membership function. The membership function can be linear or non-linear. Commonly adopted functions are: *trapezoid*, *triangle*, *Gaussian* and *sigmoid*.

Fuzzy rule-based systems have been successfully applied to pattern recognition problems (e.g., [3]).

Each fuzzy classifier includes a set of fuzzy IF-THEN rules, fuzzy inputs and outputs. The fuzzy IF-THEN rules may be represented in different ways. A typical fuzzy rule for a fuzzy classifier is shown below [4]:

IF x_1 is A_1 AND x_2 is A_2 AND ... AND x_N is A_N THEN Class C_j with CF_j

where $X = [x_1, x_2, ..., x_N]$ is a feature vector in an N-dimensional feature space, $A_i, \quad i = 1, 2, \ldots, N$ are the antecedent fuzzy sets, C_j is the consequent class (i.e., j^{th} class of the M classes), and CF_j is the degree of certainty of the fuzzy IF-THEN rule.

The performance of the fuzzy classifiers highly depends on the structure of the derived fuzzy structure (number of fuzzy rules, antecedents, consequents, membership function types, fuzzy operators, and so on). Thus, despite of comprehensibility of fuzzy classifiers, their robustness is not considerable.

4.2 Application of Multi-objective Swarm Intelligence on Data Classification

In this section, we explain the way in which multi-objective PSO can be used for designing novel classifiers. First, we investigate the usefulness of multi-objective swarm intelligence algorithms for data classification, mainly because they can overcome some of the main weaknesses of conventional classifiers. Then, we propose a taxonomy to classify the current applications of multi-objective (or single-objective) swarm intelligence on data classification into *indirect applications* and *direct applications* with a review on some of the research done around each of these tasks.

4.2.1 Why Multi-objective Swarm Intelligence?

All the commonly used conventional classifiers, which were introduced in Section 4.1.2 have some weaknesses.

Bayesian classifiers, which are optimum classifiers, need *a priori* probabilities ($P(C_i) \quad i = 1, 2, \ldots, M$) and class-conditional probability density functions ($p(x|C_i) \quad i = 1, 2, \ldots, M$). We know that this important information is not available in practical problems.

The performance of other classifiers strongly depends on their structural parameters. For fuzzy classifiers these parameters are: the number of fuzzy rules, antecedents, consequents, and membership function types. In neural network classifiers the number of hidden layers, the weights of the connections, the number of nodes in each layer, and activation functions types are important factors which affect their performance. The value of k and the type of distance metrics adopted are the parameters that affect the performance of a k-NN classifier.

As we explained in Section 4.1.1, for designing effective and powerful classifiers we need to optimize multi-performance parameters simultaneously and such optimization should be a part of classifier design and not an aspect that is considered separately, as it often is (in conventional classifiers). The powerfulness of multi-objective optimization based on swarm intelligence techniques is our main motivation to use such algorithms for this purpose.

In fact, swarm intelligence algorithms can be used for data classification in two different ways:

1. The first category of swarm intelligence applications on data classification is to design a novel group of non-parametric classifiers. In this approach, swarm intelligence techniques are directly utilized in order to obtain the optimum decision functions or decision rules. The powerfulness of swarm intelligence algorithms for obtaining the optimum solution in complex search spaces is a good motivation to construct these types of classifiers. After defining the desired fitness function (such as recognition score, reliability, and so on), swarm intelligence algorithms search within the feature space in order to estimate the decision functions which optimize the fitness function values. We call this type of application of swarm intelligence on data classification a *direct application*. Swarm intelligence algorithms play a major role here, since they are responsible for the basic classification task.
2. In the second category, swarm intelligence is employed to optimize the structural parameters of conventional classifiers. In other words, data classification is executed by conventional classifiers and their optimum parameters are estimated using swarm intelligence algorithms. In this approach, swarm intelligence techniques serve as an aid for improving the performance parameters of conventional classifiers, but are not directly responsible for the basic classification task. We call this sort of application of swarm intelligence on data classification an *indirect application*.

4.2.2 Classifiers Based on Multi-objective Swarm Intelligence (Direct Applications)

In this section, we introduce some novel classifiers in which their basic cores contain multi-objective swarm intelligence techniques, with emphasis on multi-objective particle swarm optimization. Generally speaking, multi-objective optimization can be defined as follows:

$$\textbf{Optimize} \quad F(Y) = [f_1(Y), f_2(Y), \dots, f_k(Y)],$$

Subject to $g_j(Y) \leq 0$ for $j = 1, 2, \dots, p$ and $h_j(Y) = 0$ for $j = 1, 2, \dots, p$
Where $Y = (y_1, y_2, \dots, y_n) \in \Re^n$ is the vector of decision variables, n is dimension of the solution space, $f_i : \Re^n \longrightarrow \Re$ $i = 1, 2, \dots, k$ are the objective functions and $g_i, h_i : \Re^n \longrightarrow \Re$, are the constraint functions of the problem.

For multi-objective optimization problems, the objective functions may be optimized separately from each other such that the best solution for each objective is found. However, this solution is normally unreachable, since the objective functions tend to be in conflict with each other and the best trade-offs among all the objectives need to be found in a different way.

The family of solutions to a multi-objective optimization problem is composed of all those potential solutions which cannot be improved in one objective without worsening another. Such solutions are called *Pareto optimal.* Y_o is called Pareto optimal if:

$$\forall Y \in \Omega \quad \text{and} \quad \forall i \in I, \quad f_i(Y_o) = f_i(Y) \quad \text{or} \ \exists i \in I, f_i(Y_o) > f_i(Y) \tag{4.6}$$

where Ω is the feasible solution space and I is the objective dimension. In the above definition, the "optimization" problem is considered as one of "maximization". In words, this definition says that $Y_o$is a Pareto optimal decision vector if there is no feasible vector Y that would increase some objective values without causing a simultaneous decrease in at least one other objective value. In the last few years, many multi-objective optimization techniques based on evolutionary algorithms and swarm intelligence approaches have been developed [5–8].

Data classification and multi-objective particle swarm optimization may seem to have few properties in common. However, PSO can be used to form novel classification methods which often lead to a suitable result, even in situations in which other methods would be too expensive or difficult to implement. Next, we introduce some multi-objective PSO classifiers reported in the literature:

1. MOPS-classifier

The multi-objective PSO (MOPSO) algorithm (proposed by Coello and Lechuga [9]) has been used to develop a multi-objective particle swarm classifier (called MOPS-classifier) [10]. The MOPS-classifier is able to approximate the decision hyperplanes in such a way that two performance aspects (recognition score and reliability) are *simultaneously* maximized. In the proposed method, each particle contains the weight vector of a set of decision hyperplanes and unlike the conventional PSO, the length of a particle is not fixed. Thus, the number of decision hyperplanes may be changed during the training phase and overfitting may be avoided.

In fact, the MOPS-classifier estimates the Pareto front for two objectives: *recognition score* and *reliability*. Thus, appropriate tools are prepared for the user to setup his/her desired performance aspects.

In the following, we briefly review the proposed MOPSO technique:

- Multi-objective particle swarm optimization (MOPSO)

MOPSO has been introduced by Coello and Lechuga [9] based on the idea of having a global repository in which every particle will deposit its flight experiences after each flight cycle. This technique is inspired on the external file used with the Pareto Archived Evolution Strategy (PAES) [34]. The repository is used by the particles to identify a leader that will guide the search. The mechanism of MOPSO is based on the generation of hypercubes which are produced by dividing the search space under exploration.

For designing our MOPS-classifier we modified MOPSO by using a PSO with a constriction coefficient. This algorithm is as follows:

i) Initialization the *Swarm* and the velocity of each particle of the *Swarm*:

For i=1 to *Pop* (*Pop* is the initial *Swarm_size*)
  $Y[i]$ = Random_generate();
  $V[i] = 0$;
End For;

ii) Evaluate each particle in the *Swarm*;
iii) Store the positions of the particles that represent non-dominated vectors in the repository *REP*;
iv) Generate hypercubes of the search space explored so far, and locate the particles using these hypercubes as a coordinate system where each particle's coordinates are defined according to the values of its objective functions.
v) Initialize the memory of each particle (this memory serves as a guide to travel through the search space. This memory is also stored in the repository):

a) For i= 0 to *Pop*
b) $PBEST[i] = Y[i]$;

vi) WHILE the maximum number of cycles had not been reached DO
  a) Initialize the internal parameters of PSO which are h (neighborhood size), χ (constriction coefficient), and *Swarm_size*.
  b) Compute the speed of each particle using the following expression:

$$V^{q+1}[i] = \chi \cdot ((V^q[i] + \phi_1(P[i] - Y[i]) + \phi_2(REP[h] - Y[i])), \quad (4.7)$$

Where ϕ_1 and ϕ_2 are random numbers uniformly distributed in the range $(0, \frac{\phi}{2})$ and

$$\chi = \frac{2}{\phi - 2 + \sqrt{\phi^2 - 4\phi}}, \quad \phi > 4 \quad (4.8)$$

REP[h] is a value that is taken from the repository; the index h is selected in the following way: those hypercubes containing more than one particle are assigned a fitness value equal to the result of dividing any number $x > 1$ by the number of particles that they contain. This aims to decrease the fitness of those hypercubes that contain more particles and it can be seen as a form of fitness sharing. Then, roulette-wheel selection is applied using these fitness values to select the hypercube from which we will take the corresponding particle. Once the hypercube has been selected, we randomly select a particle within such hypercube.

c) Compute the new positions of the particles adding the speed produced from the previous step, using the following equation:

$$Y^{q+1}[i] = Y^{q}[i] + V^{q+1}[i] \tag{4.9}$$

d) Maintain the particles within the search space in case they go beyond its boundaries (avoid generating solutions that do not lie on valid search space).
e) Evaluate each of the particles in the *Swarm*.
f) Update the contents of *REP* within the hypercubes. This update consists of inserting all the currently non-dominated locations into the repository. Any dominated locations from the repository are eliminated in the process.
g) When the current position of a particle is better than the position contained in its memory, the particle's position is updated. The criterion to decide what position from memory should be retained is simply obtained by applying Pareto dominance (i.e., if the current positions dominated by the position in memory, then the position in memory is kept; otherwise, the current position replaces the one in memory; if neither of them is dominated by the other, then we randomly select one of them).
h) Increment the loop counter.

vii) END WHILE

- Particle representation in the MOPS-classifier

The basic aim of the MOPS-classifier is to approximate the weight vectors of the decision hyperplanes (see Fig. 4.1). In fact, the solution space is the hyperplanes space and MOPSO tries to obtain optimal decision hyperplanes using appropriate fitness functions.

According to the above descriptions, a particle's representation is in the form of $P = [W_1, W_2, \ldots, W_i, \ldots, W_H]$ where $W_i = (w_{i1}, w_{i2}, \ldots, w_{iN}, w_{iN+1})$, is the weight vector of the ith hyperplane in the N dimensional feature space and H is the number of hyperplanes that may be changed from a minimum value to a predefined maximum value ($\log_2 M$, is the minimum value, where M is the number of classes). This means that the lengths of the particles are changed. If some mathematical operations are needed between two particles with different lengths, a proper number of zeros are inserted into the smaller particle.

This type of particle representation provides the possibility of avoiding overfitting. When a particle which includes a lower number of hyperplanes has a comparable performance with another particle which consists of a larger number of hyperplanes, the MOPS-classifier selects the particle with lower number of hyperplanes as the better position to avoid overfitting.

According to the utilized version of the PSO algorithm with the constriction coefficient, the velocity update is executed using equation (4.7). In this relation, the constriction coefficient (χ) is controlled by the parameter φ using equation (4.8). In [40], a study was presented which indicates how the particle swarm optimization algorithm works. Specifically, it has been proved that the application of a constriction coefficient allows control over the dynamical characteristics of the particle swarm, including its exploration versus exploitation behavior. In fact, the constriction coefficient prevents a build-up of the velocity because of the effect of particle's inertia. Without the constriction coefficient, particles with built-up velocities might explore the search space, but lose the ability to fine-tune a result.

- Objective function definition

The most important aims while designing the MOPS-classifier, are to simultaneously optimize two performance criteria: *error rate* and *reliability*. Thus, it is necessary to express these criteria as two objective functions. In the MOPS-classifier, two fitness functions are defined as below:

$$fit_1(P_i) = \frac{T - Miss(P_i)}{T} \tag{4.10}$$

$$fit_2(P_i) = \prod_{j=1}^{M} R_j \tag{4.11}$$

$fit_1(P_i)$ gives a value for the recognition score obtained by the i-th particle (i.e., P_i). In equation (4.10) $Miss(P_i)$ is the number of misclassified training points by P_i. Clearly, trying to maximize the objective function of fit_1 directs the classifier to approximate a set of decision hyperplanes which maximize the *recognition score*.

$fit_2(P_i)$ is defined for evaluating the reliability value, obtained by the set of hyperplanes of particle P_i. M is the number of reference classes, R_j is the reliability of the j-th class that is computed by $R_j = \frac{T_j}{T}$, where T_j is the number of correctly classified training points in class j and T is the total number of training points. In fact, this measure shows how precise the decisions of a classifier (see Section 4.1.1). There are some cases in which a classifier is able to classify many training points successfully (i.e., its recognition score is high), but its decision does not have the desired validity (its reliability is low). The major reason for this conflict is the overlapping of the classes.

In Fig. 4.3, based on the above fitness function definitions, the recognition score indicated with "white circles" is 100% but its reliability is 83.3% and the recognition score indicated by "black circles" is 80%, but its reliability is 100%.

The MOPS-classifier searches the space of weight vectors of hyperplanes to maximize fit_1 and fit_2.

Extensive experimental results on well-known data sets show the effectiveness and powerfulness of the MOPS-classifier in estimating the Pareto front for the above defined fitness functions.

- Implementation and results

Four data sets, with the number of dimensions varying from 4 to 18, nonlinear, overlapping class boundaries, and different number of reference classes (2 and 3), were used to show the effectiveness of the proposed MOPS-classifier. Each data set was divided into two different groups (one for training and the other for testing). Each group comprised 70% (for testing) and 30% (for training) of the original data set, respectively.

A description of the data sets is provided next:

A. Data Sets

Iris: The Iris data contains 50 measurements of four features from three species: Iris setosa, Iris versicolor, and Iris virginica. Features are sepal length, sepal width, petal length, and petal width.*

Cancer: The breast cancer database, obtained from the University of Wisconsin Hospital, Madison, has 683 breast mass samples belonging to two classes (Benign and Malignant), in a nine dimensional feature space.†

Mango: The Mango data set consists of eighteen measurements taken of the leaves of three different kinds of mango trees. It contains 166 samples in three different classes of three kinds of mango. The features set consists of measurements such as Z-value, area, perimeter, maximum length, maximum breadth, petiole, K-value, S-value, shape index, upper midrib/lower midrib, and perimeter upper half/perimeter lower half. The terms upper and lower are used with respect to the maximum breadth position [11].

Crude Oil: This data set has 56 samples with three classes and five features. Three classes, consisting of 7, 11, and 38 patterns, respectively, correspond to three types of oil. The five features are vanadium, iron, beryllium, saturated hydrocarbons, and aromatic hydrocarbons [12].

B. Comparison with Existing Methods

The performance of the proposed MOPS-classifier is compared with the performance of MLP, the k-NN classifier, the Bayesian classifier, and the PS-classifier. The PS-classifier is a PSO based classifier which only has one objective function: fit_1. Thus, the aim of the PS-classifier is only maximizing the recognition score [13].

---

*This data set is available at the University of California, Irvine, via anonymous ftp from `ftp.ics.uci.edu/pub/machine-learning-databases`

†This data set is available from the same site indicated before.

Different structures are used for MLP for each data set and the best obtained results are reported.

The k-NN classifier is executed making k equal to $\sqrt{T}$, where T is the number of training samples (it is known that as the number of training patterns T goes to infinity, and if the value of k and k/T can be made to approach infinity and 0, respectively, then the k-NN classifier approaches the optimal Bayesian classifier [14]. One such value of k for which the limiting conditions are satisfied is $\sqrt{T}$).

For the Bayesian classifier, the Parzen windows algorithm is used to estimate the density function of the data sets. Also, *a priori* probabilities of $\frac{T_i}{T}$ for a total of T training samples and T_i patterns from class i, are considered.

The swarm size of the MOPS-classifier and the PS-classifier is considered to be 40. All the simulations were executed in MATLAB 7.0.

C. Experimental Results

Tables 4.1 to 4.4 show the obtained recognition score and reliability values for four benchmark problems and three different algorithms during the training phase. These tables indicate the maximum, minimum, and average recognition score, as well as reliability corresponding to ten independent runs (experiments). In these tables, H is the number of hyperplanes for the revealed results.

For the *Iris* data (Table 4.1), the proposed MOPS-classifier and the Bayesian classifier provided the best reliabilities and the best average recognition score, respectively. Regarding the average score of recognition, the difference between these two classifiers is small (1.4%). Also, the single-objective PS-classifier was able to estimate the decision hyperplanes for which its average recognition scores are of 91.6%, which is better than those of the more conventional MLP and k-NN classifiers. However, a considerable gap of 13.9% can be seen for its average amount of reliability with respect to the MOPS-classifier. Since the performance aspect of reliability is not considered as an objective function in the PS-classifier, the most important drawback of its performance is related to the reliability measures.

With regard to the *Cancer* data, our MOPS-classifier provided the best reliability scores, followed by the reliability scores of other classifiers, as shown

Table 4.1. Comparative recognition scores and reliabilities for *Iris* data during training.

	Recognition score (%)			Reliability (%)		
	Max.	*Min.*	*Ave.*	*Max.*	*Min.*	*Ave.*
MLP	78.8	66.8	74.2	56.7	33.1	47.4
k-NN	87.7	79.8	83.6	53.5	35.1	48.6
Bayes	97.1	92.3	95.0	88.5	78.3	83.1
PS-classifier (H=3)	93.2	90.9	91.6	74.3	69.0	72.2
MOPS-classifier (H=3)	95.6	89.2	93.6	91.7	83.4	86.1

Table 4.2. Comparative recognition scores and reliabilities for *Cancer* data during training.

	Recognition score (%)			Reliability (%)		
	Max.	*Min.*	*Ave.*	*Max.*	*Min.*	*Ave.*
MLP	90.5	78.9	85.6	67.3	45.5	54.4
k-NN	88.4	75.3	81.0	70.4	51.0	61.7
Bayes	100	90.1	94.4	90.0	79.6	84.7
PS-classifier(H=3)	96.0	87.8	91.7	83.4	74.7	78.0
MOPS-classifier(H=2)	97.0	91.6	94.0	95.7	91.5	93.2

in Table 4.2. The results in Table 4.2 indicate improvements of 38.8%, 31.5%, 8.5%, and 15.2% for the average reliability scores with respect to MLP, k-NN, Bayesian classifier and PS-classifier, respectively. Note that the Bayesian classifier provides the best average recognition score. Overall, the performance of the MOPS-classifier is comparable to that of the Bayesian classifier for the *Cancer* data (similar to the results obtained for the *Iris* data). Again, the worst reliability scores of MLP, k-NN, Bayesian classifier, and PS-classifier are remarkable in comparison to those of the MOPS-classifier. This is because, with the exception of our MOPS-classifier, the other methods try to maximize only the recognition score, and no attempt is made to optimize reliability.

Table 4.3 shows that for the *Mango* data, the MOPS-classifier outperforms the other classifiers regarding average reliability values. Also, it can be seen that the PS-classifier outperforms the MOPS-classifier with respect to the recognition scores. However, it should be mentioned that the PS-classifier utilizes an additional number of hyperplanes to reach these recognition scores. In general, the use of a large number of hyperplanes for approximating the class boundaries is likely to lead to overfitting/overlearning of the data, which thereby leads to better performance during the training process (better abstraction), but also to a poorer generalization capability. This is supported by the results provided in Table 4.5, where it can be seen that its performance during the testing phase is worse than that of the MOPS-classifier.

Table 4.4 shows the effectiveness of the proposed MOPS-classifier, since it had the best performances (both average recognition scores and reliabilities) from

Table 4.3. Comparative recognition scores and reliabilities for *Mango* data during training.

	Recognition score (%)			Reliability (%)		
	Max.	*Min.*	*Ave.*	*Max.*	*Min.*	*Ave.*
MLP	77.4	56.0	67.2	55.6	38.5	45.9
k-NN	77.5	60.3	67.5	53.0	40.1	44.2
Bayes	93.0	86.9	89.9	62.5	42.3	50.8
PS-classifier(H=7)	90.0	83.5	86.4	57.3	39.6	48.5
MOPS-classifier(H=5)	89.8	81.5	85.1	62.0	51.7	58.0

Table 4.4. Comparative recognition scores and reliabilities for *Crude Oil* data during training.

	Recognition score (%)			Reliability (%)		
	Max.	*Min.*	*Ave.*	*Max.*	*Min.*	*Ave.*
MLP	89.1	80.6	84.5	63.6	44.9	52.8
k-NN	91.1	84.2	87.9	61.6	46.4	53.7
Bayes	100	90.0	93.8	67.3	54.9	60.4
PS-classifier(H=4)	94.5	88.9	91.1	66.3	57.0	61.2
MOPS-classifier(H=4)	100	89.8	94.0	74.0	66.1	69.3

any of the classifiers for the *Crude Oil* data classification. Regarding average recognition scores, the MOPS-classifier outperformed the PS-classifier by 2.9% and to the Bayesian classifier by 0.2%. With regard to the average reliability values, the MOPS-classifier outperformed the PS-classifier by 8.1% and to the Bayesian classifier by 8.9%.

Table 4.5 presents a comparison of results of the five classifies on different data sets during the testing stage. In this table, *S.R.* means score of recognition (%) and *Rel.* means reliability (%). The results of Table 4.5 show that two PSO-based classifiers (the PS-classifier and the MOPS-classifier) outperform the conventional classifiers: MLP, and k-NN.

The results of Table 4.5 reinforce the claim that the performance aspects (here, score of recognition and reliability) should really be a part of the classifier design and not aspects that are considered separately. For example, the PS-classifier that considers only the recognition score as the performance criterion, has an undesired performance regarding the reliability index. Its reliability in the testing phase is lower than that of the MOPS-classifier by 18.6%, 21.7%, 11.2%, and 11.8% for the *Iris*, *Cancer*, *Mango*, and *Crude Oil* data sets, respectively.

Regarding the score of recognition, the performance of the Bayesian classifier is better than that of the PS-classifier. However, it is comparable and sometimes it is worse than the scores of recognition obtained by the MOPS-classifier for the four data sets in the testing stage. Overall, the comparison of results during the testing and training stages (Tables 4.1 to 4.5) reveal that the performances of the proposed MOPS-classifier are comparable, and sometimes better than those of the Bayesian classifier. Note that the Bayesian classifier is an optimal classifier when the class distributions and the *a priori* probabilities are known.

Table 4.5. The averages of test results (%) on various data sets for five classifiers.

	MLP		*k-NN*		*Bayes*		*PS-Classifier*		*MOPS-classifier*	
	S.R.	Rel.	S.R.	Rel.	S.R.	Rel.	S.R.	Rel.	S.R.	Rel.
Iris	70.1	40.5	78.2	50.1	93.0	79.3	87.3	64.7	91.1	83.3
Cancer	80.3	55.7	75.0	56.3	90.5	79.8	83.1	67.1	90.9	88.8
Mango	61.1	43.0	61.8	46.9	80.3	51.7	70.5	48.9	80.4	60.1
Crude Oil	79.5	48.0	82.3	50.3	90.8	55.2	82.6	56.9	90.2	68.7

Fig. 4.6 shows the estimated Pareto front for the *Mango* data, which was obtained by the MOPS-classifier during one of the experiments. Since the number of non-dominated solutions for the *Mango* data set is larger than that of the other benchmarks, it is possible and more meaningful to show the Pareto front for this data set in graphical form.

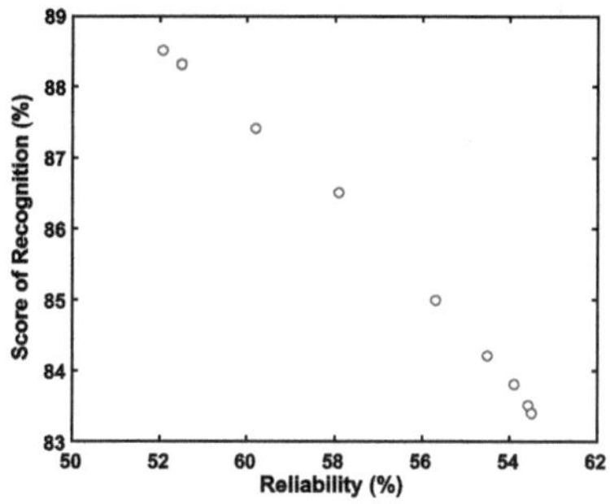

Fig. 4.6. Non-dominated solutions obtained by the *MOPS-classifier* for the *Mango* data set.

D. Sensitivity analysis of the MOPS-classifier parameters

-Constriction Coefficient

As mentioned at the sub-section entitled "Particle representation in the MOPS-classifier", the parameter φ controls the constriction coefficient value (χ) using equation 4.8. The constriction coefficient prevents a build up of the velocity due to the effect of the particle's inertia [40]. Without the constriction coefficient, particles with built up velocities might explore the search space, but lose the ability to fine-tune a result. On the other hand, an excessive constriction of the particle's speed might damage the search space exploration. Thus, the value of the constriction coefficient affects the global versus local abilities of the PSO algorithm. It can be concluded from [40] that φ determines the attraction value of the particles with respect to the best positions previously found by such particles and their neighbors. This means that the convergence properties of PSO can be controlled by φ.

As the fitness value of a particle becomes better and better, the portion of the search space that the particle explores should be smaller and smaller. This means that φ should be increased above 4 in order to decrease χ and $\chi \cdot \varphi$. The result is a decrease in the inertia of the particle, in order to emphasize a local instead of a global search. A smaller improvement in the particle's fitness has as a consequence, that a larger search space is explored. This means that decrease should happen to the value of φ, in order to increase the inertia of the particle. This has a a result an emphasis on a global search instead of a local one.

Based on our experiments, we recommend starting the algorithm with low values of φ and then increase it above 4 at the final cycles of the algorithm.

-Swarm Size

We changed the swarm size from 10 to 80. It was seen that a large value of the swarm size considerably reduces the convergence rate and slows down the execution of the algorithm. In contrast, a small value of the swarm size causes convergence to local minima and reduces the ability of the MOPS-classifier for obtaining the Pareto front. We recommend using a value between 30 and 50 for the swarm size.

-Repository Size

This parameter is used to predefine the maximum number of non-dominated solutions that can be stored in the repository [9]. Based on the characteristics of our problems and the low number of obtained non-dominated solutions, we saw that a large variation of this parameter had no meaningful effects on the performance of the MOPS-classifier.

In the following sub-section, a theorem is stated and proved regarding the performance of the MOPS-classifier, indicating that it approaches that of the Bayesian classifier, which is the optimal classifier when the class distributions and the *a priori* probabilities are known. This property and its proof was inspired by the genetic algorithm-based classifier (GA-classifier [15]).

-Relationship of the performance of the MOPS-classifier with the Bayesian error probability

Let there be k reference classes $C_1, C_2, \ldots, C_k$ with *a priori* probabilities $P(C_1)$, $P(C_2), \ldots, P(C_k)$ (easily shown by $P_1, P_2, \ldots, P_k$ and class-conditional density functions $p(x|C_1), p(x|C_2), \ldots, p(x|C_k)$ (easily shown by $p_1(x), p_2(x), \ldots, p_k(x)$. Also, let the mixture density be $p(x) = \sum_{i=1}^{k} P_i p_i(x)$. As mentioned in Section 4.1.2, according to the Bayesian classifier, a point is classified into class i iff

$$P_i p_i(x) \geq P_j p_j(x), \quad \forall j = 1, \ldots, k \quad \text{and } j \neq i \tag{4.12}$$

Suppose that $X_1, X_2, \ldots, X_T$ are independent and identically distributed (i.i.d) N-dimensional random vectors with density $p(x)$ (T is the number of training points). This indicates that there is a probability space (Ω, F, Q), where $\mathfrak{F}$ is σ field of subsets of Ω, Q is a probability measure on $\mathfrak{F}$ and

$$X_i : (\Omega, \mathfrak{F}, Q) \longrightarrow (R^N, B(R^N), P), \quad \forall i = 1, 2, \ldots \tag{4.13}$$

such that:

$$P(A) = Q(X_i^{-1}(A)) = \int_A p(x)dx \quad \forall A \in B(R^N) \quad i = 1, 2, \ldots$$

Here, $B(R^N)$ is the Borel σ field of R^N.

Suppose that:

$$\mathcal{E} = \{E : E = (S_1, S_2, \ldots, S_k),\ S_i \subseteq R^n,\ S_i \neq \phi\ \ \forall i = 1, 2, \ldots, k,$$
$$\bigcup_{i=1}^{k} S_i = R^n,\ S_i \cap S_j = \phi, \forall i \neq j\}$$

$\mathcal{E}$ provides the set of all partitions of R^N into k sets as well as their permutations, i.e., if $E_1 = (S_1, S_2, \ldots, S_k) \in \mathcal{E}$, and $E_2 = (S_2, S_1, \ldots, S_k) \in \mathcal{E}$, then $E_1 \neq E_2$. Note that $E_i = (S_{i_1}, S_{i_2}, \ldots, S_{i_k})$ implies that each S_{i_j}, $1 \leq j \leq k$, is the region corresponding to class C_j.

Let $E_o = (S_{o1}, S_{o2}, \ldots, S_{ok}) \in \mathcal{E}$ be such that each S_{oi} is the region corresponding to the class C_i in R^N and these are obtained using Bayes' decision rule. Then

$$a = \sum_{i=1}^{k} P_i \int_{S_{Oi}^C} p_i(x)d(x) \leq \sum_{i=1}^{k} P_i \int_{S_{1i}^C} p_i(x)d(x)\ \ \forall E_1 = (S_{11}, S_{12}, \ldots, S_{1k}) \in \mathcal{E}$$

Here, a is the error probability obtained using Bayes' decision rule.

It is known from the literature that such an E_o exists and that it belongs to $\mathcal{E}$ because Bayes' decision rule provides an optimal partition of R^N and for every such $E_1 = (S_{11}, S_{12}, \ldots, S_{1k}) \in \mathcal{E}$, $\sum_{i=1}^{k} P_i \int_{S_{1i}^c} p_i(x)dx$ provides the error probability for $E_1 \in \mathcal{E}$. Note that E_o does not need to be unique.

Let H_o be a positive integer and let there exist H_o hyperplanes in R^N which can provide the regions $S_{o1}, S_{o2}, \ldots, S_{ok}$. Let H_o be known *a priori*. Let the multi-objective PSO be allowed to be executed for a sufficiently large number of iterations for the generation of class boundaries.

Also, let $\mathcal{A} = \{A : A$ is a set consisting of H_o hyperplanes in $R^N\}$ and let $A_o \in \mathcal{A}$ be such that it provides the regions $S_{o1}, S_{o2}, \ldots, S_{ok}$ in R^N, i.e., A_o provides the regions which are also obtained using Bayes' decision rule. Note that $A \in \mathcal{A}$ generates several elements of $\mathcal{E}$ which result from different permutations of the list of k regions. Let $\mathcal{E}_A \subseteq \mathcal{E}$ denote all possible $E_1 = (S_1, S_2, \ldots, S_k) \in \mathcal{E}$ that can be generated from A. Let $G = \bigcup_A \mathcal{E}_A$; let

$$Z_{iE}(\omega) = \begin{cases} 1 \text{ if } X_i(\omega) \text{ is missclassified when } E \text{ is used as a decision rule} \\ \quad \text{where } E \in G, \forall \omega \in \Omega, \\ 0 \text{ otherwise} \end{cases}$$

Let
$f_{TE}(\omega) = \frac{1}{T}\sum_{i=1}^{T} Z_{iE}(\omega)$, when $E \in G$ is used as a decision rule,

and let

$$f_T(\omega) = Inf\{f_{TE}(\omega) : E \in G\}.$$

Note that the MOPS-classifier minimizes $T \cdot f_{TE}(\omega)$, the total number of misclassified points, when it attempts to maximize fit_1 (equation 4.10). This is equivalent to searching for a suitable $E \in G$ such that the term $f_{TE}(\omega)$ is minimized. It is known that for infinitely many iterations, the PSO algorithm will certainly be able to obtain such an E.

Theorem. For sufficiently large T, $f_T(\omega)$ cannot be greater than a, almost everywhere.

Proof. Let

$$Y_i(\omega) = \begin{cases} 1 \text{ if } X_i(\omega) \text{ is misclassified according to Bayes' rule } \forall \omega \in \Omega \\ 0 \text{ otherwise} \end{cases}$$

Note that $Y_1, Y_2, \ldots, Y_T$ are i.i.d. random variables. Now

$$\begin{aligned} PROB(Y_i = 1) &= \textstyle\sum_{j=1}^{k} PROB(Y_i = 1/X_i \in C_j)\ P(X_i \in C_j) \\ &= \textstyle\sum_{j=1}^{k} PROB(\omega : X_i(\omega) \in S_{oj}^c \text{ and } \omega \in C_j)\ P_j \\ &= \textstyle\sum_{j=1}^{k} \{\int_{S_{oj}^c} p_j(x) d(x)\} P_j = a \end{aligned}$$

Hence the expectation of Y_i, $E(Y_i)$ is given by $E(Y_i) = a\ \forall i$.

Then, by using String Law of Large Numbers [16], $(1/T)\sum_{i=1}^{T} Y_i(\omega) \to a$, almost everywhere, i.e., $PROB(\omega : (1/T)\sum_{i=1}^{T} Y_i(\omega) \mapsto a) = 0$ ($\mapsto$ means does not tend). Let $B = \{\omega : (1/T)\sum_{i=1}^{T} Y_i(\omega) \to a\} \subseteq \Omega$. Then $Q(B) = 1$.

Note that $f_T(\omega) \le \left(\frac{1}{T}\sum_{i=1}^{T} Y_i(\omega)\right)$, $\forall T$, and, $\forall \omega$, since the set of regions $(S_{o1}, S_{o2}, \ldots, S_{ok})$ obtained by Bayes' decision rule is also provided by some $A \in \mathcal{A}$ and, consequently, it will be included in G. Note that $0 \le f_T(\omega) \le 1\ \forall T$, and, $\forall \omega$. Let $\omega \in B$. For every $\omega \in B$, $U(\omega) = \{f_T(\omega), T = 1, 2, \ldots\}$ is a bounded, infinite set. Then by Bolzano-Weierstrass theorem [17], there is an accumulation point of $U(\omega)$.

Let $y = \text{Sup}\{y_0 : y_0 \text{ is an accumulation point of } U(\omega)\}$. From elementary mathematical analysis we can conclude that $y \le a$, since $(1/T)\sum_{i=1}^{T} Y_i(\omega) \to a$ almost everywhere and $f_T(\omega) \le \left(\frac{1}{T}\sum_{i=1}^{T} Y_i(\omega)\right)$, $\forall T$, and $\forall \omega$. Thus it is proved that for sufficiently large T, $f_T(\omega)$ can not be larger than a for $\omega \in B$. ■

The above theorem is supported experimentally by our other research where we studied the effect of the number of training points on the performance of PSO-based classifiers [18].

Note that the proof established that the number of points misclassified by the *MOPS-classifier* will always be less than or equal to the number of points misclassified by Bayes' decision rule for a sufficiently large number of training data points and iterations.

However, there are some cases in Tables 4.1 to 4.5 in which the performance of the MOPS-classifier is better than that of the Bayesian classifier. Note that $f_T(\omega) < a$, is true for only a finite number of training data points. This is due to the fact that a small number of identical training points can be generated by different statistical distributions. Consequently, each distribution will result in different error probabilities of the Bayesian classifier. The PSO-based classifier, on the other hand, will always find the decision surface yielding the smallest number of misclassified points, irrespective of their statistical properties.

As the number of points increases, the number of possible distributions that can produce them decreases. In the limiting case when $T \to \infty$, only one distribution will produce all the training points. The Bayesian classifier designed over this distribution will provide the optimal decision boundary. The PSO-based classifier, in that case, will also yield a decision boundary which is the same as that of the Bayesian decision boundary provided that the number of surfaces is known *a priori*.

2. Using PSO with aggregation function for data classification

De Falco et al. [19] have proposed a classification technique using multi-objective PSO with a weighted aggregation approach. According to this approach, all the objectives are added up to a weighted combination as shown below:

$$F_a(P) = \sum_{i=1}^{k} a_i f_i(P) \tag{4.14}$$

where a_i for $i = 1, 2, \ldots, k$ are non-negative weights and $\sum_{i=1}^{k} a_i = 1$. $F_a(P)$ is called *aggregation function*. The authors assumed that these weights are fixed. This approach is named **Conventional Weighted Aggregation** (CWA). In this case, only a single Pareto optimal solution can be found per optimization run. Next, we briefly explain their method:

Given a database with C classes and N parameters, the classification problem can be seen as that of finding the optimal positions of C centroids in an N-dimensional space, i.e., the problem consists of determining, for any centroid, its N coordinates, each of which can take on, in general, real values. With these assumptions, the i-th individual of the population is encoded as follows:

$$P_i = \left(\overrightarrow{p_i^1}, \ldots, \overrightarrow{p_i^C}, \overrightarrow{v_i^1}, \ldots, \overrightarrow{v_i^C} \right) \tag{4.15}$$

where the position of the j-th centroid is constituted by N real numbers representing its N coordinates in the problem space:

$$\overrightarrow{p_i^j} = \{p_{1,i}^j, \ldots, p_{N,i}^j\} \tag{4.16}$$

and, similarly, the velocity of the j-th centroid is made up of N real numbers representing its N velocity components in the problem space:

$$\overrightarrow{v_i^j} = \{v_{1,i}^j, \ldots, v_{N,i}^j\} \tag{4.17}$$

Then, each individual in the population is composed by $2 \times C \times N$ components, each of which is represented by a real value.

The aggregation fitness function is defined as shown below:

$$F_a(P_i) = 0.5\,(f_1(P_i) + f_2(P_i)) \tag{4.18}$$

where $f_1(P_i) = \frac{\sum_{j=1}^{T} \delta(\overrightarrow{x_j})}{T}$ and $f_2(P_i) = \frac{\sum_{j=1}^{T} d\left(\overrightarrow{x_j}, \overrightarrow{p_i}^{CL_{known}(\overrightarrow{x_j})}\right)}{T}$. In these relations,

$\delta(\overrightarrow{x_j}) = \begin{cases} 1 \text{ if } CL(\overrightarrow{x_j}) \neq CL_{known}(\overrightarrow{x_j}) \\ 0 \text{ otherwise} \end{cases}$, T is the number of training points, $d(.,.)$ denotes the Euclidean distance.

In fact, f_1 is a fitness function that estimates the incorrectly assigned instances on the training set, i.e., it takes into account all the cases in which the class $CL(\overrightarrow{x_j})$ assigned to instance $\overrightarrow{x_j}$ is different from its class $CL_{known}(\overrightarrow{x_j})$ as known from the database. f_2 is the sum on all the training set points of the Euclidean distance in an N-dimensional space between a generic instance $\overrightarrow{x_j}$ and the centroid of the class to which it belongs, according to the database $\left(\overrightarrow{p_i}^{CL_{known}(\overrightarrow{x_j})}\right)$.

The proposed classifier has been tested on 13 typical problems and compared to other techniques widely used in this field (Bayesian classifier, MLP, rule-based classifier, decision tree, and so on). Experiments have shown that the proposed method is competitive with respect to the chosen (special-purpose) classification techniques, and, for some test problems, it turns out to be better than all the others. The execution times of the proposed classifier are of the same order of magnitude as those of the other nine techniques used.

3. MOPS-classifier based on goal programming

Tsai and Yeh [20] have used multi-objective PSO to develop a classifier for inventory classification. In their approach, inventory items are classified based on a specific objective or on multiple objectives, such as minimizing costs, maximizing inventory turnover ratios, and maximizing inventory correlation. This approach determines the best number of inventory classes and how items should be categorized for the desired objectives at the same time.

To combine the three objectives of cost function, demand correlation, and inventory turnover ratio, they used the *goal programming method*, which was proposed by Charnes and Cooper in 1961. In this method, each objective value is first normalized to be between 0 and 1. The aim of the goal programming method is to minimize the difference between the objective values and the constraint values. The constraint value of each single objective is set as the best objective value obtained in the single objective problem and the objective value is the value of that single-objective obtained in the multi-objective problem.

A summary of the classification algorithm is the following. When performing a classification of items, the algorithm first sorts all the items according to their weight scores and then determines cut-off points along the sorting list. Items between two adjacent cut-off points are classified into the same group. Item properties (criteria) such as demand rate and unit holding cost are put into consideration when determining appropriate cut-off points. To incorporate different classification objectives, each property is assigned a criteria weight in order to reflect the degree of relevance of the objective. The weights are also designed as part of the search space so that proper weights will be determined. Hence, the particle in the search space is defined as:

$$P = \big(w_1, w_2, \ldots, w_k, x_{(1,2)}, x_{(2,3)}, \ldots, x_{(M-1,M)}\big) \tag{4.19}$$

where w_k is the criteria weight of property k, $x_{(i-1,i)}$ is the cut-off point between group $i-1$ and the classification group i ($x_{(1,2)} \leq x_{(2,3)} \leq \ldots \leq x_{(M-1,M)}$, M is the maximum number of classification groups).

For the criteria weight part of the search space, the value of each initial position and initial velocity is generated randomly between 0 and 1. As for the cut-off point part, the value of the initial velocity is also generated randomly between 0 and 1. However, the value of the initial position of each cut-off point has to be set such that $x_{(1,2)} \leq x_{(2,3)} \leq \ldots \leq x_{(M-1,M)}$.

The weighted scores of the items are calculated following the simple rule proposed in Guvenir and Erel [21]. First, the values of the item properties are normalized to be between 0 and 1 using the relation $\frac{i_k - \min_k}{\max_k - \min_k}$, where i_k is the property k's value of item I, and $\min_k$ and $\max_k$ are the minimum and maximum values of property k for all items. Then, the weight score of item i in particle P can be calculated as:

$$ws(P,i) = \sum_{k=1}^{K} w_k \frac{i_k - \min_k}{\max_k - \min_k} \tag{4.20}$$

Comparing the weight score of an item to the cut-off points determines the group within which that item is categorized. That is:

$$\text{classification}(P,i) = \begin{cases} 1 & \text{if } ws(P,i) \leq x_{(1,2)} \\ 2 & \text{if } x_{(1,2)} \leq ws(P,i) \leq x_{(2,3)} \\ 3 & \text{if } x_{(2,3)} \leq ws(P,i) \leq x_{(3,4)} \\ \vdots & \\ M & \text{if } x_{(M-1,M)} \leq ws(P,i) \end{cases} \tag{4.21}$$

The classification result represented by a particle is determined by the values of the criteria weights and the cut-off points within that particle, in the way described above. The fitness value of that particle is the value of the objective function corresponding to that classification result.

Extensive numerical studies were conducted, and the multi-objective PSO algorithm was compared to other classification approaches. All outcomes led to the conclusion that this algorithm performs comparatively well with respect to those schemes that are commonly used in practice.

4. Unsupervised classification based on PSO

There are some researches including the use of *single-objective* PSO to develop unsupervised classification (or clustering) techniques. Since in this chapter we are interested in the application of *multi-objective* PSO to data classification, we only address efforts of this sort.

Van der Merwe and Engelbrecht [22], Yang and Kamel [23], Omran [24, 25], Cui and Potok [26], Fun and Chen [27], used PSO for data clustering. Sousa et al. [28, 29] proposed a rule discovery technique based on PSO. In order to evaluate the usefulness of PSO for data mining, they provided an empirical comparison of the performance of three variants of PSO with respect to another evolutionary algorithm (a genetic algorithm).

4.2.3 Classifiers Based on Multi-objective Swarm Intelligence (Indirect Applications)

As mentioned in Section 4.1.2, the performance of conventional classifiers strongly depends on their structural parameters. Indirect applications of swarm intelligence on data classification referred to optimizing those structural parameters in order to achieve optimal performance. Fig. 4.7 shows the interaction of swarm intelligence techniques and conventional classifiers in indirect applications.

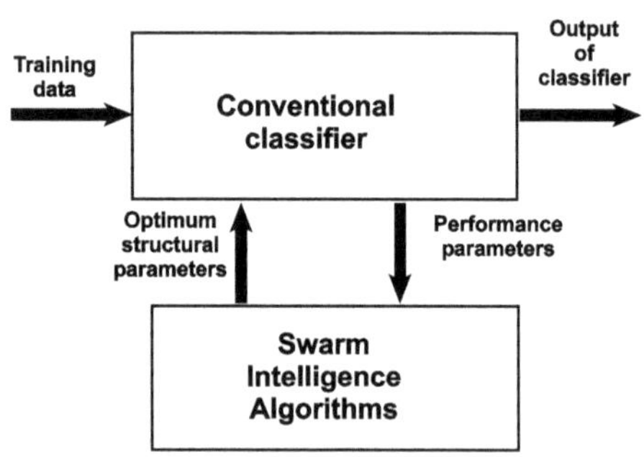

Fig. 4.7. Block diagram of indirect applications of swarm intelligence algorithms on designing conventional classifiers.

To the best of our knowledge, indirect applications include a lot of research in the literature which utilizes single-objective PSO for optimizing the performance of conventional classifiers. Some of the past and ongoing related research has been introduced in [35].

Since the basic aim of this chapter is to study the application *of multi-objective* swarm intelligence techniques on data classification, we only briefly address this research and we intentionally avoid entering into their corresponding details.

Nabavi et al. [30] presented a way of computing the weights for combining multiple neural network classifiers based on particle swarm optimization. The weights were obtained so that they minimized the total classification error rate of the ensemble system.

In [31], PSO is applied to find good values for the weights of the connections in a MLP. Settles and Rylander [32] have proposed a PSO method for training neural network classifiers.

Chen and Abraham [33] reported a preliminary study to detect breast cancer using a Flexible Neural Tree (FNT), a Neural Network (NN), a Wavelet Neural Network (WNN) and their ensemble combination. For the FNT model, a tree-structure based evolutionary algorithm and PSO were used to find an optimal FNT. For the NN and WNN, PSO was employed to optimize the free parameters. The performance of each approach was evaluated using the breast cancer data set. Simulation results showed that the obtained FNT model had a lower number of variables with a reduced number of input features and without a significant reduction in the detection accuracy.

Chen and Abraham [31] investigated how the seemingly chaotic behavior of stock markets could be well represented using several soft computing techniques. The authors considered the flexible neural tree algorithm, a wavelet neural network, local linear wavelet neural network and, finally, a feed-forward artificial neural network. The parameters of these different learning techniques were optimized by the PSO approach. Experiment results revealed that PSO could play an important role to fine tune the parameters for optimal performance.

In the task of optimizing the parameters of fuzzy classifiers, an evolutionary procedure has been addressed in [35] to design hierarchical or multilevel Takagi-Sugeno Fuzzy Systems (TS-FS). The hierarchical structure is evolved using Probabilistic Incremental Program Evolution (PIPE) with specific instructions. The fine tuning of the *if-then* rules parameters encoded in the structure is accomplished using PSO. The proposed method interleaves both PIPE and PSO optimizations. The new method results in a smaller rule-base and has a good learning ability. The proposed hierarchical TS-FS is evaluated using some forecasting problems. When compared to other hierarchical TS-FS approaches, the proposed hybrid approach exhibited competing results with high accuracy and smaller size of the hierarchical architecture.

Fun and Chen [27] introduced an evolutionary PSO learning-based method to optimally cluster N data points into K clusters. The hybrid PSO and K-means, with a novel alternative metric algorithm was called *Alternative KPSO-clustering* (AKPSO) method. This method was developed to automatically detect the cluster centers of geometrically structured data sets. In the AKPSO algorithm, an special alternative metric is considered to improve the traditional K-means clustering algorithm in order to deal with several structure data sets.

An important task in data classification is optimum feature set selection. In fact, selecting more effective feature vectors leads to better performance for the classifier. In this task, Marinakis et al. [36] proposed a classification algorithm combining nearest neighbor techniques and particle swarm optimization principles for solving the optimal feature subset selection problem. Pedrycz et al. [37] proposed a two-phase feature selection method that uses particle swarm optimization to form an initial core set of discriminatory features from the original feature space. This core set is then expanded by searching for additional discriminatory features.

Another very important application of PSO is in the domain of *cascading classifiers*. Cascading classifiers have been used to solve pattern recognition problems in recent years. The main motivations behind such a strategy are the improvement of classification accuracy and the reduction of the complexity that can be achieved with such type of classifier. The issue of class-related reject thresholds for cascading classifier systems is an important problem. It has been shown in the literature that class-related reject thresholds provide an error-reject trade-off better than a single global threshold. Oliveira et al. [38] proposed the use of PSO for finding thresholds in order to improve the error-reject trade-off yielded by class-related reject thresholds. This approach has been found to be very effective in solving real valued global optimization problems.

4.3 Conclusions

In this chapter, we have introduced both *direct* and *indirect* usages of multi-objective swarm intelligence algorithms for designing novel classifiers with a particular focus on particle swarm optimization. Furthermore, a theoretical investigation was made in this chapter in order to find the relationship between the multi-objective PSO based classifier and the Bayesian classifier. Also, we briefly reviewed some research related to applications of single-objective PSO on data classification.

References

[1] Baldi, P., Brunak, S.: Bioinformatics, The Machine Learning Approach, 2nd edn. The MIT Press, Cambridge (2001)
[2] Webb, A.R.: Statistical Pattern Recognition, 2nd edn. John Wiley and Sons, Chichester (2002)
[3] Ishibuchi, H., Nakashima, T.: Voting in Fuzzy Rule-based Systems for Pattern Classification Problems. Fuzzy Sets & Systems 103, 223–238 (1999)
[4] Ishibuchi, H., Nakashima, T., Murata, T.: Performance Evaluation of Fuzzy Classifier Systems for Multidimensional Pattern Classification Problems. IEEE Transactions on Systems, Man, and Cybernetics Part B: Cybernetics 29(5), 601–618 (1999)
[5] Fonseca, C.M., Fleming, P.J.: An Overview of Evolutionary Algorithms in Multiobjective Optimization. Evolutionary Computation 3, 1–16 (1995)
[6] Coello, C.A.C.: A Comprehensive Survey of Evolutionary-based Multiobjective Optimization Techniques. Knowledge and Information Systems 1, 269–308 (1999)
[7] Zitzler, E.: Evolutionary Algorithms for Multiobjective Optimization: Methods and Applications, Ph.D. Thesis, Swiss Federal Institute of Technology, Zürich, Switzlerland (1999)
[8] Reyes-Sierra, M., Coello, C.A.C.: Multi-objective Particle Swarm Optimizers: A Survey of the State-of-the-Art. International Journal of Computational Intelligence Research 2(3), 287–308 (2006)

[9] Coello, C.A.C., Lechuga, M.S.: MOPSO: A Proposal for Multiple Objective Particle Swarm Optimization. In: Proceedings of the 2002 Congress on Evolutionary Computation (CEC 2002), vol. 2, pp. 1051–1056 (2002)
[10] Zahiri, S.H.: Designing Multi-objective Classifier Based on the Multi-objective Particle Swarm Optimization. Iranian Journal of Electrical and Computer Engineering 4(2), 91–98 (2007) (in persian)
[11] Pal, S.K.: Fuzzy Set Theoretic Measures for Automatic Feature Evaluation–II. Information Sciences 64, 165–179 (1992)
[12] Johnson, R.A., Wichern, D.W.: Applied Multivariate Statistical Analysis. Prentice-Hall, Englewood Cliffs (1982)
[13] Zahiri, S.H., Seyedin, S.A.: Swarm Intelligence Based Classifiers. Journal of the Franklin Institute 344, 362–376 (2007)
[14] Fukunaga, K.: Introduction to Statistical Pattern Recognition. Academic Press, New York (1972)
[15] Bandyopadhyay, S., Murthy, C.A., Pal, S.A.: Theoretical Performance of Genetic Pattern Classifier. Journal of the Franklin Institute 336, 387–422 (1999)
[16] Duda, R.O., Hart, P.E.: Pattern Classification and Scene Analysis. Wiley, New York (1973)
[17] Apostol, T.M.: Mathematical Analysis. Narosa Publishing House, New Dehli (1985)
[18] Zahiri, S.H.: Theoretical Performance Investigation of PSO based Classifiers. In: Proceedings of 11th International CSI Computer Conference, CSICC 2006, Tehran, Iran, pp. 1–7 (2006) (in Persian)
[19] De Falco, I., Della Cioppa, A., Tarantino, E.: Facing Classification Problems With Particle Swarm Optimization. Applied Soft Computing 7, 652–658 (2007)
[20] Tsai, C.-Y., Yeh, S.-W.: A multiple Objective Particle Swarm Optimization Approach for Inventory Classification. International Journal of Production Economics 114(2), 656–666 (2008)
[21] Guvenir, H.A., Erel, E.: Multicriteria Inventory Classification Using a Genetic Algorithm. European Journal of Operational Research 105, 29–37 (1998)
[22] Van der Merwe, D.W., Engelbrecht, A.P.: Data Clustering Using Particle Swarm Optimization. In: Proceedings of the 2003 IEEE Congress on Evolutionary Computation, pp. 215–220 (2003)
[23] Yang, Y., Kamel, M.: Clustering Ensemble Using Swarm Intelligence. In: Proceedings of the 2003 IEEE Swarm Intelligence Symposium (SIS 2003), pp. 65–71 (2003)
[24] Omran, M.: Particle Swarm Optimization Methods for Pattern Recognition and Image Processing, Ph.D. Thesis, University of Pretoria, South Africa (2005)
[25] Omran, M., Salman, A., Engelbrecht, A.P.: Image Classification Using Particle Swarm Optimization. In: Proceedings of the 4th Asia-Pacific Conference on Simulated Evolution and Learning 2002 (SEAL 2002), Singapore, pp. 370–374 (2002)
[26] Cui, X., Potok, T.E.: Document Clustering Analysis Based on Hybrid PSO+K-means Algorithm. Journal of Computer Sciences, 27–33 (2005) ISSN 1549-3636
[27] Fun, Y., Chen, C.Y.: Alternative KPSO-Clustering Algorithm. Tamkang Journal of Science and Engineering 8(2), 165–174 (2005)
[28] Sousa, T., Neves, A., Silva, A.: Swarm Optimisation as a New Tool for Data Mining. In: International Parallel and Distributed Processing Symposium (IPDPS 2003), pp. 48–53 (2003)
[29] Sousa, T., Silva, A., Neves, A.: Particle Swarm based Data Mining Algorithms for Classification Tasks. Parallel Computing 30(5-6), 767–783 (2004)

[30] Nabavi-Kerizi, S.H., Abadi, M., Kabir, E.: A PSO-based Weighting Method for Linear Combination of Neural Networks. In: Computers and Electrical Engineering (in press, 2009) doi:10.1016/j.compeleceng.2008.04.006
[31] Chen, Y., Abraham, A.: Hybrid Learning Methods for Stock Index Modeling. In: Kamruzzaman, J., Begg, R.K., Sarker, R.A. (eds.) Artificial Neural Networks in Finance, Health and Manufacturing: Potential and Challenges. Idea Group Inc. Publishers, USA (2006)
[32] Settles, M., Rylander, B.: Neural Network Learning Using Particle Swarm Optimizers. Advances in Information Science and Soft Computing, 224–226 (2002)
[33] Chen, Y., Abraham, A.: Hybrid Neurocomputing for Detection of Breast Cancer. In: The Fourth IEEE International Workshop on Soft Computing as Transdisciplinary Science and Technology (WSTST 2005), Japan, pp. 884–892. Springer, Germany (2005)
[34] Knowles, J.D., Corne, D.W.: Approximating the Nondominated Front Using the Pareto Archived Evolution Strategy. Evolutionary Computation 8(2), 149–172 (2000)
[35] Abraham, A., Grosan, C., Ramos, V.: Swarm Intelligence in Data Mining. Springer, Heidelberg (2006)
[36] Marinakis, Y., Marinaki, M., Dounias, G.: Particle Swarm Optimization for Pap-smear Diagnosis. Expert Systems with Applications 25(4), 1645–1656 (2008)
[37] Pedrycz, W., Park, B.J., Pizzi, N.J.: Identifying Core Sets of Discriminatory Features Using Particle Swarm Optimization. Expert Systems with Applications, Part 1 36(3), 4610–4616 (2009)
[38] Oliveira, L.S., Britto, A.S., Sabourin, R.: Improving Cascading Classifiers with Particle Swarm Optimization. In: International Conference on Document Analysis and Recognition (ICDAR 2005), Seoul, South Korea, pp. 570–574 (2005)
[39] Tou, J.T., Gonzalez, R.C.: Pattern Recognition Principles. Addison-Wesley, Reading (1992)
[40] Clerc, M., Kennedy, J.: The particle swarm–explosion, stability, and convergence in a multidimensional complex space. IEEE Transactions on Evolutionary Computation 6(1), 58–73 (2002)

5

Optimizing Decision Trees Using Multi-objective Particle Swarm Optimization

Jonathan E. Fieldsend

School of Engineering, Computing and Mathematics
University of Exeter
Harrison Building, North Park Road, Exeter, EX4 4QF, UK
J.E.Fieldsend@exeter.ac.uk

Summary. Although conceptually quite simple, decision trees are still among the most popular classifiers applied to real-world problems. Their popularity is due to a number of factors – core among these is their ease of comprehension, robust performance and fast data processing capabilities. Additionally feature selection is implicit within the decision tree structure.

This chapter introduces the basic ideas behind decision trees, focusing on decision trees which only consider a rule relating to a single feature at a node (therefore making recursive axis-parallel slices in feature space to form their classification boundaries). The use of particle swarm optimization (PSO) to train near optimal decision trees is discussed, and PSO is applied both in a single objective formulation (minimizing misclassification cost), and multi-objective formulation (trading off misclassification rates across classes).

Empirical results are presented on popular classification data sets from the well-known UCI machine learning repository, and PSO is demonstrated as being fully capable of acting as an optimizer for trees on these problems. Results additionally support the argument that multi-objectification of a problem can improve uni-objective search in classification problems.

5.1 Introduction

The problem of classification is a popular and widely confronted one in data-mining, drawing heavily from the fields of machine learning and pattern recognition. Classification, most simply put, is the assignment of a class $\mathcal{C}_i$ to some observed datam $\mathbf{x}$, based on some functional transformation of $\mathbf{x}$, $\hat{p}(\mathcal{C}_i|\mathbf{x}) = f(\mathbf{x}, \mathbf{s}, \mathbf{D})$, where $\hat{p}(\mathcal{C}_i|\mathbf{x})$ is an estimate of the underlying probability of observation $\mathbf{x}$ belonging to class i (the class typically assigned by the classifier to $\mathbf{x}$ being that class with the highest estimated probability), and $\mathbf{D}$ is some set of pre-labelled data used in the selection of model parameters $\mathbf{s}$. Depending on the classifier used, it may produce the probability directly, a score that can be converted into a probability, or a hard classification (that is, assigning all the probability to a single class). The learning (or optimization) aspect in classifiers

C.A. Coello Coello et al. (Eds.): Swarm Intel. for Multi-objective Prob., SCI 242, pp. 93–114.
springerlink.com

relates to $\mathbf{s}$, the tunable parameters of classifier function $f()$.* These are adjusted so as to minimize the difference between the estimated probabilities assigned to data, and the underlying true probabilities. Often the latter are not known, and this has to be approximated using the corpus of *training* data $\mathbf{D}$, and possibly some prior.

The range of classification problems is extensive, with applications as diverse economics and finance, biology, engineering and safety systems, medicine, etc. One of the most popular classifiers (if not *the* most popular) is the decision tree [2]. A decision tree consists of a sequence of connected nodes, each of which act as a discriminator. The edges between the nodes are uni-directional, and travelling down from the root node, they act to sequentially partition the data space until a terminus node (also known as a leaf) is reached. The leaf assigns a class (or infrequently a pseudo class probability) to the unique volume of feature space covered by the leaf. The internal nodes themselves partition the data by applying a rule to the data, typically these are of the form 'if-then-else'. For instance, the rule may be 'if feature 2 is greater than 2.5, go to child node 1, else go to child node 2'. This feature may for instance be the age of an individual or the amount of money applied for in a loan – and the child nodes may contain further rules, or be leaf nodes relating to actual decisions.

Decision trees derive a lot of their popularity from their ease of comprehension. The assignment of class can be traced back to a sequence of rules, which make it easy to explain to end users; this also aids tremendously in their application to e.g. safety critical systems [3] or medical applications [4], where 'black box' classifiers have difficulty passing regulatory hurdles due to their opaque processing nature. Additionally their computational complexity is relatively low and are ideally specified for batch processing, meaning they are widely used for real-time classification tasks, and problems requiring a large throughput of data.

The chapter will proceed as follows, in Sect. 5.2 decision trees will be discussed in further depth along with their properties. In Sect. 5.3 the basic particle swarm optimization (PSO) algorithm is introduced, followed by Sect. 5.4 which discusses the representation of decision trees to enable optimization by PSO. Sect. 5.5 discusses various multi-objective problems related to decision trees, and introduces a multi-objective PSO variant to optimize decision trees. Sect. 5.6 presents empirical results using the most popular data sets from the widely used UCI machine learning repository [5], with a chapter summary presented in Sect. 5.7.

5.2 Decision Trees

As mentioned in the preceding section, the nodes in a decision tree act as rules, recursively partitioning the decision space. If the rule covers a single feature, as

*It sometimes also relates to choosing $f()$, or to selecting the most informative members of $\mathbf{D}$ to learn from (data which can also form a part of $\mathbf{s}$ in some classifiers, e.g. k-nearest neighbors and support vector machines [1]).

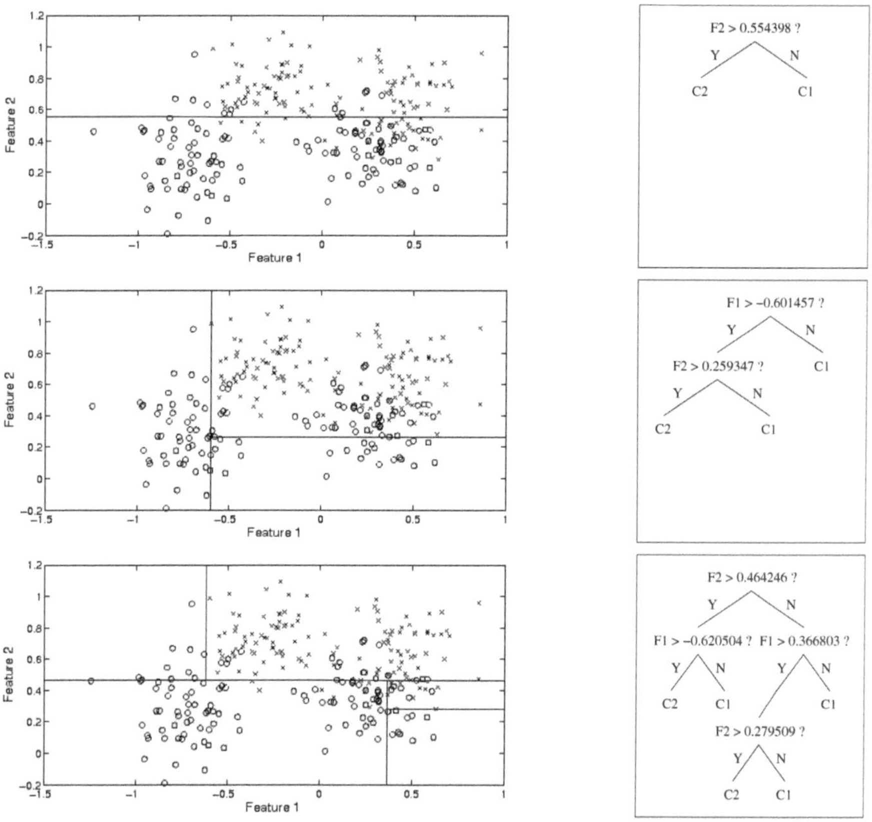

Fig. 5.1. *Left*: Decision tree partitioning of the Ripley data feature space, minimizing total misclassification (class 1 data denoted by circles, class 2 by crosses). *Right*: Corresponding trees.

the example given above, then the partitions are axis parallel*, although this generates quite a simple decision boundary, as the tree depth is increased the feature space can be partitioned into more and more different sections, and the resulting decision boundary becomes more complex (though piecewise linear, and axis parallel in these sections).

Fig. 5.1 shows example decision tree decision boundaries (and corresponding trees) for the two dimensional synthetic data from [6]†. Note that the trees shown are of various depths, illustrating the way the decision boundary complexity can

*Some decision trees also combine features in single rules, enabling non-axis-parallel partitions of feature space. This form will not be covered further here however.

†This data is generated by sampling from four two-dimensional Gaussians, with centers of class 1 instances at $\mu_{11} = (-0.7, 0.3)$, $\mu_{12} = (0.3, 0.3)$ and centers of class 2 instances at $\mu_{21} = (-0.3, 0.7)$, $\mu_{22} = (0.4, 0.7)$, with identical covariances of $0.03I$.

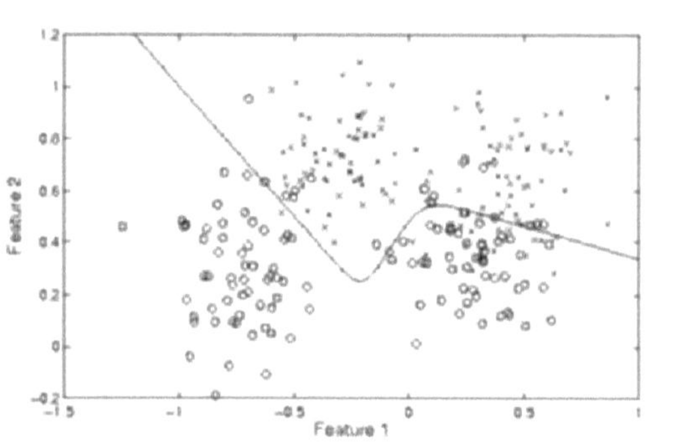

Fig. 5.2. Bayes rule decision boundary on the synthetic Ripley data (the best possible decision boundary, assuming equal misclassification costs, generated from the underlying data model).

increase with tree depth. The corresponding Bayes rule decision boundary for this problem is shown in Fig. 5.2 for completeness.

Among the most popular traditional learning algorithms for decision trees are CART (classification and regression trees) [7], ID3 (iterative dichotomizer 3) [8] and C4.5 [9]. The search through all the possible symbol choices for a symbol (the rule or class to contain in a node), along with all possible thresholds is typically infeasible (irrespective of the computational cost of varying the depth of the tree). As such tree learning algorithms typically employ some form of greedy search, performing a local exhaustive search at the data covered by a node when selecting its symbol and threshold. The issue of when to stop growing the tree (make a node a terminal) is confronted in a number of different ways. The most popular are by stopping the growth when the reduction in prediction error (typically entropy is used) falls below a certain threshold, when the number of points covered by a node falls below a certain threshold, or letting a tree grow large and then prune back (remove and recombine) leafs, based on some trade-off of accuracy (error) and complexity (tree size).

Given the nature of the tree learning/optimization (its size and complexity), and that the learning methods currently used are only locally optimal, evolutionary optimization algorithms have also gained popularity as decision tree parameter optimizers.

5.3 Particle Swarm Optimization

The PSO heuristic was initially proposed for the optimization of continuous nonlinear functions [10]. Subsequent work in the field has developed some methods for its operation in discrete domains (e.g. [11]) however the continuous domain remains its principle field of deployment.

In standard PSO a fixed population of M potential solutions, $\{\mathbf{s}_i\}_{i=1}^{M}$, is maintained, where each of these solutions (or particles) is represented by a point in P-dimensional space (where P is the number of parameters to be optimized).

Each of these solutions maintains knowledge of its 'best' previously evaluated position (its personal best) $\mathbf{p}_i$, and also has access to the 'best' solution found so far by the population as a whole, $\mathbf{g}$, which by definition is also one of the swarm member's personal best. The rate of position change of a particle/solution from one iteration/generation to the next depends upon its previous local best position, the global best position, and its previous trajectory (its velocity, $\mathbf{v}_i$). The general formula for adjusting the jth parameter of the ith particle's velocity is:

$$v_{j,i} := wv_{j,i} + c_1r_1(p_{j,i} - s_{j,i}) + c_2r_2(g_j - s_{j,i}) \tag{5.1}$$

$$s_{j,i} := s_{j,i} + \chi v_{j,i}. \tag{5.2}$$

Where w, c_1, c_2, $\chi \geq 0$. w is the inertia of a particle (how much its previous velocity affects its next trajectory), c_1 and c_2 are constraints on the velocity toward the global and local best and χ is a constraint on the overall shift in position (often a maximum absolute velocity, $V_{\max}$ is also applied). r_1 and r_2 are random draws from the continuous uniform distribution, i.e. $r_1, r_2 \sim U(0,1)$. In [10] the final model presented has w and χ fixed at 1, and c_1 and c_2 fixed at 2. Later work has tended toward varying the inertia term downward during the search to aid final convergence.

The PSO heuristic has proved to be an extremely popular optimization technique, with reputation for relatively fast convergence, and as such is its application to decision tree optimization has recently gained interest.

5.4 Representation

As you may have noted from the description above, decision trees may be variable in size, with their parameters a mixture of unordered discrete (i.e. which features and rules to include in nodes, and which class to assign to a leaf) and ordered continuous (i.e. which thresholds to use in rules). As such the evolutionary algorithm of choice for optimizing them has tended to be genetic algorithms (see e.g. [12] for a recent discussion of these methods in a multi-objective setting). Heuristic tree growing and pruning methods are among the more traditional forms of decision tree construction [2], as discussed in Sect. 5.2, and advanced methods from the machine learning literature like Bayesian averaging have also been applied [13] (although it is an open problem to effectively sample from the posterior). Recently Veenhuis *et al.* [14] introduced a general PSO-based 'tree swarming algorithm' for optimizing generic tree structures (i.e. decision trees, parse tress, program trees, etc.) with respect to a single quality (objective) measure. A slightly modified version of their tree representation is used here, and is described below (with variations from [14] highlighted).

Fig. 5.3 illustrates a full ordered 2-ary tree with a single root, 4 layers and directed edges (denoted by $T_{4,2}$). As the tree is full, the final layer (nodes 7-14) are terminal nodes (leafs), having no children of their own. The size of a general tree with L layers and arity of A, $T_{L,A}$, in terms of the number of internal and terminal nodes is

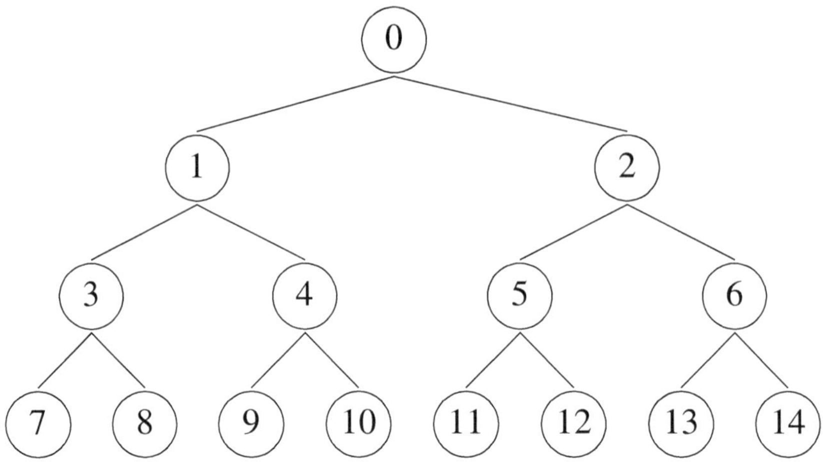

Fig. 5.3. Illustration of a full A-ary tree with L layers, where A=2 and L=3.

$$\texttt{size}(T_{L,A}) = \sum_{i=0}^{L-1} A^i. \tag{5.3}$$

(Note also there are $\texttt{size}(T_{L,A}) - 1$ edges in a full $T_{L,A}$ tree.)

Although trees may be mapped to a continuous valued vector for use in Equation 5.1, it is easier for comprehension to describe the mapping in terms of a continuous valued matrix (the final transformation from a matrix to a vector representation being trivial). First consider the mapping of the nodes in a tree to an array. As laid out in [14], and illustrated in Fig. 5.3, the nodes may be numbered in the tree, starting at 0 at the root, and counting from left to right at each subsequent level. The index of a child node c of any particular parent node p can be calculated as:

$$\texttt{child\_index}(p, c, A) = Ap + c. \tag{5.4}$$

Where c denotes the cth child of parent p (i.e. $1 \leq c \leq A$), and $0 \leq p \leq \texttt{size}(T_{L,A}) - 1$. Using Equation 5.4 one can travel down the tree until the appropriate leaf is reached. Fig. 5.4 shows the node array constructed in this fashion corresponding to the tree in Fig. 5.3.

The representation of node traversal has now been covered, however the key problem is in the transformation of the rules (symbols) to continuous values for use in PSO, which can impose a (variable) order on the symbols. This is confronted by [14] with the use of a symbol vector, whose length is equal to the total number of symbols possible in a node (in the case of decision trees, this would be the number of different rules plus the number of different classes). Each element in the symbol vector is a score (here on the range [0,1]), with the symbol used being determined by the index of the element in the symbol vector with the maximum score. Consider for instance a classification problem with five features and three classes, which we want to describe using a tree exclusively

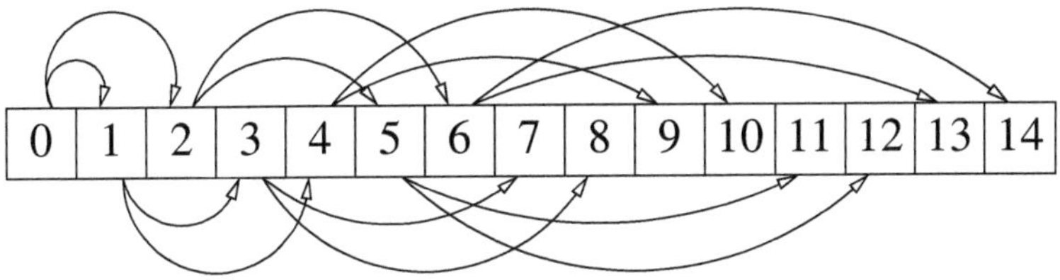

Fig. 5.4. Illustrative node array.

using rules of the 'if feature greater than else' form. This would lead to eight distinct symbols, the first five relating to which feature to use in the rule, and the last three to which class to assign, i.e., {'if feature 1 is greater than, else','if feature 2 is greater than, else', 'if feature 3 is greater than, else', 'if feature 4 is greater than, else','if feature 5 is greater than, else', 'class 1', 'class 2', 'class 3'}. A node with a symbol vector whose maximum element was 1, 2, 3, 4, or 5 would be an internal node, whereas one with a maximum element 6, 7 or 8 would be a leaf (note this representation allows the mapping to be used for sparser trees, if the node belongs to a layer$< L$ and is assigned a leaf symbol, then none of that node's subsequent children will be evaluated).

Using the symbol vector notation, the decision tree can be represented as a matrix of symbol scores, **M**, with each column denoting a node and each row denoting a symbol, as shown below for a $T_{3,2}$ tree with five symbols.

$$\begin{array}{c} \\ S_1 \\ S_2 \\ S_3 \\ S_4 \\ S_5 \end{array} \begin{array}{c} \begin{array}{ccccccc} 0 & 1 & 2 & 3 & 4 & 5 & 6 \end{array} \\ \begin{pmatrix} M_{1,1}, & M_{1,2}, & M_{1,3}, & M_{1,4}, & M_{1,5}, & M_{1,6}, & M_{1,7} \\ M_{2,1}, & M_{2,2}, & M_{2,3}, & M_{2,4}, & M_{2,5}, & M_{2,6}, & M_{2,7} \\ M_{3,1}, & M_{3,2}, & M_{3,3}, & M_{3,4}, & M_{3,5}, & M_{3,6}, & M_{3,7} \\ M_{4,1}, & M_{4,2}, & M_{4,3}, & M_{4,4}, & M_{4,5}, & M_{4,6}, & M_{4,7} \\ M_{5,1}, & M_{5,2}, & M_{5,3}, & M_{5,4}, & M_{5,5}, & M_{5,6}, & M_{5,7} \end{pmatrix} \end{array} .$$

The symbol S_i to use for a particular node $j + 1$ being the determined by the maximum element of the jth column of **M**. The issue of the threshold is resolved in [14] by making the decision trees 3-ary, with the first element of the symbol vector of the first child of an internal node determining the threshold (by rescaling the value contained from $[0, 1]$ to $[\mathtt{min}(F_i), \mathtt{max}(F_i)]$, where F_i denotes the feature used in the rule, and $\mathtt{min}(F_i)$ and $\mathtt{max}(F_i)$ returning the minimum and maximum respectively of feature i in the training data. This is a somewhat wasteful representation as only the first symbol of the first child node will ever be used, and none of its subsequent children will ever be accessed (although they will be represented) as it is treated as a terminal node. The same effect may be implemented with much less space required by adding an extra row on the bottom of matrix M to hold the threshold to be used if the node is internal. An arguably even better approach, is to add z extra rows on the bottom of M (where z is the number of features), so that there are different threshold values represented for each different feature. This allows the thresholds at nodes to be learnt in parallel and prevents the problems that may arise when changing the feature potentially makes the single threshold stored inappropriate. This does increase the number of dimensions of the problem, but should act to improve the smoothness of the search space.

It is worth noting that whereas the matrix representation is easier to interpret, there is no reason for it not to be represented in a program as a vector. Conversion between the two representations is trivial, and if a preferred optimizer is already implemented to deal exclusively with vector represented solutions, this intermediate conversion may be used as an interface between the solution and evaluation.

5.5 Multi-objective PSO

As discussed above, PSO has previously been applied to single objective decision tree optimization (where total misclassification was the objective to be minimized). There are however situations where one might want to optimize a decision tree with respect to multiple objectives. Three specific situations are covered here. Firstly when also minimizing the size of the tree. This may be important due to processing time – the larger the tree the longer the processing time for a particular query, which in certain situations may be a critical factor (e.g., real time fraud detection). Secondly when also minimizing the number of features used by the tree. Initial and subsequent feature selection for a classification system has a cost associated with it (sometimes quite significant, e.g. chemical/biological measurements), minimizing the number of features used acts to lower this cost, and also remove any redundancy across features. Thirdly when multiple error measures need to be optimized – for instance in binary classification problems the overall misclassification rate may be less important than the relative true and false positive rates, where misclassification costs are not equal (e.g. cancer detection). A brief outline of these types of multi-objective problem is given below.

5.5.1 Multiple Objectives

Structure

As mentioned in the introduction, one of the attractions of decision trees as classifiers is the fast computation time when classifying new data. The computational cost is directly proportional to the number of internal nodes in a tree, and therefore the smaller the tree the faster the processing ability. Additionally there tends to be a trade-off between a decision tree's *generalization* ability and size (a tree that is too big – too flexible – may overfit to the data being trained on). This is usually apparent when the number of data samples covered by each leaf is very small (hence the use of pruning in some tree learning algorithms, as mentioned in Sect. 5.2). A natural additional objective is therefore to minimize the size of the tree (see for example [12]).

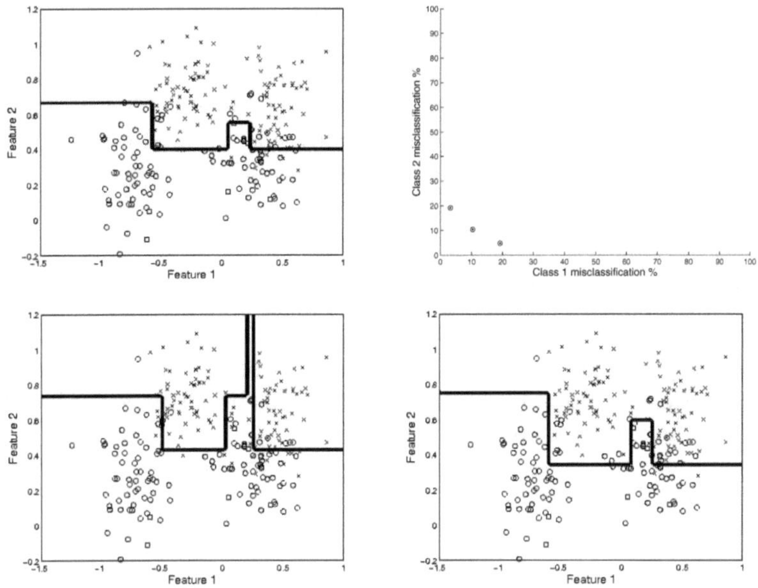

Fig. 5.5. a,c,d) Decision tree partitioning of the Ripley data feature space, trading off misclassifiction rates between the two classes. b) Corresponding points in misclassification rate space.

Feature space

In many real world situations data collection costs time and money. Features which don't contribute to the performance of a classifier can be detrimental if included, and others may duplicate information or have only a marginal effect. As such feature selection, and feature minimization, are also of concern when constructing decision trees, and can also be cast as an additional objective, both as a means of improving generalization performance, and to decrease the cost of future data collection.

Multiple error terms

By minimizing the total misclassification error one is implicitly stating that the misclassification costs across classes are equivalent. Often this is not the case (for example when screening for cancers, or in safety related classification problems). Where the costs are unknown *a priori* and/or the shape of the trade-off front is unknown, it is appropriate to trade-off the different misclassification rates in parallel [15, 16]. An illustration of this is provided in Fig. 5.5 using the synthetic data described earlier. The upper right plot shows the objective space mapping of three decision trees, where the objectives are minimizing the class 1 misclassification rate and minimizing the class 2 misclassification rate. Note this is equivalent to the widely used Receiver Operating Characteristics (ROC) curve representation used for binary classification tasks – by representing the curve in terms of both misclassification rates, instead of focusing on a single class (correct assignment and incorrect assignment rates to that class), the problem is more easily extended to *multiple* (i.e. > 2) class problems [15, 16]. The class mapping on feature space caused by the three mutually non-dominating trees are also plotted in Fig. 5.5, and arranged in the order they are plotted in the trade-off front.*

5.5.2 Dominance and Pareto Optimality

The vast majority of recent multi-objective optimization algorithms (MOAs) rely on the properties of dominance and Pareto optimality to compare and judge potential solutions to a problem, as such these will be briefly reviewed here before discussing the multi-objective PSO algorithm.

The multi-objective optimization problem is concerned with the simultaneous extremization of D objectives:

$$y_i = f_i(\mathbf{s}), \quad i = 1, \ldots, D \tag{5.5}$$

where each objective evaluation depends upon the parameter vector $\mathbf{s} = \{s_1, s_2, \ldots, s_P\}$. These parameters may also be subject to various inequality and equality constraints:

$$e_j(\mathbf{s}) \geq 0, \quad j = 1, \ldots, J, \tag{5.6}$$
$$g_k(\mathbf{s}) = 0, \quad k = 1, \ldots, K. \tag{5.7}$$

Without loss of generality it can be assumed that the objectives are to be minimized, thus the problem can be expressed as:

$$\text{Minimize} \quad \mathbf{y} = \mathbf{f}(\mathbf{s}) = \{f_1(\mathbf{s}), f_2(\mathbf{s}), \ldots, f_P(\mathbf{s})\}, \tag{5.8}$$
$$\text{subject to} \quad \mathbf{e}(\mathbf{s}) = \{e_1(\mathbf{s}), e_2(\mathbf{s}), \ldots, e_J(\mathbf{s})\} \geq 0, \tag{5.9}$$
$$\mathbf{g}(\mathbf{s}) = \{g_1(\mathbf{s}), g_2(\mathbf{s}), \ldots, g_K(\mathbf{s})\} = 0. \tag{5.10}$$

When concerned with a single objective, an optimal solution is one which minimizes the objective subject to any constraints within the model. In the situation where there is more than one objective, then it is often the case that solutions exist for which performance cannot be improved on one objective without sacrificing performance on at least one other. Such solutions are said to be *Pareto optimal*, with the set of all such solutions being the *Pareto set*, and their image in objective space known as the *Pareto front*.

The notion of *dominance* is crucial to understanding Pareto optimality, and is relied upon heavily in most modern multi-objective optimizers. A decision

*Note that as there is no assessment on a test set of data of the generalization ability of these trees (visually we would be concerned of the overfitting of the bottom left tree for instance, given our knowledge of the underlying data generation process).

Algorithm 2. A multi-objective PSO algorithm.

Require: I — Number of PSO iterations
Require: M — Number of particles/solutions
Require: N — Dimension of search problem
Require: w — Inertia value
Require: c_1 — Global search weight
Require: c_2 — Local search weight
Require: χ — Constriction variable
Require: $V_{\max}$ — Absolute maximum velocity
$\{\mathbf{S},\mathbf{V}\}$:= `initialize_population`(M,N) — See Algorithm 3
$i := 0$
$\mathbf{Y}$:= `evaluate`$(\mathbf{S})$ — Assess particle
$\mathbf{G}$:= `initialize_gbest`$(\mathbf{S},M)$ — See Algorithm 5
$\mathbf{P}$:= `initialize_pbest`$(\mathbf{S},M)$ — See Algorithm 4
while $i < I$: **do**
  $\mathbf{V}$:= `update_velocity`$(\mathbf{S},\mathbf{P},\mathbf{G},w,c_1,c_2,M,N)$ — See Algorithm 6
  $\mathbf{V}$:= `restrict_velocity`$(\mathbf{V},V_{\max})$ — Ensure no velocity element exceeds $V_{\max}$
  $\mathbf{P} := \mathbf{S} + \chi\mathbf{V}$
  $\mathbf{Y}$:= `evaluate`$(\mathbf{S})$ — Assess particle
  $\mathbf{G}$:= `update_gbest`$(\mathbf{G},\mathbf{S},M)$ — See Algorithm 8
  $\mathbf{P}$:= `update_pbest`$(\mathbf{P},\mathbf{S},M)$ — See Algorithm 7
  w := `update_inertia`(w,i) — Decrease inertia
  $i := i+1$
end while

vector (also known as a solution/parameter vector) $\mathbf{s}$ is said to *strictly dominate* another $\mathbf{v}$ (denoted $\mathbf{s} \prec \mathbf{v}$) iff

$$f_i(\mathbf{s}) \leq f_i(\mathbf{v}), \quad \forall i = 1,\ldots,D \quad \text{and} \tag{5.11}$$

$$f_i(\mathbf{s}) < f_i(\mathbf{v}), \quad \text{for at least one } i. \tag{5.12}$$

Note that the dominance relationship $\mathbf{s} \prec \mathbf{v}$ is denoted in the parameter space domain, whereas the calculation is in the objective space mapping of the parameters. As such $\mathbf{f}(\mathbf{s}) \prec \mathbf{f}(\mathbf{v})$ is perhaps more accurate, however the accepted shorthand will be used throughout the rest of the chapter.

A set of M decision vectors $\{\mathbf{w}_i\}_{i=1}^{M}$ is said to be a *non-dominated* set (i.e. and estimate of the Pareto set) if no member of the set is dominated by any other member:

$$\mathbf{w}_i \not\prec \mathbf{w}_j \quad \forall i,j = 1,\ldots,M. \tag{5.13}$$

The aim of most MOAs is to find such a non-dominated set – whose image in objective space is well converged to, and spread across, the true Pareto front, and is therefore a good approximation of the underlying Pareto set.*

*Note that due to the non-linear mappings involved, closeness in objective space may not relate to closeness in decision space. As such, even if the Pareto front is known a *priori*, closeness to it is not a guarantee that the set members are "close" in parameter space to the Pareto set.

Algorithm 3. `initialize_population`(M, N).

```
S := ∅                                        Create empty set
V := ∅                                        Create empty set
i := 1
while i ≤ M do
    j := 1
    while j ≤ N do
        S_{i,j} := U(0,1)                     Insert a uniform sample from (0,1)
        V_{i,j} := 0                          Initialize velocity – random samples also possible
        j := j + 1
    end while
    i := i + 1
end while
return {S, V}                                 RETURN initial search population and velocity
```

Algorithm 4. `initialize_pbest`$(\mathbf{S}, M)$.

```
i := 1
while i ≤ M do
    P^i := ∅                                  Create empty personal best set for search particle
    P^i := P^i ∪ S_i                          insert current position as initial set
    i := i + 1
end while
return P                                      RETURN personal best sets
```

5.5.3 The Optimizer

There are a number of different approaches to extending PSO to multi-objective problems (see e.g. [17, 18] for reviews). Building on the previous work of Alvarez-Benitez *et al.* [19], the implementation used here relies solely on dominance to select the guides for individual particles (although the use of distance measures on *search* space is also investigated). This circumvents the issue of bias and appropriate objective weighting that other alternative selection processes can lead to [17].

Using the decision tree representation presented earlier, a general multi-objective PSO algorithm is presented in Algorithm 2.

As detailed in Algorithm 2, the principle inputs to the optimizer are the coefficients from Equation 5.1, along with the number of iterations the optimizer is to be run for (alternatively convergence measures could be used instead [20]), the number of particles in the search population, I, and the solution size, N (which can be calculated from the Equations given in Sect. 5.4). The `evaluate()` procedure (lines 3 and 9) relies on the transformation of the solution vector to a tree (as described in Sect. 5.4), and the subsequent evaluation on a set of data $\mathbf{D}$ and calculation of errors (examples of which are given in Sect. 5.5.1). The search population $\mathbf{S}$, and associated velocities $\mathbf{V}$ are initialized using draws from the uniform distribution (line 1, and Algorithm 3), and after evaluating the search

Algorithm 5. `initialize_gbest`$(\mathbf{S}, M)$.

```
G := ∅                                   Create empty global best set
i := 1
while i ≤ M do
    if S_j ⊀ S_i,  ∀S_j ∈ S, i ≠ j then
        G := G ∪ S_i          (IF particle non-dominated) add to global best set
    end if
    i := i + 1
end while
return G                                 RETURN global best set
```

Algorithm 6. `update_velocity`$(\mathbf{S}, \mathbf{P}, \mathbf{G}, w, c_1, c_2, M, N)$.

```
i := 1
while i ≤ N do
    j := 1
    while j ≤ M do
        r_1 := U(0,1)                Draw from a continuous uniform distribution
        r_2 := U(0,1)                Draw from a continuous uniform distribution
        V_{i,j} := w V_{i,j} + r_1 c_1 (get(G, S_i) − S_{i,j}) + r_2 c_2 (get(P^i, S_i, i) − S_{i,j})
        j := j + 1       (Alter velocity according to Equation 5.1, see Algorithm 9)
    end while
    i := i + 1
end while
return V                                 RETURN updated velocity
```

Algorithm 7. `update_pbest`$(\mathbf{P}, \mathbf{S}, M)$.

```
i := 1
while i ≤ M do
    if P^i_j ⊀ S_i   ∀P^i_j ∈ P^i then
        P^i := P^i ∪ S_i          (IF particle non-dominated) add to personal best set
        j := 1
        while j ≤ |P^i| do
            if S_i ≺ P^i_j then
                P^i := P^i \ P^i_j  (IF member is dominated) remove from personal best set
            end if
            j := j + 1
        end while
    end if
    i = i + 1
end while
return P                                 RETURN personal best sets
```

population the global best and personal best vectors are initialized. As discussed in Sect. 5.5.2, there is usually no single 'best' solution when optimizing with respect to more than one objective, and as such a set of mutually non-dominated solutions, which are the best so far encountered by the search population, are

Algorithm 8. update_gbest($\mathbf{G}, \mathbf{S}, M$).

$i := 1$
while $i \leq M$ **do**
  if $\mathbf{G}_j \nprec \mathbf{S}_i \quad \forall \mathbf{G}_j \in \mathbf{G}$ **then**
    $\mathbf{G} := \mathbf{G} \cup \mathbf{S}_i$ (IF particle non-dominated) add to global best set
    $j := 1$
    while $j \leq |\mathbf{G}|$ **do**
      if $\mathbf{S}_i \prec \mathbf{G}_j$ **then**
        $\mathbf{G} := \mathbf{G} \setminus \mathbf{G}_j$ (IF member is dominated) remove from global best set
      end if
    end while
    $j := j + 1$
  end if
  $i = i + 1$
end while
return $\mathbf{G}$ RETURN global best set

Algorithm 9. get($\mathbf{A}, \mathbf{s}$), implementation relying solely on dominance.

$\mathbf{A}^{\mathbf{s}} = \emptyset$ Initialize empty set of dominating individuals
$i := 1$
while $i \leq |\mathbf{A}|$ **do**
  if $\mathbf{A}_i \preceq \mathbf{s}$ **then**
    $\mathbf{A}^{\mathbf{s}} := \mathbf{A}^{\mathbf{s}} \cup \mathbf{A}_i$ Add set member that weakly dominates $\mathbf{s}$
  end if
  $i := i + 1$
end while
$r := \mathcal{U}(1, |\mathbf{A}^{\mathbf{S}}|)$ Draw from a discrete uniform distribution
return $\mathbf{A}^{\mathbf{S}}_r$ RETURN random weakly dominating individual

Algorithm 10. get($\mathbf{A}, \mathbf{s}$), implementation relying solely on distance in search space. Five closest members used in implementation here, though can vary this parameter.

$\mathbf{d} = \{\}_{i=1}^{|\mathbf{A}|}$ Initialize empty vector of distances
$i := 1$
while $i \leq |\mathbf{A}|$ **do**
  $d_i := ||\mathbf{A}_i - \mathbf{s}||^2$ Add Euclidean distance between solutions $\mathbf{s}$
  $i := i + 1$
end while
$\{\mathbf{d}, I\} = \texttt{sort}(\mathbf{d})$ Sort distances in ascending order, and index
$r := \mathcal{U}(1, \mathbf{min}(5, |\mathbf{A}|))$ Draw from a discrete uniform distribution
return $\mathbf{A}^{\mathbf{S}}_{I_r}$ RETURN random *close* individual

maintained in $\mathbf{G}$ (lines 4 and 11 of Algorithm 2 and Algorithms 5 and 8). Likewise there is likely no single personal best, and a set of sets $\mathbf{P}$ is maintained, with $\mathbf{P}^i$ containing the best set of mutually non-dominating solutions found by particle

$\mathbf{S}_i$ in the search so far (lines 5 and 12 of Algorithm 2 and Algorithms 4 and 7), as used previous in e.g. [21].

The basic PSO update algorithm in Equation 5.1 is implemented in lines 7-9 of Algorithm 2, with line 7 laid out more fully in Algorithm 6. The key part in Algorithm 6 is line 7 where the `get()` function is called to return a global best individual and personal best individual respectively. Two different implementations of these methods are implemented here. The first method, relying solely on dominance, is laid out in Algorithm 9. Here the selected global best for a solution, $\mathbf{S}_i$, is a member of $\mathbf{G}$ which dominates $\mathbf{S}_i$ (selected at random from the subset of $\mathbf{G}$ which weakly dominates $\mathbf{S}_i$). Likewise the personal best guide of $\mathbf{S}_i$ is chosen at random from the subset of $\mathbf{P}^i$ which weakly dominates $\mathbf{S}_i$. An alternative implementation of `get()` is presented in Algorithm 10, which uses a distance measure (Euclidean) in search space to determine the five 'closest' members of $\mathbf{G}$ to $\mathbf{P}_i$ and uses one of them at random as a guide (and similarly from $\mathbf{P}^i$ for choosing a personal best guide).

5.6 Empirical Results

In this empirical section, certain inputs are kept fixed across all experiments. The number of search particles $M = 20$, The global and local search weights, c_1 and c_2 are fixed at 2.0 (a common choice in the literature) and the inertia weight w is initially set at 1.0, and decreased linearly throughout the search until it reaches 0.1 at the termination of the algorithm. χ is fixed at 1.0, and $V_{\max}$ at 0.1 (one tenth of the range of the elements). The dimensionality N of the search problem is determined by the number of features, classes of the particular classification problem, and maximum tree size used (as previously discussed in Sect. 5.4). Likewise the number of iterations the optimizer is run for is varied with the size of the problem.

The first experiment will be to confirm the performance of the optimizer compared to previously publish results, namely those of the single objective PSO tree representation drawn from [14]. The implementation here varies slightly (as laid out in Sect. 5.4) and local search in [14] is implemented via selecting from the closest (in parameter space) search particles as opposed to from a stored personal best; so it would be useful to quantify the effect of these changes. Algorithm 2 can be used with just a single objective – the effect is that $\mathbf{G}$ will only ever contain a single particle, and likewise each $\mathbf{P}^i$ will only ever contain a single solution. A number of papers in the literature have also suggested that transforming a uni-objective problem to a multi-objective one can actually increase performance with respect to a single objective (see e.g. [22–25]), due to the effect on the mapping from search-space to objective space (making it smoother, but adding gradient, therefore making it easier to traverse). As such, as well as running Algorithm 2 to optimize the single overall misclassification rate, it is also run with respect to minimizing the individual misclassfication rates, but keeping note of the member in $\mathbf{G}$ which minimizes the overall misclassification.

Table 5.1. Mean and median total misclassification error results on the Iris data set, over 100 runs after 500 iterations (see Fig. 5.6 and 5.7 for plots versus iteration). Values for TSO, GP and C4.5 taken from [14].

Average	Uni-PSO	$MOPSO_1$	$MOPSO_2$	$MOPSO_3$	TSO	GP	C4.5
Mean	5.27	3.24	5.91	2.90	5.84	5.6	5.9
Median	4.00	2.67	4.67	2.00	–	–	–

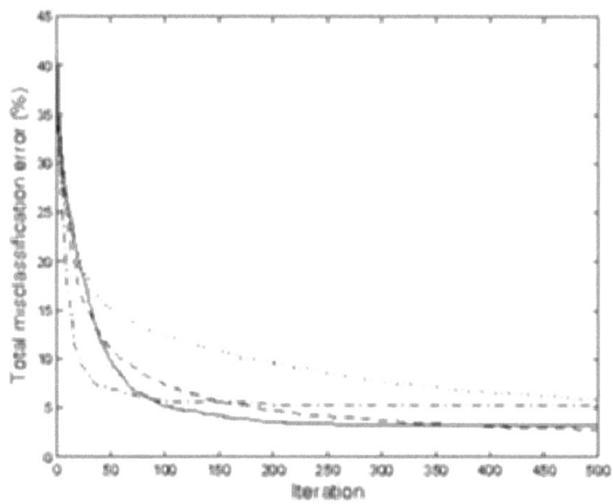

Fig. 5.6. Total misclassification error on Iris data versus PSO generation. Dash-dotted line denotes mean of the global individual of 100 runs using a single objective, solid line $MOPSO_1$, dotted line $MOPSO_2$ and dashed line $MOPSO_3$. (For MOPSO optimizers, global individual selected from the trade-off set based on minimizing total error.)

The experiments in [14] use the Iris data set from the UCI Machine Learning repository [5].* The same PSO meta-parameters are used here, with 500 iterations performed and all 150 examples of the dataset used. The algorithms are run 100 times and Table 5.1 shows the resulting mean and median total misclassification error after 500 iterations (identical classifier evaluations) of the different optimizers used. The first four are implementations of Algorithm 2, the first when minimizing a single objective (uni-PSO); when minimizing the individual misclassification rates† with the dominance guide selection method ($MOPSO_1$); minimizing individual misclassification rates with the distance guide selection method ($MOPSO_2$); and minimizing the individual misclassification rates with a (random) 50/50 use of the two different guide methods ($MOPSO_3$). The last three columns give the mean result reported in [14] of their single objective PSO

*The most popular data set in the repository, with 28,209 hits since 2007 at 30-07-2008.

†As it is a 3 class problem this results in 6 different objectives, i.e. the off-diagonal elements of the confusion rate matrix.

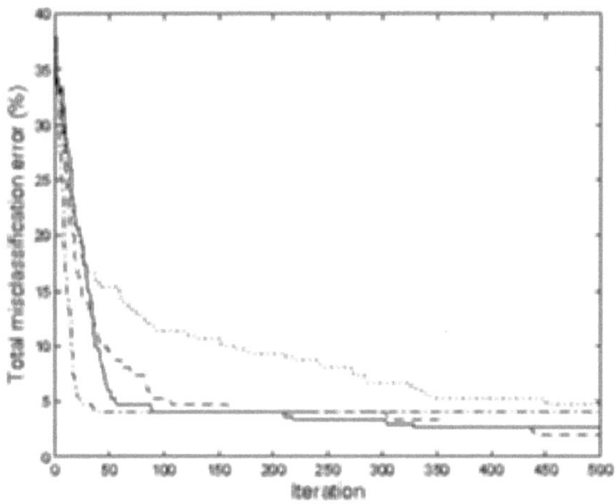

Fig. 5.7. Total misclassification error on Iris data versus PSO generation. Dash-dotted line denotes median of the global individual of 100 runs using a single objective, solid line $MOPSO_1$, dotted line $MOPSO_2$ and dashed line $MOPSO_3$. (For MOPSO optimizers, global individual selected from the trade-off set based on minimizing total error.)

Table 5.2. Properties of classification problems (sample number refers to number of complete data points in set – i.e. without missing values) [5].

Data set	# classes	# features	# samples
Adult	2	14	48842
Breast cancer Wisconsin (original)	2	10	683
Breast cancer	2	9	277
Iris	3	4	150
Statlog (Australian credit approval)	2	14	690

Table 5.3. Mean and median results of uni-objective PSO and $MOPSO_3$ over 30 runs.

	uni-obj		$MOPSO_3$	
Data set	mean	median	mean	median
Adult	17.72	17.56	18.13	18.02
Breast cancer Wisconsin (original)	3.26	3.22	2.76	2.63
Breast cancer	21.31	21.30	21.16	21.30
Statlog (Australian credit approval)	13.53	13.62	12.68	12.75

optimizer (TSO), that of a genetic programming (GP) optimizer, and C4.5. It is encouraging to note that the optimizers introduced here (except for $MOPSO_2$) outperform the results published in [14].

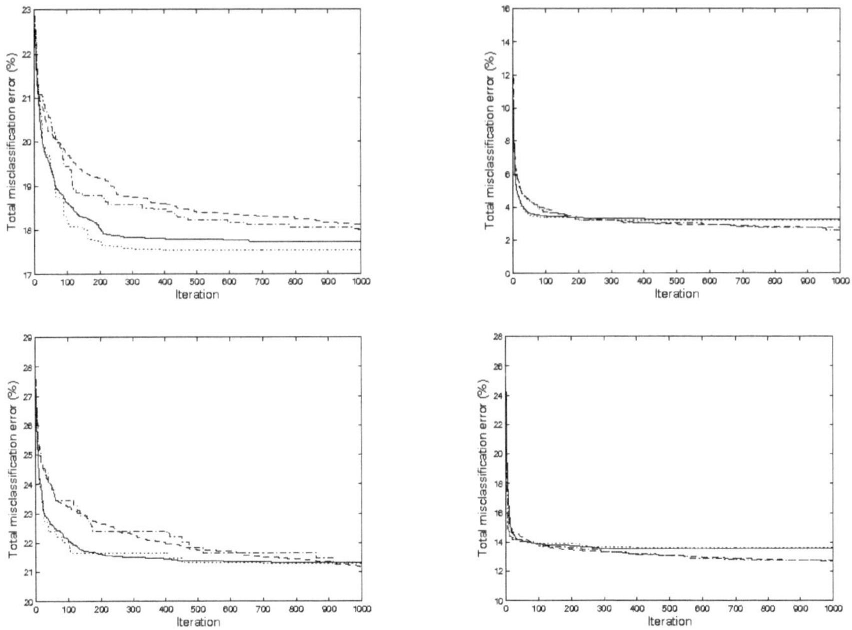

Fig. 5.8. Average total misclassification error over 30 runs. Solid line, mean uni-objective PSO, dotted line median uni-objective PSO, dashed line mean $MOPSO_3$, dash-dotted line median $MOPSO_3$. Top left plot Adult dataset, top right Breast cancer Wisconsin (original) data set, bottom left Breast cancer data set, bottom right Statlog (Australian credit approval) data set.

Fig. 5.6 shows how the mean total error varies with iteration for the four PSO implementations, and Fig. 5.7 does the same for the median. Both plots tell the same underlying story – though it should be noted that the median is a more robust statistic. Between 10 and 60 iterations the uni-objective PSO finds significantly better solutions (with respect to total error) compared to the multi-objective optimizers. This is most likely due to its focused search. The performance of the uni-objective PSO plateaus at around 200 iterations (mean) and 50 iterations (median) – indicating it has converged. The multi-objective optimizers by comparison keep improving the total misclassification error throughout the search, with $MOPSO_1$ and $MOPSO_3$ overtaking the uni-objective optimizer, with a significant out performance after 200-300 iterations (depending on the optimizer). $MOPSO_2$ (with guides selected based on distance) performs less well comparatively, however it is interesting to note that it is still improving its performance throughout the search and looks set to overtake the uni-objective optimizer if run for more iterations.

The optimizers which use dominance to select their guides, exclusively or in tandem with using a distance measure, perform much better than the one which uses distance exclusively. Interestingly it is only toward the very end of the runs

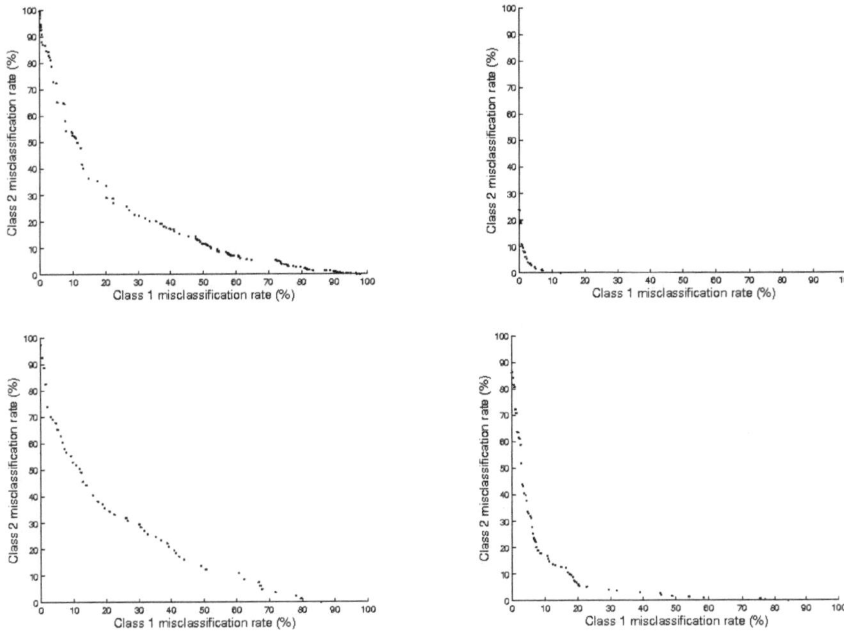

Fig. 5.9. Example trade-off fronts. Top left plot Adult data set, top right Breast cancer Wisconsin (original) data set, bottom left Breast cancer data set, bottom right Statlog (Australian credit approval) data set.

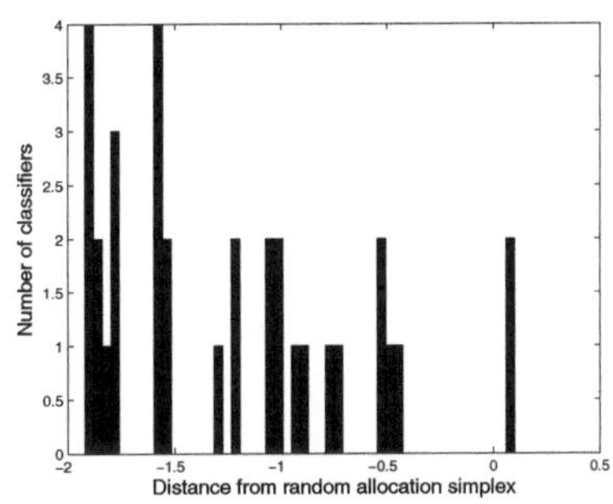

Fig. 5.10. Distance of global best points from the random allocation simplex, taken from a single run of $MOPSO_3$ on the Iris data set. The distance value of -2 relates to the optimal (typically not achievable) origin. A point with a distance value greater than zero is worse than random.

that the optimizer that uses a mixture overtakes that using strictly dominance. It seems likely that this is the effect of the distance selection promoting greater search. Toward the end of the run as the optimizer converges the inertia is very small, so if the number of global and personal best points which dominate a particle $\rightarrow 1$ (or even are identical to the particle) the variation/search aspect may shrink prematurely. Selecting at random from a subset of the closest global and personal bests to act as guides means the search aspect is maintained, whilst at the same time convergence is promoted by selecting 50% of the time using dominance.

The optimizers were run on a number of other data sets taken from the UCI machine learning repository, whose details are described in Table 5.2. Numerical results are presented in Table 5.3 for the uni-objective PSO and $MOPSO_3$ variant, running for 1000 iterations with 6 layers for these larger problems.

As the results in Table 5.3 show, on these problems (except for the Adult data set) the multi-objective search discovers better *overall* (equal cost) solutions than the optimizer that is designed specifically for that objective (again in the same number of function evaluations). Not only this but at the end of the run an estimated optimal trade-off set is returned, Fig. 5.9 shows trade-off solutions returned by the multi-objective optimizer from a single randomly selected run on the 2-class problems. The Adult data set seems more difficult than the others for the multi-objective optimizer to push forward. Looking at Table 5.2, this data set has considerably more data samples than the others, meaning the possible objective combinations is also much larger. In this case MOPSO variants such as the sigma method [26] may fair better at pushing the front forward as opposed to filling it out.

The Iris problem having 6 objectives is less easy to visualize in this format, so an alternative representation is provided in Fig. 5.10, providing a histogram of the distances of points on the trade-off front from the random allocation simplex [15, 16]. The random allocation simplex is the plane in misclassification rate space which denotes classifiers which assign classes to data points at random (with some probability). For the 3-class (6 objective) problem, the optimal point at the origin is an absolute distance of 2 (number of classes-1) from the random allocation simplex. In keeping with the minimization representation, points between the origin and random allocation simplex are given negative distances and points *behind* (worse than) the random allocation simplex are given positive distances. As is clear, the vast majority of points represent classifiers that perform better than random, with a number a great distance in front of the random allocation simplex.

5.7 Summary

This chapter has covered the optimization of decision trees, both with single and multiple objectives, utilizing the representation originally introduced in [14] for single objective optimization, and drawing on the extensive work in the literature on applying PSO to multi-objective problems. Key results not only include the success in general of applying multi-objective PSO to decision tree classification problems, able to find a set of decision trees which trade-off the different misclassification error rates, but that it also tends to find better single objective solutions compared to uni-objective PSO.

The multi-objective PSO experienced slower convergence and worse performance with the larger Adult dataset. It would be useful to investigate this further in future to see if there actually is a correlation between size and comparative performance. This may be a result of the fact that as the data set size increases the *resolution* in objective space also increases, making the potential number of points on any front increase and thereby impede convergence. Data subsampling or algorithms with greater convergence pressure make be more appropriate for these types of problem.

The multi-objective PSO variant with an equal mix of dominance based guide selection and distance based guide selection performed best of the three variants compared, with the assessment that this was due to its mixing of convergence properties and search properties. A similar effect may well be found by having a different weighting between the guide selection (as the dominance approach does tend to converge faster, and not plateau like the uni-objective variant), or alternatively using a turbulence (mutation) term (see e.g. [21]).

A final point of note is that the size of representation (and therefore size of the search space) is influenced by the maximum number of layers in the decision tree, the number of features and the number of classes. There is a potential if these are high for the search landscape to become excessive, with large flat (uninformative) sections, even with respect to multiple objectives. As such, investigation into smaller representations for continuous optimizers, or for discrete PSO, would be worth investigating.

References

[1] Bishop, C.: Pattern Recognition and Machine Learning. Information Science and Statistics. Springer, Heidelberg (2006)
[2] Duda, R., Hart, P.: Pattern Classification and Scene Analysis, 2nd edn. Wiley, Chichester (2001)
[3] Everson, R., Fieldsend, J.: Multi-objective optimization of safety related systems: An application to short term conflict alert. IEEE Transactions on Evolutionary Computation 10, 187–198 (2006)
[4] Schetinin, V., Fieldsend, J., Partridge, D., Coats, T., Krzanowski, W., Everson, R., Bailey, T., Hernandez, A.: Confident interpretation of bayesian decision tree ensembles for clinical applications. IEEE Transactions on Information Technology in Biomedicine 11, 312–319 (2007)
[5] Asuncion, A., Newman, D.: UCI machine learning repository (2007)
[6] Ripley, B.: Neural networks and related methods for classification (with discussion). Journal of the Royal Statistical Society Series B 56, 409–456 (1994)
[7] Brieman, L., Friedman, J., Olshen, R., Stone, C.: Classification and Regression Trees. Chapman & Hall/CRC (1984)
[8] Quinlan, J.: Induction of decision trees. Machine Learning 1, 86–106 (1986)
[9] Quinlan, J.: C4.5 Programs for Machine Learning. Morgan Kaufmann, San Francisco (1993)
[10] Kennedy, J., Eberhart, R.: Particle swarm optimization. In: IEEE International Conference on Neural Networks, Perth, Australia, pp. 1942–1948. IEEE Service Center (1995)

[11] Kennedy, J., Eberhart, R.: A discrete binary version of the particle swarm algorithm. In: Proceedings of the IEEE Conference on Systems, Man and Cybernetics, pp. 4104–4109. IEEE Press, Los Alamitos (1997)
[12] Kim, D.: Minimizing structural risk on decision tree classification. In: Jin, Y. (ed.) Multi-Objective Machine Learning. Studies in Computational Intelligence, vol. 16, pp. 241–260. Springer, Heidelberg (2006)
[13] Denison, D., Holmes, C., Mallick, B., Smith, A.: Bayesian Methods for Nonlinear Classification and Regression. Probability and Statistics. Wiley, Chichester (2002)
[14] Veenhuis, C., Köppen, M., Krüger, J., Nickolay, B.: Tree swarm optimization: An approach to pso-based tree discovery. In: 2005 IEEE Congress on Evolutionary Computation, vol. 2, pp. 1238–1245 (2005)
[15] Everson, R., Fieldsend, J.: Multi-class roc analysis from a multi-objective optimisation perspective. Pattern Recognition Letters 27, 918–927 (2006)
[16] Everson, R., Fieldsend, J.: Multi-objective optimisation for receiver operating characteristic analysis. In: Jin, Y. (ed.) Multi-Objective Machine Learning. Studies in Computational Intelligence, vol. 16, pp. 531–556. Springer, Heidelberg (2006)
[17] Fieldsend, J.: Multi-objective particle swarm optimisation methods. Tech. Rep. 419, Department of Computer Science, University of Exeter (2004)
[18] Coello Coello, C., Pulido, G., Lechuga, M.: Handling multiple objectives with particle swarm optimization. IEEE Transactions on Evolutionary Computation 8, 256–279 (2004)
[19] Alvarez-Benitez, J., Everson, R., Fieldsend, J.: A mopso algorithm based exclusively on pareto dominance concepts. In: The Third International Conference on Evolutionary Mutli-Criterion Optimization, pp. 459–473 (2005)
[20] Fieldsend, J., Everson, R., Singh, S.: Using unconstrained elite archives for multiobjective optimization. IEEE Transactions on Evolutionary Computation 7, 305–323 (2003)
[21] Fieldsend, J., Singh, S.: A multi-objective algorithm based upon particle swarm optimisation, an efficient data structure and turbulence. In: 2002 UK Workshop on Computational Intelligence (UKCI 2002), Birmingham, UK, pp. 37–44 (2002)
[22] Fieldsend, J., Singh, S.: Pareto evolutionary neural networks. IEEE Transactions on Neural Networks 16, 338–354 (2005)
[23] Knowles, J., Watson, R., Corne, D.: Reducing local optima in single-objective problems by multi-objectivization. In: Zitzler, E., Deb, K., Thiele, L., Coello Coello, C.A., Corne, D.W. (eds.) EMO 2001. LNCS, vol. 1993, pp. 269–283. Springer, Heidelberg (2001)
[24] Jensen, M.: Guiding single-objective optimization using multiobjective methods. In: Raidl, G.R., Cagnoni, S., Cardalda, J.J.R., Corne, D.W., Gottlieb, J., Guillot, A., Hart, E., Johnson, C.G., Marchiori, E., Meyer, J.-A., Middendorf, M. (eds.) EvoIASP 2003, EvoWorkshops 2003, EvoSTIM 2003, EvoROB/EvoRobot 2003, EvoCOP 2003, EvoBIO 2003, and EvoMUSART 2003. LNCS, vol. 2611, pp. 268–279. Springer, Heidelberg (2003)
[25] Abbass, H., Deb, K.: Searching under multi-evolutionary pressures. In: Fonseca, C.M., Fleming, P.J., Zitzler, E., Deb, K., Thiele, L. (eds.) EMO 2003. LNCS, vol. 2632, pp. 391–404. Springer, Heidelberg (2003)
[26] Mostaghim, S., Teich, J.: Strategies for finding good local guides in multi-objective particle swarm optimization (mopso). In: IEEE 2003 Swarm Intelligence Symposium (2003)

6

A Discrete Particle Swarm for Multi-objective Problems in Polynomial Neural Networks used for Classification: A Data Mining Perspective

Satchidananda Dehuri[1], Carlos A. Coello Coello[2], Sung-Bae Cho[1], and Ashish Ghosh[3]

[1] Soft Computing Laboratory, Department of Computer Science, Yonsei University, 134 Shinchon-dong, Seodaemun-gu, Seoul 120-749, South Korea
{satchi,sbcho}@sclab.yonsei.ac.kr
[2] CINVESTAV-IPN, Depto. de Computación (Evolutionary Computation Group), Av. IPN No. 2508, Col. San Pedro Zacatenco Mexico, D.F. 07360, Mexico
ccoello@cs.cinvestav.mx
[3] Machine Intelligence Unit, Indian Statistical Institute, 203 B. T. Road, Kolkata-700 108, India
ash@isical.ac.in

Summary. Approximating decision boundaries of large datasets to classify an unknown sample has been recognized by many researchers within the data mining community as a very promising research topic. The application of polynomial neural networks (PNNs) for the approximation of decision boundaries can be considered as a multiple criteria problem rather than as one involving a single criterion. Classification accuracy and architectural complexity can be thought of as two different conflicting objectives when using PNNs for classification tasks. Using these two metrics as the objectives for finding decision boundaries, this chapter adopts a Discrete Pareto Particle Swarm Optimization (DPPSO) method. DPPSO guides the evolution of the swarm by using the two aforementioned objectives: classification accuracy and architectural complexity. The effectiveness of this method is shown on real life datasets having non-linear class boundaries. Empirical results indicate that the performance of the proposed method is encouraging.

6.1 Introduction

In this chapter, an attempt is made, from a data mining [19, 40, 52] perspective to study the application of polynomial neural networks (PNNs) [46] in classification. Classification is a problem consisting of generating decision boundaries that can effectively distinguish various classes in feature space. For solving this problem, classical algorithms such as PNN [3, 57] and its variants try to measure the performance by considering only one evaluation criterion, e.g., classification accuracy. However, a more important criterion such as the architectural complexity embedded in the PNN is normally completely ignored. The PNN's architecture involves more computational time as the partial descriptions (PDs)

C.A. Coello Coello et al. (Eds.): Swarm Intel. for Multi-objective Prob., SCI 242, pp. 115–155.
springerlink.com

grow over the training period layer-by-layer making the network more complex. Furthermore, the complexity of the PNN is based on various parameters such as *number of input variables*, *order of the polynomial*, *number of layers of the PNN* and *number of PDs in a layer*. Altogether, these parameters constitute the complexity of the PNN architecture. This implicit criterion (i.e., architectural complexity) is one of the main bottlenecks of PNNs which prevents data mining researchers from using them for harvesting knowledge form large datasets. Therefore, in order to transform this sort of network into a valuable tool for large datasets, we need to minimize the complexity of their architecture without compromising classification accuracy.

Decision-makers aim for solutions that simultaneously optimize multiple objectives such as architectural complexity and predictive accuracy, such that they can obtain the best possible trade-offs among them. Multi-criteria problems are normally characterized by a set of solutions, none of which is better than the others with respect to all the objectives. Such solutions constitute the so-called Pareto optimal set, whose corresponding vectors are called non-dominated, and their image in objective space is called the Pareto front. Multi-Objective Evolutionary Algorithms (MOEAs) [11, 13] are an increasingly popular approach to confront this type of problem. The use of evolutionary algorithms (EAs) [23, 24, 28, 44] as a search and optimization engine is motivated by the inherent complexity of many real-world problems (either single- or multi-objective), having a large number of decision variables and huge and accidented search spaces. EAs are population-based techniques, which helps them to converge to different elements of the Pareto optimal set in a single run. Also, because of their heuristic nature, EAs are less susceptible to the shape or continuity of the Pareto front, which are major concerns when using traditional mathematical programming techniques [45]. Particle swarm optimization (PSO) [35] is another heuristic approach having similar characteristics than EAs, and whose use has been very successful in a variety of application domains [54]. Additionally, PSO is a relatively simpler than EAs, both in its algorithmic structure and in its use. These were the main reasons that motivated us to adopt it for the work reported in this chapter.

Recently, researchers are emphasizing more the use of PSO to solve multi-objective optimization problems (a detailed discussion on this topic is provided in Section 6.4). Adapting the original (single-objective) PSO algorithm so that it can deal with multi-objective problems (i.e., to make it a multi-objective particle swarm optimizer (MOPSO)) requires the redefinition of the notion of global and personal best in order to make them appropriate for a multi-objective context. As indicated before, in multi-objective optimization there is no absolute global or personal best but rather a set of non-dominated solutions, all of which are equally good, in the absence of any preference information. Thus, it turns out to be a non-trivial task to choose both the personal and global best for guiding each of the particles in a swarm [7].

Architectural complexity, which is an important design criterion for PNNs, can be reduced based on the following assumption. The proposed PNN model is a three-layer architecture: the input layer contains only the input features, the hidden layer contains PDs along with the set of original features and the output layer contains only one neuron. We can select an optimal set from the PDs generated in the hidden layer along with the input features using the proposed DPPSO as a driving force to further decrease the architectural complexity. This optimal set of features is to be fed to the output layer. Further, the proposed DPPSO approach implicitly optimizes the weights between the hidden layer and the output layer.

The remainder of this chapter is organized as follows. Section 6.2 discusses the classification problem of data mining and knowledge discovery (DM and KDD). Section 6.3 describes the basic architecture and algorithmic view of PNN. In Section 6.4 we provide some basic concepts of multi-objective optimization and PSO for multi-objective problems. The proposed method is formulated and discussed in Section 6.5. In Section 6.6 we show some experimental results of the proposed method. Section 6.7 draws the conclusions with a possible path of future research followed by an extensive set of domain related references.

6.2 Classification and Data Mining

Nowadays, the data mining community is mostly concerned with large collections of data, usually owned by large organizations who can afford expensive hardware (and software and consultants) to perform data mining tasks such as classification, clustering and association rule mining, among others. Even small databases hide information, and we can use data mining techniques to extract it. But there is no clear definition of what a large or a small database actually is. How many records or how many mega, giga, or tera-bytes of data are necessary for a database to be considered "large" and thus worthy of data mining? The answer to the question "how much data?" could lie in the representativeness of the database. Even a very large data set may be inadequate for mining purposes simply because it does not truly represent the population from which it was drawn. If a data set does not represent the population from which it is drawn, no amount of other checking, surveying, and measuring will produce a valid model. Therefore, in addition to the data mining tasks, we have to also give emphasis to some pre and post processing of the collected data. These all together constitute the field known as *knowledge discovery in databases* (KDD).

KDD has evolved, and continues to evolve, from the intersection of research in fields such as databases, machine learning [61], pattern recognition [42], statistics [6], computational intelligence and reasoning with uncertainty, knowledge acquisition for expert systems, data visualization, machine discovery [27], scientific discovery, information retrieval, swarm intelligence and high-performance computing. KDD software systems incorporate theories, algorithms, and methods from all of these fields. The goal of this section is to provide an overview of the classification task and how it relates to KDD. The KDD process can be defined as follows:

Definition: *The nontrivial process of identifying valid, novel, potentially useful, and ultimately understandable patterns in data.*

The term *process* in the above definition implies that there are many steps including *creating target datasets*, *data cleaning and preprocessing*, *choosing the function of data mining*, *choosing the data mining algorithms*, *interpretation*, *using the discovered knowledge and refinement* all repeated in multiple iterations. The process is assumed to be nontrivial in that it goes beyond computing closed-form quantities. Fig. 6.1 illustrates the KDD process.

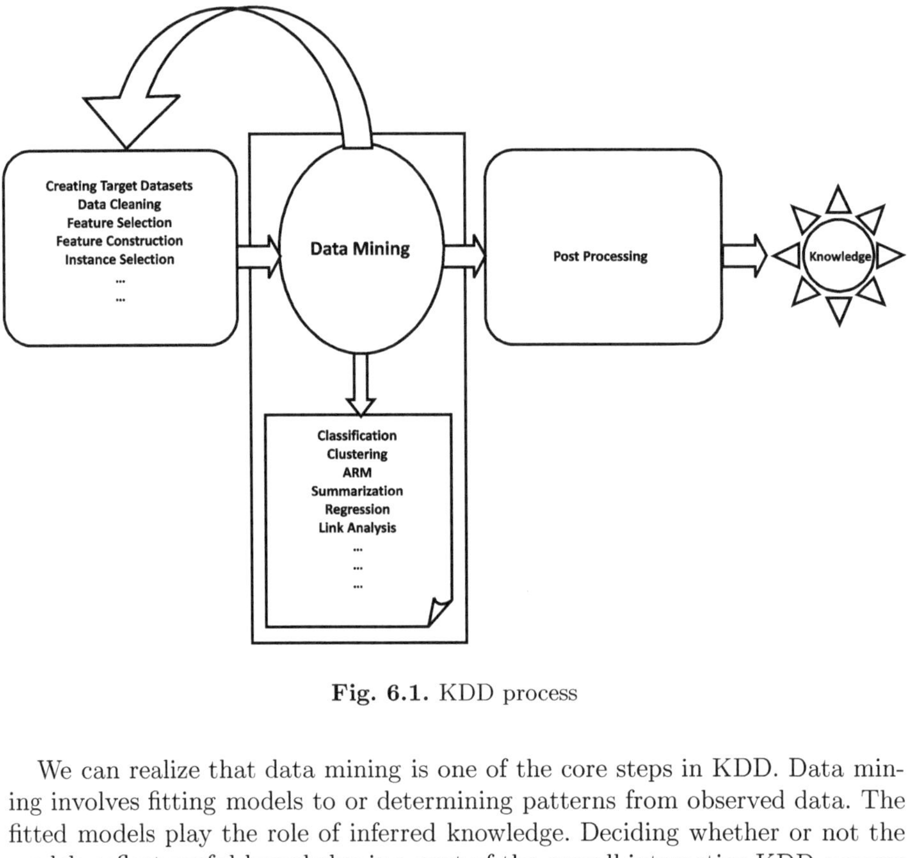

Fig. 6.1. KDD process

We can realize that data mining is one of the core steps in KDD. Data mining involves fitting models to or determining patterns from observed data. The fitted models play the role of inferred knowledge. Deciding whether or not the models reflect useful knowledge is a part of the overall interactive KDD process for which subjective human judgment is usually required. A wide variety and number of data mining algorithms are described in the literature from the fields of statistics, pattern recognition, machine learning, databases and, recently, from swarm intelligence. Most data mining algorithms can be viewed as compositions of a few basic techniques and principles. In particular, data mining algorithms consist largely of some specific combinations of three components:

Model: There are two relevant factors: the function of the model (e.g., classification and clustering) and the representational form of the model (e.g., a linear function of multiple variables and a Gaussian probability density function). A model normally contains parameters that are to be determined from the data.

The preference criterion: A basis for preference of one model or set of parameters over another, depends on the given data. The criterion is usually some form of goodness-of-fit function of the model to the data, perhaps tempered by a smoothing term to avoid over-fitting, or generating a model with too many degrees of freedom to be constrained by the given data.

The search algorithm: The specification of an algorithm for finding particular models and parameters, given data, a model (or family of models), and a preference criterion. A particular data mining algorithm is usually an instantiation of the model/preference/search components (e.g., a classification model based on a decision tree representation, model preference based on data likelihood, determined by greedy search using a particular heuristic. Algorithms often differ largely in terms of the model representation (e.g., linear and hierarchical), and model preference or search methods are often similar across different algorithms. The literature on learning algorithms frequently does not state clearly the model representation, preference criterion, or search method used; these are often mixed up in a description of a particular algorithm.

The most common model functions in current data mining practice include:

Classification: maps (or classifies) unknown patterns into one of a predefined set of classes.

Regression: maps a data item to a real-value prediction variable.

Clustering: maps a pattern into one of several sets of predefined classes (or clusters). Clusters are defined by finding natural groupings of data items based on similarity metrics or probability density models.

Summarization: provides a compact description for a subset of data. A simple example would be the mean and standard deviations for all fields. More sophisticated functions involve summary rules, multivariate visualization techniques, and functional relationships among variables. Summarization functions are often used in interactive exploratory data analysis and automated report generation.

Dependency modeling: describes significant dependencies among variables. Dependency models exist at two levels: structured and quantitative. The structural level of the model specifies (often in graphical form) which variables are locally dependent; the quantitative level specifies the strengths of the dependencies using some numerical scale.

Link analysis: determines relations between fields in the database (e.g., association rules [61] to describe which items are commonly purchased with other items in grocery stores). The focus is on deriving multi-field correlations satisfying support and confidence thresholds.

Sequence analysis: models sequential patterns (e.g., in data with time dependence, such as time-series analysis). The goal is to model the states of the process generating the sequence or to extract and report deviation and trends over time.

In this chapter our main emphasis is on the classification task of data mining. The task of classification is the problem of finding a good strategy to assign classes to samples based on past observations of sample-class pairs. We shall only assume that all samples x are contained in the set X, often referred to as the *input space*. Let Y be a finite set of classes called the *output space*. If not otherwise stated, we will only consider the two element output space $+1, -1$, in which case the classification problem is called **binary classification task**. Suppose we are give a sample of m training patterns, $x = [x_1, x_2, ..., x_m] \in X^m$, together with a sample of corresponding classes, $y = [y_1, y_2, ..., y_m \in Y^m]$. Let the training set be denoted as $z = (x, y) \in (X \times Y)^m = Z^m$, and assume that z is a sample drawn identically and independently distributed according to some unknown probability measure P_Z.

Definition: *The problem is to find the unknown (decision functional) relationship $h \in Y^X$ between patterns $x \in X$ and targets $y \in Y$ based solely on a sample $z \in (X \times Y)^m$ of size $m \in N$ drawn identically and independently distributed from an unknown distribution P_{XY}. If the output space Y contains a finite number $|Y|$ of elements then the task is called classification.*

Of course, having knowledge of $P_{XY} = P_Z$ is sufficient for identifying this relationship as for all patterns x,

$$P_{Y|X=x}(y) = \frac{P_Z((x, y))}{\sum_{\tilde{y} \in Y} P_Z((x, \tilde{y}))}. \tag{6.1}$$

Thus, for a given pattern $x \in X$ we could evaluate the distribution $P_{Y|X=x}$ over classes and decide on the class $\widehat{y} \in Y$ with the largest probability $P_{Y|X=x(\widehat{y})}$. Estimating P_Z based on the given sample z, however poses a nontrivial problem.

Over the years, we have witnessed many classification techniques from fields such as statistics, machine learning, and pattern recognition, among others. However, traditional statistical classification procedures such as discriminant analysis are built on Bayesian decision theory [66]. In these procedures, an underlying probability distribution must be assumed in order to calculate the posterior probability upon which the classification decision is made. One major limitation of these statistical models is that they work well only when the underlying assumptions are satisfied. The efficiency of these methods depends to a large extent on the various assumptions or conditions under which the models are developed. Users must have a good knowledge of both, data properties and model capabilities before the models can be successfully applied to the domain of interest.

Artificial neural networks (ANNs) [26, 63] have emerged as an important tool for classification. The recent (vast) research activities in neural classification have established that neural networks are a promising alternative to various conventional classification methods. ANNs are capable of generating complex mapping between the input and the output space and thus these networks can form arbitrarily complex nonlinear decision boundaries. This does not mean that finding such type of networks is easy. On the contrary, problems such as local minima trapping, saturation, weight interference, initial weight dependence, and over-fitting make neural networks training a difficult task. Moreover, most neural learning methods, being based on gradient descent, cannot search the non-differentiable landscape of multi-layer architectures. This is a key issue, since it can be proved that if a network is allowed to adapt its architecture, it can solve any learnable problem in polynomial time. The next section will provide the foundations to our proposed method.

6.3 Polynomial Neural Networks

This Section is divided into two subsections. Subsection 6.3.1 discusses the general characteristics and architecture of polynomial neural networks (PNNs). An algorithmic framework commonly adopted with PNNs is provided in subsection 6.3.2.

6.3.1 Architecture of PNNs

The PNNs architecture is based on the Group Method of Data Handling (GMDH) [31, 33]. GMDH [32] was developed by Ivakhnenko in the late 1960s for identifying non-linear relations between input and output variables. Ivakhnenko was inspired by the form of Kolmogorov-Gabor polynomials, which is the discrete analogue of Volterra functional series. The discrete analogue of Volterra functional series can be expressed as:

$$y = f(x_1, x_2, ..., x_d) = A_0 + \sum_{i=1}^{d} A_i x_i + \sum_{i=1}^{d}\sum_{j=1}^{d} A_{ij} x_i x_j + \dots, \tag{6.2}$$

where $(x_1, x_2, ..., x_d)$ is the input variable vector and $A_i's$ are the weight vector. Viewing computation time and other associated problems, Ivakhnenko attempted to resemble the Kolmogorov-Gabor polynomial by using low order polynomials for every pair of the input variables. He proved that a second order polynomial i.e.,

$$f(x_i, x_j) = A_0 + A_1 x_i + A_2 x_j + A_3 x_i x_j + A_4 x_i^2 + A_5 x_j^2, \tag{6.3}$$

can reconstruct the complete Kolmogorov-Gabor polynomial through an iterative perceptron type procedure. This approach offers a better accuracy, requires a fewer number of observations and, therefore, reduces the computational time.

However, there are several drawbacks associated with the GMDH such as its limited generic structure, overly complex network, etc., and hence prompted a new class of polynomial neural networks (PNNs). In summary, these networks come with a high level of flexibility as each node can have a different number of input variables as well as exploit a different order of polynomial (e.g., linear, quadratic, cubic, etc). Unlike neural networks whose topologies are commonly decided prior to all detailed (parametric) learning, the PNN architecture is not fixed in advance but becomes fully optimized (both parametrically and structurally).

Various types of topologies of the PNN have been proposed [51] for different applications on the basis of the number of input variables and the order of partial description (PD) in each layer. Depending on the number of input variables, two kinds of PNN structure are encountered in most practical situations such as basic PNNs and modified PNNs. Furthermore, these two are classified into two classes such as generic and advanced based on the order of the polynomial. Fig. 6.2 summarizes their differences and classifications. Here we will quickly cover the basic architecture of the PNNs as our proposed method is entirely built on the concept of it. The PNN architecture utilizes different types of polynomials such as linear, quadratic, cubic, etc. Table 6.1 enumerates some of the polynomials most commonly adopted with PNNs.

Table 6.1. Some of the polynomials most commonly adopted with PNNs

Name	Polynomials with 2 Variables
Linear	$A_0 + A_1x_1 + A_2x_2$
Quadratic	$A_0 + A_1x_1 + A_2x_2 + A_3x_1^2 + A_4x_2^2 + A_5x_1x_2$
Cubic	$Quadratic + A_6x_1^3 + A_7x_2^3 + A_8x_1^2x_2 + A_9x_1x_2^2$

By choosing the most significant number of variables and an order for the polynomial, we can obtain the best ones from the extracted PDs according to the selected nodes of each layer. Additional layers are generated until the best performance of the extended model is obtained. Such methodology leads to an optimal PNN structure. Let us assume that the input-output of the data is given in the following form: $[X_i, y_i] =<< x_{i1}, x_{i2}, ..., x_{id} >, y_i >$, where $i = 1, 2, 3, \ldots, n$, n is the number of instances and d is the number of attributes. The following matrix represents the complete set of n instances with their corresponding class label.

$$\begin{pmatrix} x_{11} & x_{12} & \cdot\cdot & x_{1d} & : & y_1 \\ x_{21} & x_{22} & \cdot\cdot & x_{2d} & : & y_2 \\ \cdot & \cdot & \cdot\cdot & \cdot & \cdot & \cdot \\ x_{n1} & x_{n2} & \cdot\cdot & x_{nd} & : & y_n \end{pmatrix}$$

The input-output relationship of the above data by PNN model can be described in the following manner: $\widetilde{y} =< x_{i1}, x_{i2}, \ldots, x_{id} >, i = 1(1)n$. In case of an insufficient amount of data samples and regression coefficients that are hard to estimate, PNNs suggest an iterative procedure to overcome this difficulty. The

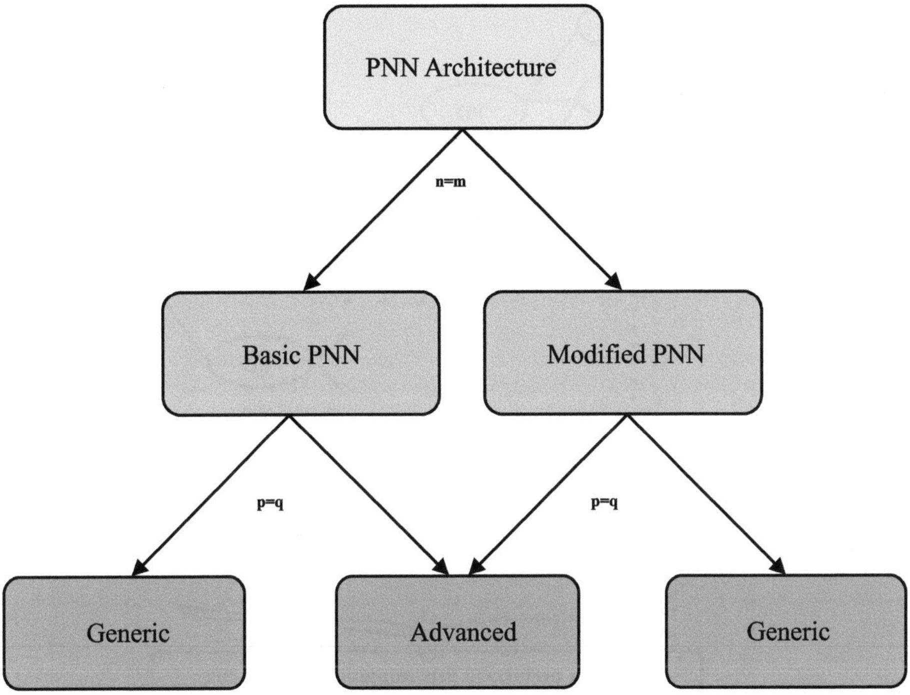

Fig. 6.2. Classifications of PNNs depending on the number of input variables (denoted as n and m) and order of the polynomial (denoted as p and q, respectively).

architecture of the basic PNN is shown in Fig. 6.3. A set of partial descriptions (PDs) are generated as the basic building block of the PNN model and is shown in Fig. 6.4.

To compute the estimated output y, we construct a PD for each possible pair of independent variables. For example, if the number of independent variables is d, then the total number of possible PDs is $\frac{d!}{2! \times (d-2)!}$. Training samples are used for determining the coefficients of the PDs by the least square fit method. An optimal set of PDs can be selected to the next layer by competition among themselves. Here, choosing an optimal set of PDs is purely based on an external criterion. This operation is repeated until some stopping criterion is met. Once the final layer PD has been constructed, the node that shows the best performance is selected as the output node and the remaining PDs are discarded. Furthermore, by backtracking, the nodes of the previous layers that do not have influence on the output node PD are deleted.

6.3.2 Algorithmic Framework of the PNNs

The behavior of the PNN algorithm mimics GMDH. The best algorithm for the classification problem depends on the number of input features (i.e., this is

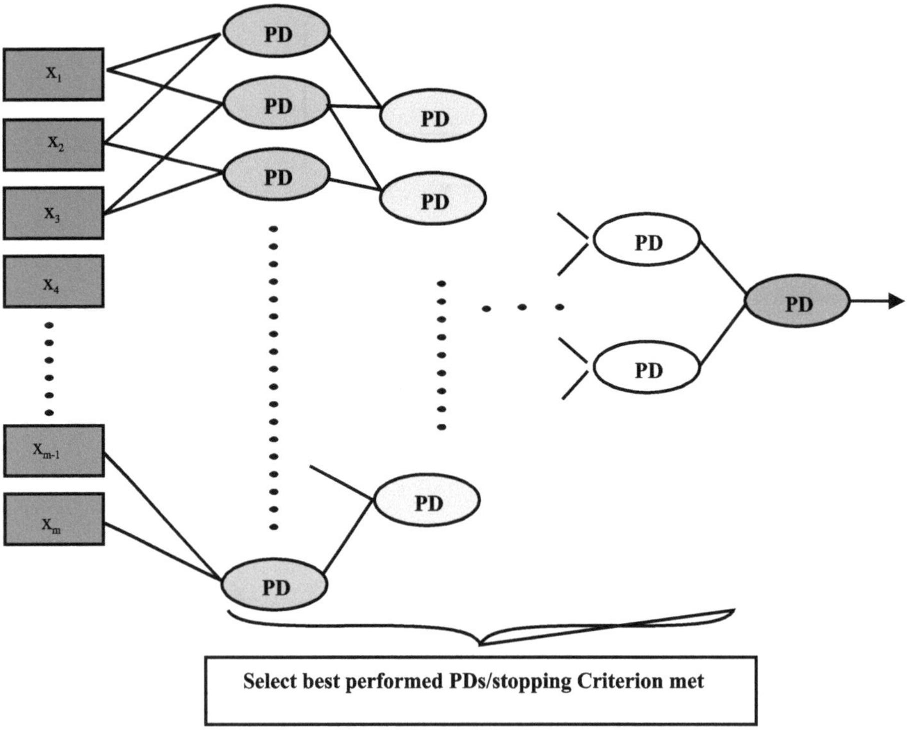

Fig. 6.3. Architecture of a basic Polynomial Neural Network (PNN)

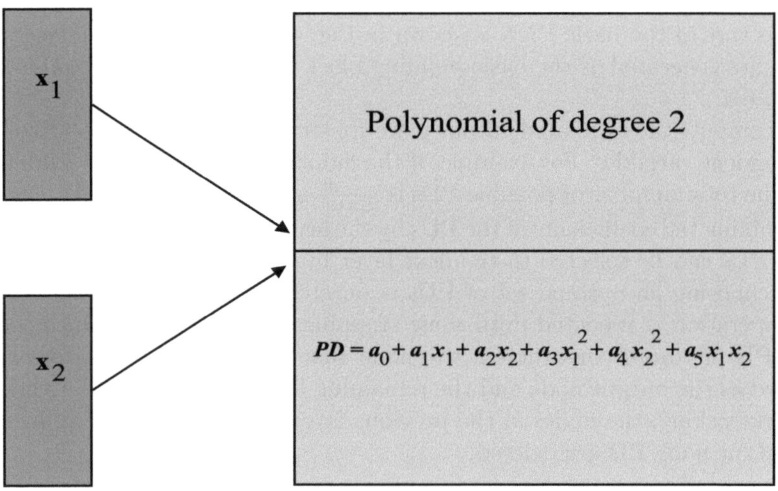

Fig. 6.4. The building block of a PNN

different from the original set of features) and the order of the polynomial. The algorithm of PNN can be described by the following sequence of steps:

1. **Determine the model's input features:** the set of attributes of the given dataset defines the models input features and, if needed, carry out the pre-processing of input data.
2. **Partition the dataset:** the input-output dataset $[X_i, y_i]$ is divided into two parts, that is a training and testing dataset i.e., $n = ntr + n_{te}$, where n, n_{tr}, and n_{te} denote the size of the dataset, the training subset, and the test subset, respectively. Using the training part, we can construct the PNN model (including an estimation of coefficients of the PDs of every layer of the PNN) and the test data are used to evaluate the estimated PNN model.
3. **Select the topological structure of the PNN:** the structure of the PNN is selected based on the number of input features and the order of the PDs in each layer. Depending on the number of features, the PNN structures can be categorized into two types, namely a basic PNN and a modified PNN. In the case of a basic PNN, the number of input variables of the PDs is the same in every layer, whereas in a modified PNN, the number of input variables of the PDs varies from layer to layer.
4. **Generate the PDs:** in particular, we select the input features of a node from d input features $x_1, x_2, \ldots, x_d$. The total number of PDs located at the current layer differs according to the number of the selected input features from the nodes of the preceding layer. This results in $n_{PD} = \frac{d!}{(d-r)! \times r!}$ nodes, where r is the number of chosen input variables. The choice of the input variables and the order of a PD is very important to select the best model with respect to the characteristics of the data, model design strategy, non-linearity and predictive capability.
5. **Estimate the coefficient of the PD:** The vector of coefficients $c_i = c_0, c_1, c_2, c_3, c_4, c_5$ is derived by minimizing the mean square error between y_i and $\widehat{y_{ji}}$,
$$e = \frac{1}{n_{tr}} \sum_{i=1}^{n_{tr}} (y_i - \widehat{y_{ji}})^2 \qquad (6.4)$$
where $\widehat{y_{ji}} = c_0 + c_1 x_k + c_2 x_l + c_3 x_k x_l + c_4 x_k^2 + c_5 x_l^2$, $1 \leq k,l \leq d, j = 1, 2, 3, \ldots, \frac{d(d-1)}{2}$. In order to find out the coefficients, we need to minimize the error criterion e. Differentiating e with respect to all the coefficients, we get a set of linear equations. Upon solving this set of linear equations, we can get the value of the coefficients. This procedure is implemented repeatedly for all nodes of the layer and also for all layers of the PNN starting from the input layer and moving to the output layer.
6. **Select the PDs with the best predictive capability:** Each PD is estimated and evaluated using both the training and the testing data sets. Using the evaluated values, we choose the PDs which give the best predictive performance for the output variable. Normally we use a prespecified cutoff value of the performance for all the PDs. In order to be retained at

the next generation, the PD has to exhibit its performance above the cutoff value.

7. **Check the stopping criterion:** Two termination methods can be exploited:
   - The following stopping condition indicates that an optimal PNN model has been accomplished at the previous layer, and the modeling can be terminated. This condition reads as $e_c \geq e_p$, where e_c is a minimal identification error of the current layer, and e_p denotes a minimal identification error that occurred at the previous layer.
   - The PNN algorithm terminates when the number of iterations (predetermined by the designer) is reached. When setting up a stopping (termination) criterion, one should be prudent in achieving a balance between model accuracy and an overall computational complexity associated with the development of the model.
8. **Determine new input features for the next layer:** if any of the two above criteria fails, then the model has to be expanded by repeating steps 4-8 of the algorithm.

6.4 Multi-objective Particle Swarm Optimization

In this section, we will discuss both continuous and discrete versions of the particle swarm optimization algorithm. Furthermore, an overview of multi-objective particle swarm optimization in continuous domains is provided. In order to make the flow of this section cohesive, we divided it into five subsections. Subsection 6.4.1 provides an overview of multi-objective problems. Subsection 6.4.2 gives some insights regarding multi-objective problem solving approaches. Continuous particle swarm optimization and its extension to multi-objective problems is discussed in Subsections 6.4.3 and 6.4.4, respectively. Subsection 6.4.5 discusses discrete particle swarm optimization.

6.4.1 Multi-objective Problems

Multi-objective optimization, also known as multi-criteria optimization or, multi-performance or/multi-attribute or vector optimization is defined as the problem of finding a vector of decision variables which satisfies and optimizes a vector function whose elements represent the objective functions. These functions form a mathematical description of performance criteria, which are usually in conflict with each other. Hence, the term "optimize" means finding a solution which gives values for all the objective functions that are acceptable to the user.

In most real-world optimization problems, particularly those in design, it is required to optimizate simultaneously more than one objective function. Let us consider the following examples.

For the construction of bridges, a good design is characterized by the minimization of the total mass and the maximization of the stiffness. In chemical plant design, or in the design of a groundwater remediation facility, the objectives to be considered include total investment and net operating costs. Aircraft design requires the simultaneous optimization of fuel efficiency, payload, and weight. A good sunroof design in a car could aim to minimize the noise the driver hears and maximize the ventilation. The traditional portfolio optimization problem attempts to simultaneously minimize the risk and maximize the fiscal return. In these and most other cases, it is unlikely that the different objectives will be optimized by a single set of parameter choices. Hence, some trade-off between the criteria is needed to ensure a satisfactory design.

Multicriteria optimization has its roots in late-nineteenth-century welfare economics, in the works of Francis Ysidro Edgeworth and Vilfredo Pareto [49, 50]. Multi-objective optimization methods as the name suggests, deal with finding optimal solutions to problems having multiple objectives. So, in this type of problems, the aim is to produce trade-offs among the objectives, rather than solutions that only satisfy one crieterion. Mathematically, this can be stated and visualized as follows:

Minimize

$$\overrightarrow{f}(x) = [f_1(x), f_2(x), \ldots, f_n(x)]^T, \tag{6.5}$$

where $n \geq 2$ and $C = \{x : h(x) = 0, g(x) \leq 0,$ and $l_x \leq x \leq u_x\}$ denotes the feasible set constrained by equality and inequality constraints, as well as by explicit variable bounds.

In a multi-objective setting, there is no single optimal solution, but instead, the goal is to generate a set of solutions that represent the best possible trade-off among the objectives. This is known as the *Pareto optimal set*. Essentially, a vector $x^{'} \in C$ is said to be *Pareto optimal* if all other vectors $x \in C$ have a worse value for at least one of the objective functions $f(.)$, or else have the same value for all objectives. Formally, we have the following definition:

A point $x^{'} \in C$ is said to be *global Pareto optimal* or a global efficient solution for a multi-objective optimization problem (MOP) if and only if there is no $x \in C$ such that $f_i(x) \leq f_i(x'))$ for all $i \in 1, 2, ..., n$, with at least one strict inequality.

We can also have *local Pareto optimal* points, for which the definition is the same as the one just given, except that we restrict attention to a feasible neighborhood of $x^{'}$. That is, if $B(x^{'}, \gamma)$ denotes a sphere of radius γ around the point $x^{'}$, we require that for some $\gamma > 0$, there is no $x \in C \cap B(x^{'}, \gamma)$ such that $f_i(x) \leq f_i(x^{'})$, $\forall i = 1, 2, ..., n$ with at least one strict inequality.

The vectors corresponding to the solutions included in the Pareto optimal set are called non-dominated or non-inferior. The plot of the objective functions whose non-dominated vectors are in the Pareto optimal set is called the *Pareto front.*

6.4.2 Multi-objective Problem Solving Approaches

Broadly we can categorize the multi-objective problem solving approaches into three types:

1. The conventional weighted-sum-approach: transforming the original multi-objective problem into a single-objective problem by using an aggregating formula.
2. The lexicographical approach, where the objectives are ranked in order of priority.
3. The Pareto-based approach, which consists of finding as many non-dominated solution as possible and returning a set of non-dominated solutions to the user.

In a conventional weighted-sum-approach, a multi-objective problem is transformed into a single-objective problem. Basically, a numerical weight is assigned to each objective (evaluation criterion) and then the values of the weighted criteria are combined into a single value by either adding or multiplying all the weighted criteria.

Thus the fitness function of a given candidate solution can be measured by either the two following types of formulas:

$$f(x) = \sum_{i=1}^{n} w_i.f_i(x) \tag{6.6}$$

or

$$f(x) = \prod_{i=1}^{n} f_i(x)^{w_i} \tag{6.7}$$

where $w_i, i = 1, ..., n$, denotes the weight assigned to criteria $f_i(x)$ and n is the number of evaluation criteria.

This method is popular and more usable because of its simplicity. However, there are several drawbacks associated with this method. First, the setting of the weights in these formulas is *ad-hoc*. The second problem with the weights is that once a formula with precise values for the weights has been defined and given to a data mining algorithm, it will be effectively trying to find the best model for that particular setting of weights, missing the opportunity to find other models that might be actually more interesting to the user, since they could represent a better trade-off among the different quality criteria. Third, weighted formulas involving a linear combination of different quality criteria cannot generate solutions in a non-convex region of the solution space.

In the lexicographic approach, different priorities are assigned to different objectives, and then the objectives are processed (i.e., optimized separately) according to their priority. So, when two or more candidate solutions are compared with each other, in order to choose the best one, their performance measures are compared with respect to the highest-priority objective. If one candidate solution is significantly better than the other with respect to that objective, the former is chosen. Otherwise, the performance measure of the two candidate solutions is compared with respect to the second objective. Again, if one candidate solution is significantly better than the other with respect to that objective, the former is chosen, otherwise the performance measure of the two candidate solutions is compared with respect to the third criterion. This process is repeated until one finds a clear winner or until one has used all the criteria. In the latter case, if there was no clear winner, one can simply select the solution optimizing the highest-priority objective. The lexicographic approach has an advantage over conventional weighted approach, as it treats each of the criteria separately, recognizing that each criterion measures a different aspect of quality of a candidate solution. The disadvantage of this approach is how to specify a tolerance threshold for each criterion.

In the Pareto-based approach, instead of transforming a multi-objective problem into a single-objective problem and then solving it by using a single-objective search method, an algorithm that selects solutions based on the definition of Pareto optimality is adopted (see Section 6.4.1). The approach proposed here for solving classification problems using PNNs belongs to this category.

The disadvantage of the Pareto-based approach is that it is normally very difficult to choose a single (most preferred) solution from the Pareto optimal set (which could be very large), and in practice, only one of such solutions will be actually used. One of its advantages is that the Pareto-based approach is more generic than the Minimum Description Length (MDL) principle, since the latter is used only to cope with accuracy and simplicity, whereas the Pareto-based approach can cope with any kind of non-commensurable quality measure.

6.4.3 Continuous Particle Swarm Optimization

Particle swarm optimization (PSO) is a collaborative swarm intelligence approach that has been commonly used to optimize continuous functions. It was originally proposed by James Kennedy and Russell Eberhart in 1995 [16]. The algorithm, which is based on a metaphor of social interaction, probes a search space by fine-tuning the trajectories of individual particles as they are conceptualized as moving points in a multi-dimensional space. The individual particles are drawn stochastically toward the positions of their own best fitness achieved so far and the best fitness achieved so far by any of their neighbors within a small neighborhood γ (or the entire swarm). Since its inception, PSO has gained increasing popularity among researchers and practitioners as a robust and efficient technique for solving difficult optimization problems. Next, we will briefly explain how it works.

Particle Swarm Optimization Algorithm

The standard PSO algorithm maintains a population of n particles initialized with random positions $\overrightarrow{x_i}$ and velocities $\overrightarrow{v_i}$. Each particle represents a potential solution to a problem in a d-dimensional space (i.e. $\overrightarrow{x_i} = [x_{i1}, x_{i2}, \ldots, x_{id}]$, where $i = 1, 2, \ldots, n$) and undergoes evaluation by a function f. Positions and velocities are adjusted and the function is evaluated with the new coordinates at each time step. When a particle discovers a position that is better than the position found so far, it stores the coordinates in a vector $\overrightarrow{p_i} = [p_{i1}, p_{i2}, ..., p_{id}]$. The difference between $\overrightarrow{p_i}$ (the best point found by i so far) and the individual's current position is stochastically added to the current velocity, causing the trajectory to displace the position of that point. Further, each particle is defined within the context of a topological neighborhood comprising itself and some (or all of the) particles in the population. The stochastically weighted difference between the neighborhood's best position $\overrightarrow{p_g}$ and the individual's current position is also added to its velocity, adjusting it for the next time step. These adjustments to the particle's movement through the space causes it to search around the two best positions.

Without loss of generality, consider a minimization task of the objective function f. The equation for updating the dth dimension of the personal best position $\overrightarrow{p_i}$ is presented in equation (6.8), with time step t.

$$p_i^d(t+1) = \{ \begin{matrix} p_i^d(t) & if \;\; f(\overrightarrow{x_i}(t+1)) \geq f(\overrightarrow{p_i}(t)) \\ x_i^d(t+1) & if \;\; f(\overrightarrow{x_i}(t+1)) < f(\overrightarrow{p_i}(t)). \end{matrix} \tag{6.8}$$

At each iteration t, the dth dimension of particle i's velocity and position are updated using equations (6.9) and (6.10) separately, where w, c_1, and c_2 are non-negative real parameters, $r_{1i}^d(t)$ and $r_{2i}^d(t)$ are two independent uniform random numbers distributed in the range [0, 1].

$$v_i^d(t+1) = w.v_i^d(t) + c_1 r_{1i}^d(t)(p_i^d(t) - x_i^d(t)) + c_2 r_{2i}^d(t)(p_g^d(t) - x_i^d(t)), \tag{6.9}$$

$$x_i^d(t+1) = x_i^d(t) + v_i^d(t+1) \tag{6.10}$$

There exist many factors that would influence the convergence property and performance of the PSO algorithm, including the selection of w, c_1, and c_2, velocity clamping, position clamping, topology of neighborhood, etc. For more about the convergence analysis and parameters selection of PSO, interested readers can refer to [9, 34, 59].

The algorithm in pseudocode is the following:

- Initialize Population
- REPEAT
- FOR i = 1 to Population Size
- IF $f\overrightarrow{x_i} < f(\overrightarrow{p_i})$ then $\overrightarrow{p_i} = \overrightarrow{x_i}$
- $\overrightarrow{p_g} = \min(\overrightarrow{p}_{neighbors})$
- FOR d = 1 to Dimension
- $v_i^d = w.v_i^d + c_1 r_{1i}^d(p_i^d - x_i^d) + c_2 r_{2i}^d(p_g^d - x_i^d)$
- $x_i^d = x_i^d + v_i^d$
- END FOR
- END FOR
- UNTIL termination criterion is met.

In spite of their short existence, PSO has gained widespread appeal amongst researchers and has shown to offer good performance in a variety of application domains, with a great potential for hybridization and specialization. For more information on PSO, interested readers can refer to [4].

6.4.4 Continuous PSO for Multi-objective Problems

Recently a number of researchers have been working towards the development of promising algorithms for multi-objective problems using particle swarm optimization. In this section we give a basic multi-objective particle swarm optimization (MOPSO) algorithm derived from proposals that have published over the years. However, for those interested in knowing more about different MOPSO variants, a survey on this topic up to the year 2006 can be found in [53]. Although most of these studies were generated in tandem, each one implements their MOPSO in a different fashion. Unlike traditional multi-objective evolutionary algorithms (MOEAs) available in the literature, PSO imposes additional constraints that are normally relaxed when extending it to the multi-objective case. For example, in PSO, the swarm of particles is fixed in size, and each of its members cannot be replaced. Instead, the personal best (p_i) and the global best (p_g) of each particle is adjusted. However, when dealing with multi-objective problems, a set of non-dominated solutions (the best individuals found so far by the search engine) must replace the (single) global best individual in the standard single-objective PSO case. Additionally, there may be no single previous best individual for each member of the swarm. Choosing both p_g and p_i to direct a swarm member's flight is, therefore, a non-trivial problem in MOPSO.

Generally, the three main aims when solving a multi-objective problem (using either evolutionary or non-evolutionary techniques) are the following: i) maximize the number of points of the Pareto optimal set; ii) minimize the distance between the true Pareto front (assuming that its location is known to us) and the obtained one; iii) maximize the spread of the solutions generated, such that the distributions of vectors are smooth and uniform.

Based on the population-based nature of PSO, it is desirable to produce several (different) non-dominated solutions with a single run. So, as with any other evolutionary algorithm, the three main issues to be considered when using PSO to multi-objective optimization are: i) how to select the p_g particles in order to give preference to non-dominated solutions over those that are dominated; ii) how to retain the non-dominated solutions found during the search process in order to report solutions that are non-dominated with respect to all the past populations and not only with respect to the current one. Also, it is desirable that these solutions are well spread along the Pareto front; iii) how to maintain diversity in the swarm in order to avoid convergence to a single solution; and iv) how to choose the personal best p_i and also the local best p_l.

While solving single-objective optimization problems (as we have seen in subsection 6.4.3) the leader p_g that each particle uses to update its position is determined once a neighborhood topology is determined. However, in the case of multi-objective optimization problems, each particle might have a set of leaders (local/global) from which just one can be selected in order to update its position. Such set of leaders is usually stored in an external repository, which we will call EX_SWARM. This is a repository in which the non-dominated particles found so far are stored. The particles contained in EX_SWARM are used as p_g when the positions of the particles of the swarm have to be updated. Furthermore, the contents of the EX_SWARM can be visualized as our approximation of the Pareto front obtained by the algorithm. The following algorithm describes the way in which a basic MOPSO works.

Algorithm MOPSO

1. INITIALIZATION of the SWARM;
2. EVALUATE the fitness of each particle of the SWARM;
3. EX_SWARM = SELECT the non-dominated solutions from the SWARM;
4. timer = 0;
5. REPEAT
6. FOR each particle
7. SELECT the p_g from EX_SWARM
8. SELECT the p_i
9. UPDATE the Velocity and Position
10. MUTATION /* Optional */
11. EVALUATE the Particle
12. UPDATE the p_i
13. END FOR
14. UPDATE EX_SWARM
15. timer = timer+1
16. UNTIL (timer <= MAXIMUM_ITERATIONS)
17. Report Results in the EX_SWARM

The basics and major steps of the aforementioned algorithm can be explained as follows: first, the swarm is initialized either randomly or by adopting any domain specific heuristic. Then, a set of leaders is also initialized with the non-dominated particles from the swarm and stored in EX_SWARM. Later on, some sort of selection technique is used to choose (usually) one p_g for each particle of the swarm (in the next section we will discuss different strategies to select p_g that have been developed by various authors). At each iteration, for each particle, a leader is selected and the flight is performed. Most of the existing MOPSOs apply some sort of mutation operator after performing the flight. Then, the particle is evaluated and its corresponding p_i is updated. A new particle replaces its p_i usually when this particle is dominated or both are non-dominated with respect to each other. After all the particles have been updated, the set p_g is updated. This process is repeated for a certain number of iterations.

Issues Related to Continuous MOPSO

The following important issues are associated with the algorithmic design aspects of MOPSO. Even though a good number of proposals came to the forefront of resolving these issues, they needs more careful attention:

Selection and update of p_g , p_i, and p_l: How to select a single (local/global) leader out of a set of non-dominated solutions which are all equally important. Similarly, the selection and update of p_i also draws careful attention. How to maintain the external archive.

Exploring new solutions: How to promote diversity through the two main mechanisms to explore new solutions: updating of positions and mutation operator.

Selection of p_g, p_i and p_l

Selecting p_g, p_l, and p_i is identified as a non-trivial problem in designing a MOPSO. Therefore, the main challenge in MOPSO parlance is to pick a suitable (personal/ global/ local) guide to move the particles towards the true Pareto front with a proper maintenance of diversity.

Selection of p_g: A very simple and straightforward approach is to consider every non-dominated solution as a new leader and then, choose one leader based on certain criterion. Nearly all the methods developed thus far are classified into different classes depending on how they choose the leader.

Random Selection: In this uniform selection the leader is chosen randomly among the non-dominated solutions present in SWARM or chosen from EX_SWARM. The main advantage of this approach is that it is fast, i.e., $O(1)$. However it biases selection to densely represented areas of the obtained Pareto front.

Density Measures: So far many density measures have been contributed by several authors. Here we will present two measures that have been commonly used in multi-objective optimization.
i)*Nearest Neighbor Density Estimator* [15]: This estimator gives us an idea of how crowded are the closest neighbors of a given particle in objective function space. This measure estimates the perimeter of the cuboid formed by using the nearest neighbors as the vertices. Particles with a larger value of this estimator are preferred.
ii)*Kernel Density Estimator* [14, 25]: When a particle is sharing resources with neighbors within a specific radius σ_{share}, its fitness must be shared in proportion to the number and closeness of particles surrounding it. Such neighborhoods are called niches.

Preserving and Spreading of Non-dominated Solutions: Most often, it is important to preserve the non-dominated solutions obtained thus far. At the end we can report all the non-dominated solutions qualifying for the true Pareto front. This is important not only for practical reasons, but also for theoretical ones.

A very common approach is to adopt an external repository (we named here EX_SWARM) to store all the non-dominated solutions encountered from the beginning of the independent run. Adopting such type of external repository raises several questions e.g., i) when a particle will be eligible to enter into the repository?; ii) how to control the size of the repository?, and so on. Over the years many researchers have tried to answer these questions, but a lot of work still remains to be done in this regard. Next, we will discuss a few proposals that have been suggested by some researchers.

A particle will qualify to enter into the repository if: (a) it is non-dominated with respect to the particles of the repository or (b) if it dominates any of the solutions within the repository (in this case, the dominated solutions have to be deleted from the repository). This approach has, however, the drawback of increasing the size of the repository very quickly. This is an important issue because the repository has to be updated at each generation. Thus, this update may become very expensive, in terms of time. In the worst case, all members of the swarm may wish to enter into the repository, at each generation. Thus, the corresponding updating process, at each generation, has a time complexity of $(k \times n^2) \approx O(n^2)$, where n is the size of the swarm and k is the number of objectives.

In this way, the time complexity of the updating process for the complete run is of $O(mn^2)$, where m is the total number of iterations. Because of practical reasons, archives tend to be bounded [11], which makes it necessary to use an additional criterion to decide which non-dominated solutions to retain, once the archive is full. In evolutionary multi-objective optimization, researchers have adopted different techniques to prune the archive (e.g., clustering [65] and geographical-based schemes that place the non-dominated solutions in cells in order to favor less crowded cells when deleting in-excess non-dominated solutions [39]). However, the use of an archive introduces additional issues: for example, do we impose an additional criteria to enter the archive instead of just using non-dominance (e.g., do we use the distribution of solutions as an additional selection criterion)?

Besides the use of an external repository, it is also possible to use a *plus selection* in which parents compete with their offspring and those which are non-dominated (and possibly comply with some additional criterion such as providing a better distribution of solutions) are selected for the upcoming generation. In the case of PSO, a *plus selection* involves selecting from a combination of two consecutive swarms.

Recently, several other researchers have proposed the use of relaxed forms of dominance. The most commonly adopted relaxed form of dominance in PSO has been ε-dominance [41]. The objective of this concept in multi-objective PSO has been to filter out solutions from the external repository. By using ε-dominance, we define a set of boxes of size ε and only one non-dominated solution is retained for each box (e.g., the one closest to the lower lefthand corner). The use of ε-dominance, as proposed in [41] guarantees that the preserved solutions are non-dominated with respect to all solutions generated during the run. It is worth noting, however, that, when using ε-dominance, the size of the final external repository depends on the ε-value, which is normally a user-defined parameter [41]. Mostaghim and Teich [48] have found that when comparing ε-dominance against existing clustering techniques for fixing the repository size, the ε-dominance method can explore solutions much faster than the clustering technique with a comparable (and even better in some cases) convergence and diversity.

Until now in item 3 we have discussed how to manage the external repository ignoring the point of selecting p_g. Let us see a few of the frequently used proposals: i) the first proposal is based on unbiased selection from the Pareto front. The Pareto front constructed by the solutions of the external archive are partitioned into grids [39] or bins of equal width [18, 21], then one partition is selected and a leader is chosen from such partition, ii) the second method proposed in [10, 22] also uses partitioning, but biases selection towards partitions which have fewer members. This can be achieved, by using roulette wheel selection of partitions, iii) in [22] (third approach) the authors attempt to project the swarm members towards their nearest p_g using the Euclidean distance method. Here the weighted Euclidean distances between the swarm members and the archive members are calculated and the closest archive members are used as the p_g.

Selection of p_i: As we have already stated, in PSO a particle's movement is not only guided by p_g but also by p_i. Choosing the personal guide of a particle also has a significant impact on the performance of the algorithm. Next, we show some of the existing proposals regarding the maintenance and selection of p_i.

In [10] a single personal best p_i is maintained. p_i is replaced if it is dominated by the newly created particle. If the newly created particle is non-comparable with p_i then one of the two is randomly selected for new p_i.

In [47] a single personal best is maintained. If the newly created particle is better than p_i, then p_i will be replaced. Otherwise, if both are non-comparable then the new one is selected for a new p_i.

In [29] one personal best p_i is also maintained. p_i will be replaced only when the newly created solution dominates p_i.

In [22] a set of personal bests is maintained by an external archive. This set contains the non-dominated solutions that describe the estimated Pareto front found by the newly created particle during the search process. This method is computationally expensive, but it can promote a greater degree of search by the heuristics.

In [2], a set of personal bests p_i is maintained in an external archive for each particle. The authors proposed hierarchical clustering methods for selection of one personal leader from the archive. Furthermore, the authors claimed that by maintaining these personal archives the results obtained are significantly better than those obtained with their single (shared) archive counterpart. However, this sort of approach is evidently more expensive, computationally speaking.

Selection of p_l: The selection of local guide p_l and global guide p_g is indifferent if the neighborhood size spreads across the entire swarm. Otherwise, we have to carefully select the local leader for each particle to maintain diversity in the solutions and the convergence of the swarm towards the true Pareto front.

Many authors are trying to misuse the term local leader, e.g. Abido [2] has used the term local leader for maintaining and selecting a personal best p_i from the external archive. Similarly, Mostaghim et al. [48] used the term local leader for maintaining and selecting a global best p_g from an external archive. To the best of our knowledge, the consideration of local leaders in a MOPSO is still unexplored and constitutes an interesting research direction.

Exploring New Solutions

The PSO algorithm has gained popularity, among other things, because of its fast convergence. This is a desirable feature, as long as the algorithm does not have premature convergence (i.e., convergence to a local optimum). Premature convergence is caused by the rapid loss of diversity within the swarm. So, the appropriate promotion of diversity in PSO is a very important issue in order to control its premature convergence. As we have already mentioned, when adopting PSO for solving multi-objective optimization problems, it is possible to promote diversity through the selection of leaders. However, this can also be done through the two main mechanisms used for creating new solutions.

Updating of positions. The use of different neighborhood topologies determines how fast is the process of transferring the information through the swarm. Since in a fully connected topology all particles are connected with each other, the information is transferred faster than in the case of a smaller neighborhood. Under the same argument, a specified neighborhood topology also determines how fast is the diversity lost within the swarm. Since in a fully connected topology, the transfer of information is fast, when using this topology, diversity within the swarm is also lost rapidly. In this way, topologies that define neighborhoods smaller than the entire swarm for each particle can also preserve diversity within the swarm a longer time.

Diversity can also be promoted by means of an inertia weight, since the inertia weight is employed to control the impact of the previous history of velocities on the current velocity. Thus, the inertia weight influences the trade-off between global (wide-ranging) and local (nearby) exploration abilities [54]. A large inertia weight facilitates global exploration (searching new areas) while a smaller inertia weight tends to facilitate local exploration to fine-tune the current search area. The value of the inertia weight may vary during the optimization process.

Shi [55] asserted that by linearly decreasing the inertia weight from a relatively large value to a small one through the course of the PSO run, PSO tends to exhibit a higher global search ability at the beginning of the run and a higher local search ability towards the end of the run. On the other hand, Zhen et al. [64] argue that either the global or the local search ability of PSO is associated with a small inertia and that a large inertia weight provides the algorithm more chances to be stabilized. In this way, inspired on the process of the simulated annealing algorithm, the authors proposed to use an increasing inertia weight throughout the PSO run. The addition of velocity to the current position to generate the next position is similar to the mutation operator in evolutionary algorithms, except that the *mutation* operator in PSO is guided by the experience of a particle and that of its neighbors. In other words, PSO performs mutation with a *conscience* [54].

Use of a mutation (or turbulence) operator. As mentioned in the previous section, when a particle updates its position, a mutation with *conscience* occurs. Sometimes, however, some unconsciousness or *craziness*, as called by Kennedy and Eberhart in the original proposal of PSO [36], is needed. Craziness, also referred to as "turbulence", reflects the change in a particle's flight which is out of control [22]. In general, when a swarm stagnates, that is, when the velocities of the particles are almost zero, it becomes unable to generate new solutions which might lead the swarm out of this state. This behavior can lead to the whole swarm being trapped in a local optimum from which it becomes impossible to escape. Since the global best individual attracts all members of the swarm, it is possible to lead the swarm away from a current location by mutating a single particle if the mutated particle becomes the new global best. This mechanism potentially provides a means both of escaping from local optima and of speeding up the search [56]. In this way, the use of a mutation operator is very important in order to escape from local optima and to improve the exploratory capabilities of PSO. When a solution is chosen to be mutated each component is then mutated (randomly changed) or not with certain probability. Actually, different mutation operators have been proposed that mutate components of either the position or the velocity of a particle. In our experience, the choice of a good mutation operator is a difficult task that has a significant impact on performance. On the other hand, once we have selected a specific mutation operator another difficult task is to decide how much mutation to apply: with how much probability, in which moments of the process, in which specific component of a particle, etc. Several approaches have used different mutation operators. However, there are also approaches which do not use any kind of mutation operator and that show good performance. So, the use of mutation is an issue that certainly deserves a more careful study.

MOPSO in Data Mining: A Quick review

In this section we will review the application of MOPSO in data mining thus far achieved by various researchers. Fig. 6.5 shows the current trends of MOPSO-related research regarding both methodology and applications.

The x and y axis of Fig. 6.5 represent, respectively, the year and the number of publications scattered in journals, proceedings, and book chapters. The current trend is promising in terms of the development of methodologies and applications to various domains. Although we collected an important number of references regarding the application of MOPSO in various domains, we found very few applications of MOPSO in data mining. Additionally, existing references focus mainly on solving the problem of classification rule mining, and on clustering, and very few are addressing the problem of data preprocessing. We did not find a single reference that addresses association rule mining, summarization,

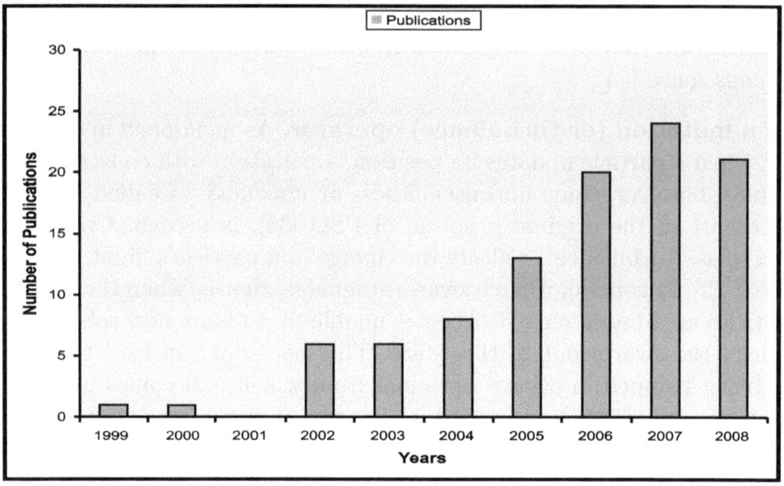

Fig. 6.5. Current research trends of MOPSO

regression, link analysis, or data post processing. Therefore, there is plenty of room for extending the use of MOPSO in data mining.

We will now briefly discuss the MOPSO methods so far developed by various authors for the data mining domain.

Feng et al. [20] developed a method for preprocessing in electric load data by identifying outliers using a multi-objective PSO based clustering algorithm. Their proposed work automatically determines the optimum number of clusters and it starts by partitioning the data domain into a relatively large number of clusters in order to reduce the effects of the initial conditions. The experimental results show that the proposed MOPSO-based clustering algorithm performs better than state-of-the-art clustering algorithms such as k-means and self organizing feature maps (SOM). The approach can also identify and justify noisy outliers from a data set in a simple but effective manner, while avoiding the danger of the outliers' influence in power system operation.

A multi-objective particle swarm optimization approach for rule discovery was proposed by Li et al. [43]. In their work *predictive accuracy* and *comprehensibility* are taken as the two conflicting objectives for optimization using MOPSO. The predictive accuracy is defined as the proportion between the number of instances classified correctly and the total number of records. Similarly, the comprehensibility is defined based on the number of rules discovered by the model. The lower the number of conditions the more comprehensible the rule is. The experimental results showed that their algorithm exhibited better performance than its single-objective counterpart.

Torácio and Pozo [58] presented a method for classification rule discovery based on multiple objective PSO. In their work, two sets of objectives were optimized by PSO. The first set contained sensitivity and specificity criteria. Similarly, set two contained positive and negative confidence. Their method finds the best non-dominated rules of a problem by selecting in the rules generating process.

Carvalho and Pozo [8] have proposed a multi-objective particle swarm optimization for non-order data mining rules from a mixed data domain. These rules can be used as an unordered classifier in such a way that, the rules are more intuitive and easier to understand. Further, they can be interpreted independently one from the other. The algorithm is evaluated by using the area under the ROC curve and comparing the performance of the induced classifiers with others obtained with well-known rule induction algorithms.

Ishida et al. [30] presents a method of classification rule discovery by exploring Multi-objective PSO and GRASP-PR. The rules are selected at the rule creation process following Pareto dominance concepts and forming unordered classifiers.

Tsai et al. [60] presented a particle swarm optimization approach for inventory classification problems where inventory items are classified based on multiple objectives, such as minimizing costs, maximizing inventory turnover ratios, and maximizing inventory correlation. In addition, their approach determines the best number of inventory classes and how items should be categorized for the desired objectives at the same time.

6.4.5 Discrete Particle Swarm Optimization

The continuous particle swarm algorithm discussed so far has been found to be robust in solving problems featuring non-linearity and non-differentiability, multiple optima, and high dimensionality through adaptation which is derived from social-psychological theory [16]. The PSO algorithm was originally developed for continuous valued problems, however many problems are defined on discrete valued space where the domain of the variables is finite. Typical examples include feature selection in data mining and bioinformatics [12], routing and scheduling among others [17]. Kennedy and Eberhart [37] introduced a discrete binary version of PSO for discrete optimization problems. Then, after many successively modified versions, binary PSO has evolved over the years [38]. Here, we will discuss the basic discrete PSO.

In binary PSO, each particle represents its position in binary values. Each particle's value can then be changed (or mutated) from one to zero or viceversa. A particle may be seen as if it moved to nearer and farther corners of a hypercube by flipping various bits; thus, the velocity of the particles may be described by the number of bits changed per iteration, or the Hamming distance between the particle at time t and at time $t+1$. A particle with zero bits flipped does not move, while it moves to the farthest corner by reversing all of its binary coordinates. This does not answer the question, however, of what corresponds to velocity in a binary function, that is, what is the velocity or rate of change of a single bit or coordinate.

The solution to this dilemma is to define trajectories, velocities, etc., in terms of changes of probabilities that a bit will be in one state or the other. Thus, a particle moves in a state space restricted to zero and one on each dimension, where each velocity component represents the probability of the corresponding component of taking the value of 1. For example, if the dth component of the velocity corresponding to the dth component of the ith particle is $v_{id} = 0.3$, then we can say that there is a thirty percent chance that the dth position of the ith particle x_{id} becomes 1 and a 70% chance that the value will be zero. Depending on the previous personal best p_{id} and the global best p_{gd} position of the particle, the following components can be calculated as follows:

p_{id}	p_{gd}	$(p_{id} - p_{gd})$	$(p_{gd} - x_{id})$
0	0	0 or -1	0 or -1
1	1	0 or 1	0 or 1

In summary, the formula for updating the velocity and position of a particle is the same as in the continuous case. However, the following modifications need to be done in the case of binary PSO.

1. The value of p_{id}, p_{gd} and x_{id} are from the domain $[0, 1]$.
2. The velocity $v_{id} \in [0, 1]$, since it is a probability.
3. The new position of the particle can be obtained as follows:

$$x_{id}(t+1) = \begin{cases} 1 \text{ if } r_{id} < sig(v_{id}(t+1)) \\ 0 \qquad\qquad \text{otherwise} \end{cases}$$

where r_{id} is a uniform random number and the function $sig(.)$ is used here to map all real valued numbers of the velocity to the range $[0, 1]$. The $sig(.)$ function is defined as:

$$v_{id}(t) = sig(v_{id}(t)) = \frac{1}{1 + e^{-v_{id}(t)}}. \tag{6.11}$$

In addition, we need to pay more attention to the issues related to the parameters of binary PSO. In the continuous version of PSO, large numbers for the maximum velocity of the particle encourage an increase of the range explored by a particle, whereas the opposite occurs in the binary version of PSO. There are also some difficulties with choosing proper values for the inertia weight w. For binary PSO, values of $w < 1$ prevent convergence. For values $-1 < w < 1$, the velocity becomes 0 over time. In this case, $sig(0) = 0.5$, so for $w < 1$ we have $\lim_{t \to \infty} sig(v_{id}(t)) = 0.5$. If $w > 1$, the velocity increases over time and $\lim_{t \to \infty} sig(v_{id}(t)) = 1$ so all bits change to 1. If $w < -1$ then $\lim_{t \to \infty} sig(v_{id}(t)) = 0$, so the probability that the bits change to 0 increases. Hence, it is very important to take into consideration the inertia weight in binary PSO. Some approaches have been proposed in [17, 38] to deal with this problem.

Although lots of improvements have been made on binary PSO for the single objective optimization domain, very few references are available for the multi-objective optimization domain. Therefore, extending binary PSO to the multi-objective domain is a promising research area. Moreover, to the best of our knowledge, the work reported in this chapter, is the first attempt to use discrete PSO for solving multi-objective problems in data mining.

6.5 A MOPSO for Optimizing Polynomial Neural Networks

Neural networks are computational models capable of learning through adjustments of topology/architecture and internal weight parameters according to a training algorithm in response to some training samples. Yao [62] describes three common approaches to neural network training: (1) for a neural network with a fixed architecture find a near-optimal set of connection weights; (2) find a near optimal neural network architecture; and (3) simultaneously find both a near optimal set of connection weights and neural network topology/architecture.

Recall that multi-objective optimization deals with the simultaneous optimization of several possibly conflicting objectives and generates the Pareto optimal set. Each solution in the Pareto optimal set represents a trade-off among the various objectives. In supervised learning, model selection involves finding a good trade-off between at least two objectives: classification accuracy and architectural complexity. The usual approach is to formulate the problem in a single objective manner by taking a weighted sum of the objectives, but Abbass [1] presented several reasons as to why this method is inefficient. Thus, a multi-objective optimization approach is suitable since the architecture with connection weights and predictive accuracy can be determined concurrently and a Pareto optimal set can be obtained in a single simulation run. This gives the user more flexibility to choose a final solution.

In data mining, the scalability is a very important issue: the scalability of the architecture can be determined from the architectural complexity. In this chapter, we are considering the architectural complexity of the polynomial neural network, which is based on parameters such as: 1) chosen input features, 2) order of the polynomial, 3) number of layers, and 4) number of partial descriptions (PDs) in a layer.

Let us assume that the chosen number of features and the total number of given features are r and m, respectively. Further, the chosen number of features remains unchanged in the entire network. The order of the polynomial and the number of layers are P and L, respectively. Based on these assumptions, Table 6.2 shows the total number of possible PDs in each of the layers of a PNN.

Table 6.2. Possible number of PDs in each layer of a PNN

Layer	Possible Number of PDs
1st	$m_1 = \frac{m!}{r!(m-r)!}$
2nd	$m_2 = \frac{m_1!}{r!(m_1-r)!}$
$\vdots$	$\vdots$
L^{th}	$m_L = \frac{m_{L-1}!}{(m_{L-1}-r)!}$

Therefore, the total possible number of PDs excluding input and output layers is

$$M = \sum_{i=1}^{L} m_i, = \frac{1}{r!}(\sum_{i=0}^{L-1} \frac{m_i!}{(m_i - r)!}), \tag{6.12}$$

where $m = m_0$.

In summary, the architectural complexity of the PNN can be reduced by minimizing equation (6.12).

Furthermore, in this work we are generating only one hidden layer; but we are considering the input variables twice (i.e., in the input layer and in the hidden layer). Therefore, the architectural complexity of the network is defined as

$$f_{AC} = 2m + \frac{m_0!}{(m_0 - r)!}, \tag{6.13}$$

where f_{AC} is the function for measuring the architectural complexity.

The measure for classification accuracy is defined based on the concept of cost matrix. Let the problem be a 2 class problem. Then, the cost matrix for the 2 class problem is defined as follows:

$$f_{CA} = \begin{pmatrix} c_{11} & c_{12} \\ c_{21} & c_{22} \end{pmatrix} \tag{6.14}$$

The entries in the cost matrix have the following meaning in the context of our study. c_{11} is the number of samples classified to class 1 having their true class to be class 1. c_{12} is the number of samples classified to class 2 having their true class to be class 1. Similarly, c_{21} is the number of samples classified to class 1 having their true class to be class 2. c_{22} is number of samples classified to class 2 having their true class to be class 2.

We can define several standard terms based on the general $c \times c$ cost matrix: The classification accuracy is the proportion of the total number of predictions that were correct, and it is determined using equation (6.15)

$$f_{CA1} = \frac{\sum_{i=1}^{c} c_{ii}}{\sum_{i=1}^{c}\sum_{j=1}^{c} c_{ij}}. \tag{6.15}$$

The correct classification accuracy corresponding to the $i^{th} < i = 1, 2, 3, \ldots, C >$ class is

$$f_{Ci} = \frac{c_{ii}}{\sum_{j=1}^{c} c_{ij}}, i = 1, 2, \ldots, c. \tag{6.16}$$

The incorrect classification accuracy corresponding to the $ith < i = 1, 2, 3, \ldots, C >$ class is

$$f_{ICi} = \frac{\sum_{j=1, j \neq i}^{c} c_{ii}}{\sum_{j=1}^{c} c_{ij}}, i = 1, 2, \ldots, c. \tag{6.17}$$

The accuracy determined using equation (6.15) might not be an adequate performance measure when the distributions of the classes are unbalanced. For example, there are 100 samples, 95 of which are class 1, 3 of which are class 2 and 2 of which are class 3. If the system classifies all of them as class 1, the accuracy would be 95% even though samples belonging to class 2 and class 3 are misclassified.

Therefore, we measure the classification accuracy by using the following equation:

$$f_{CA2} = \frac{1}{c}\left(\frac{c_{11}}{\sum_{j=1}^{c} c_{1j}} + \frac{c_{22}}{\sum_{j=1}^{c} c_{2j}} + \ldots + \frac{c_{cc}}{\sum_{j=1}^{c} c_{cj}}\right). \tag{6.18}$$

Finally, the two identifiable objective functions which are very often conflicting to each other in PNN are: classification accuracy and architectural complexity of the network. These two objectives are measured by equations (6.13) and (6.18).

6.5.1 Our Discrete MOPSO Approach

We have proposed a discrete multi-objective particle swarm optimizer for the simultaneous optimization of classification accuracy and architectural complexity of PNNs. Fig. 6.6 shows the topological structure of a typical PNN evolved by our proposed discrete particle swarm optimizer (DPSO).

In our proposed approach, the PDs are generated for only one layer (which can be treated as the only hidden layer of the network). Along with these PDs, the original set of input features are fed to the output layer. The output layer contains only one node with a very sophisticated transfer function. The PDs and

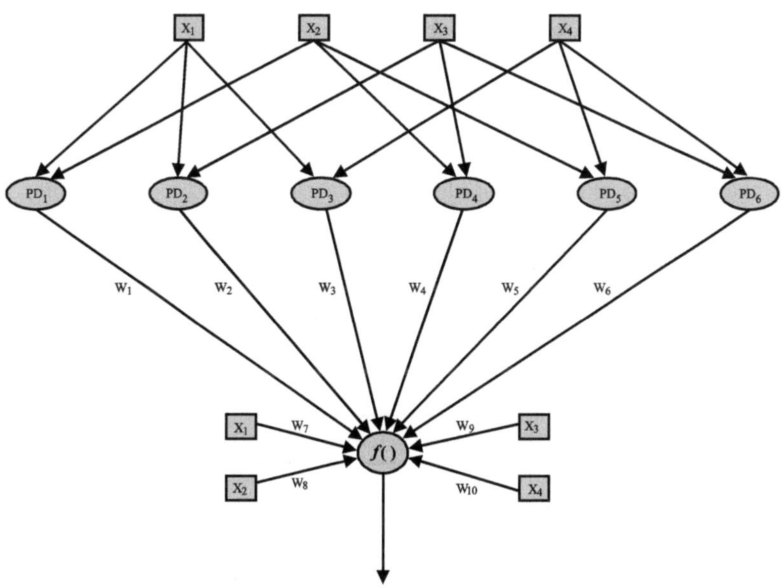

Fig. 6.6. Topological structure of a PNN evolved by our proposed DPSO

input features present in the hidden layer are connected to the output layer by a set of weights.

The task of discrete particle swarm optimization is to select an optimal set of PDs and features for the output layer such that the classification accuracy is maximized and the PDs as well as the features should be minimized (i.e., an implict minimization of architectural complexity). In addition, the proposed algorithm optimizes the weight between the hidden layer and the output layer in the course of running the algorithm. The algorithm applies direct encoding and uses classification accuracy, and architectural complexity as the performance measures that guide the evolution of the swarm.

In the above system, m represents the number of features in the dataset and $N = \frac{m!}{2!(m-2)!}$ represents the number of PDs in the hidden layer.

6.5.2 Algorithmic Framework

The algorithm deals with discrete variables (attributes and PDs) and its swarm of candidate solutions contains particles of varied length. Potential solutions to the multi-objective optimization problem at hand are represented by a swarm of particles. There are N particles in a swarm. Each particle is logically divided into two parts. The first part represents the architectural complexity and the second part represents the weights corresponding to each active bit of the first part. Therefore, the length of the first part varies from 1 to $(N+m)$. Similarly for second part. Before a detailed discussion of the proposed algorithm, let us fix some fundamental components of the multi-objective DPSO.

Representation of the Particles: The position of the particle is logically divided into two parts: i) the first part represents the PDs and number of input features and, ii) the second part represents the connection weights. Each component of the first part either contains 1 or 0 (i.e., either the PDs/attributes selected or not selected). The second part contains a set of real values. For example, given the list of PDs and attributes $< PD_1, PD_2, PD_3, f_1, f_2, f_3 >$ and size of the swarm $S = 5$, the population of particles could look like this:

$$\begin{aligned}
\vec{x}_1 &= \underbrace{0,1,0,0,1,1|-,0.3,-,-,0.29,0.53} \\
\vec{x}_2 &= \underbrace{1,1,0,0,0,1|0.2,0.3,-,-,-,0.3} \\
\vec{x}_3 &= \underbrace{1,0,0,0,1,1|0.6,0,-,-,0.4,0.5} \\
\vec{x}_4 &= \underbrace{0,1,0,1,1,1|-,0.3,-,0.8,0.29,0.53} \\
\vec{x}_5 &= \underbrace{0,1,1,0,1,0|-,0.3,0.5,-,0.29,-}
\end{aligned} \tag{6.19}$$

Note that in the particle x_1, from left to right 0 represents PD_1 which is not selected, 1 represents that PD_2 is selected, and so on. Therefore x_1 represents a candidate solution where PD_2, x_2 and x_3 are all selected. Their corresponding weights are given in the second part. Fig. 6.7 illustrates the topological mapping of the vector representations of the aforementioned particles.

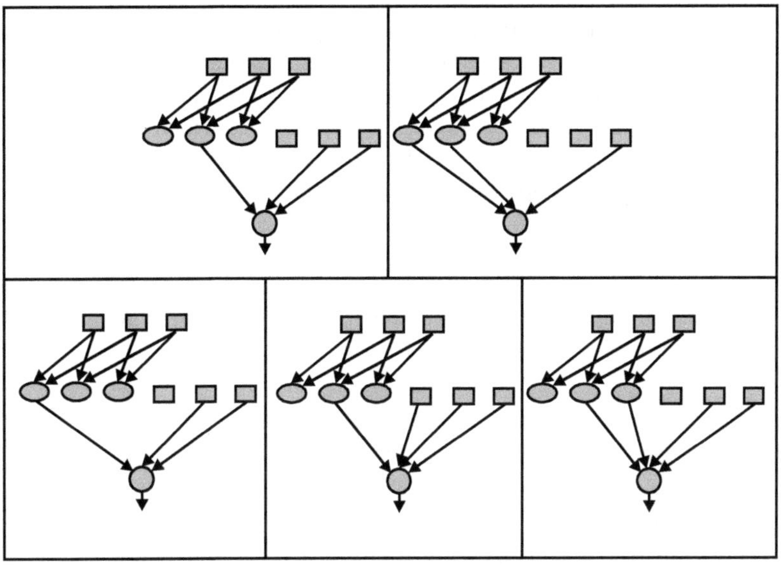

Fig. 6.7. Topological structure of the particles

Representation of the Velocity: The velocity of each particle is also logically divided into two parts. The first part represents the vector of probabilities and is constrained within the interval [0.0, 1.0]. The second part contains the real values. In the case of the first part, a map is introduced to map all real valued numbers of the velocity to the range [0.0, 1.0]. A logistic transformation such as the sigmoidal function ($sig(.)$) is used here to accomplish the task:

$$v_{id} = sig(v_{id}) = \frac{1}{1 + e^{-v_{id}}}. \tag{6.20}$$

In addition to the above encoding strategies, the following major elements are needed and are briefly described as follows:

Non-dominated personal set: It is a set that stores the non-dominated solutions obtained by the ith particle up to the current iteration. As the ith particle moves through the search space, its new position is added to this set and the set is updated to keep only the non-dominated solutions. A partitional medoid-based clustering algorithm is employed to reduce the non-dominated personal set size if it exceeds a certain pre-specified value. It is denoted as EX_PNS.

Non-dominated global best: It is a set that stores the non-dominated solutions obtained by all particle of the swarm up to the current iteration. First, we take the union of all the non-dominated local sets. Then, the non-dominated solutions out of this union are members in the non-dominated global set. The same partitional medoid-based clustering algorithm is employed to control the size of the global set. It is denoted as EX_GNS.

Selection of the personal best and the global best: The i_{th} personal best and global best are selected by using a Euclidean distance measure in objective space between the members of the non-dominated personal and global sets. The values of the non-dominated personal and global sets which are closest to the ith particle are nominated to become the personal and global bests, respectively. In the case of a tie, the proposed method randomly selects one as its personal/ global best.

In the proposed discrete multi-objective particle swarm optimization, the swarm has S particles and each particle is an $N+m$ dimensional vector, where m and N are the number of attributes and PDs, respectively. The computational steps of the proposed algorithm are the following:

Pseudocode()

1. DETERMINE the number of input features and the order of the polynomial forming the partial description (PD) of the data.
2. FOR each PDs
3. ESTIMATE the coefficients
4. INVOKE DMOPSO_PNN()
5. DECLARE a set of non-dominated solutions

The function DMOPSO_PNN() is the heart of the above algorithm. The details of this procedure are described next:

MOPSO_RUN()

1. INITIALIZE the Swarm and its Velocity.
2. FOR each particle of the swarm
3. FOR each sample
4. EVALUATE the classification accuracy by equation (6.18) and architectural complexity by equation (6.13).
5. NON-DOMINATED PERSONAL SET UPDATING: The updated position of the particle is added to EX_PNS. If any dominated solution is in EX_PNS, it will be deleted and the set will be updated accordingly. If the size exceeds the upper limit, then invoke the partitional medoid clustering algorithm.
6. NON-DOMINATED GLOBAL SET UPDATING: The union of all the non-dominated personal sets is formed and the non-dominated solutions out of this union are the members of this set. The size of this set can be controlled (if it exceeds the pre-specified value) by invoking the partitional medoid clustering algorithm.
7. SELECT the personal and global best.
8. VELOCITY UPDATING: Using the personal best and the global best, the velocity of each particle is updated using the following equation:

$$\overrightarrow{v}(t) = \overrightarrow{v}(t-1) + c_1 r_1(\overrightarrow{x}(t-1) - \overrightarrow{x_p}(t-1)) + c_2 r_2(\overrightarrow{x}(t-1) - \overrightarrow{x_g}(t-1))$$

9. POSITION UPDATING: The new position of the particle is obtained by:

$$\overrightarrow{x}(t) = \begin{cases} 1 \text{ if } \overrightarrow{r} < sig(\overrightarrow{v}(t)) \\ 0 \qquad \text{otherwise} \end{cases}$$

where r is a random number uniformly generated from the interval [0, 1].
10. STOPPING CRITERION: If the number of iterations exceeds the pre-specified maximum, then stop, else go to step 2

The standard k-medoid partitional clustering algorithm is used for controlling the size of the archives EX_PNS and EX_GNS.

6.6 Empirical Validation of the Proposed Method

The performance of the proposed method is evaluated using the benchmark classification databases IRIS, WINE, PIMA, and BUPA Liver Disorders. All these databases were taken from the UCI machine repository [5].

6.6.1 Description of the Datasets

Let us briefly discuss the datasets, which we have taken for our experimental setup.

IRIS Dataset: This is the most popular and simple classification dataset based on multivariate characteristics of a plant species (length and thickness of its petal and sepal) divided into three distinct classes (Iris Setosa, Iris Versicolor and Iris Virginica) of 50 instances each. One class is linearly separable from other two; the latter are not linearly separable from each other. In a nutshell, it has 150 instances and 5 attributes. Out of 5 attributes, four attributes are predicting attributes and one is the goal attribute. All the predicting attributes are real values.

WINE Dataset: This dataset resulted from a chemical analysis of wines grown in the same region in Italy but derived from three different cultivars. In a classification context, this is a well-posed problem with well-behaved class structures. The total number of instances is 178 and it is distributed as 59 for class 1, 71 for class 2 and 48 for class 3. The number of attributes is 14 including class attribute and all 13 are continuous in nature. There are no missing attribute values in this dataset.

PIMA Indians Diabetes Database: This database is a collection of all the female patients who are at least 21 years old and of PIMA Indian heritage. It contains 768 instances, 2 classes of positive and negative and 9 attributes including the class attribute. The attribute contains either integer or real values. There are no missing attribute values in the dataset.

Table 6.3. Summary of the datasets used for validating the proposed method

Dataset	Samples	Features	Classes	Class 1	Class 2	Class 3
IRIS	150	4	3	50	50	50
WINE	178	13	3	59	71	48
PIMA	768	8	2	268	500	-
BUPA	345	6	2	145	200	-

BUPA Liver Disorders: This dataset is related to the diagnosis of liver disorders and was created by BUPA Medical Research, Ltd. It consists of 345 records, and 7 attributes including the class attribute. The class attribute is repeated with only two class values for the entire database. The first 5 attributes are all blood tests, which are thought to be sensitive to liver disorders that might arise from excessive alcohol consumption. Each record corresponds to a single male individual. Table 6.3 presents the summary of the main features of the datasets used for our experimental study.

The experimental method involved a 2-fold cross validation. The samples of the datasets were divided into two almost equal folds. The folds were randomly generated keeping in mind a balanced distribution of the classes. This is also known as stratified cross validation. Each of the 2 folds was used once as a test set and another one for training. The training set was used twice in the algorithm. First, it was used for estimating the coefficient of the PDs and then it was used for guiding the evolution of the swarm. The division of the datasets and their class distribution is shown in Table 6.4.

Table 6.4. Classwise distribution of samples after division

Dataset	Division	Samples	Class 1	Class 2	Class 3
IRIS	Subset 1	75	25	25	25
	Subset 2	75	25	25	25
Wine	Subset 1	89	29	36	24
	Subset 2	89	30	35	24
PIMA	Subset 1	384	134	250	-
	Subset 2	384	134	250	-
BUPA	Subset 1	172	72	100	-
	Subset 2	173	73	100	-

6.6.2 Parameters and Numerical Results

As the proposed algorithm is stochastic, 10 independent runs were performed for each fold. The classification accuracy obtained, averaged over 10 runs, is reported in Fig. 6.8. The parameters used for the simulation studies are given in Table 6.5. In addition to the user-defined parameters adopted for the discrete multi-objective PSO, we set the maximum capacity of the external archives

Table 6.5. Protocols for setting the parameters of our proposed approach

Parameters	Values
Size of the swarm	100
Maximum Iterations	500
Cognitive Parameter	1.49
Social Parameter	2.0

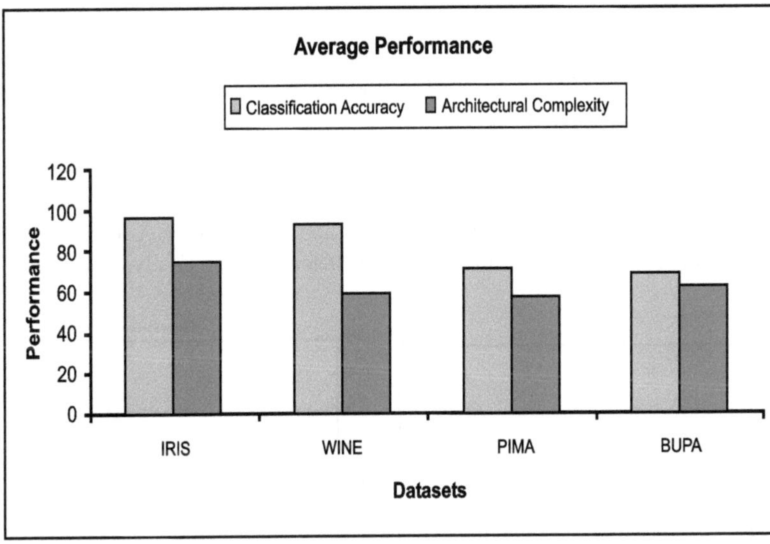

Fig. 6.8. Classification accuracy and architectural complexity obtained by the proposed method

as $EX\_PSN = 10$ and $EX\_GSN = 50$. The main objective of these external archives is to keep a historical record of the non-dominated vectors found along the search process. The size of these external archives is controlled by the k-medoid algorithm and the value of k is set to 5 for EX_PSN and 15 for EX_GSN.

First, we show some of the experimental results obtained by our discrete multi-objective PSO to illustrate the advantages of considering non-dominance. The confusion matrix shows the experimental results, which is selected from our proposed method by considering a sort of lexicographic ordering on the objectives for the entire dataset. The optimum solution is then obtained by maximizing the classification accuracy separately, starting with the most important one and proceeding according to the assigned order of importance of the objectives. Furthermore, this matrix shows the classification accuracy obtained from the run which is best among 10 independent runs. The cost matrices for IRIS and WINE are given in the following matrix:

$$\begin{pmatrix} 50 & 0 & 0 & | & 59 & 0 & 0 \\ 1 & 49 & 0 & | & 0 & 70 & 1 \\ 0 & 0 & 50 & | & 0 & 0 & 48 \end{pmatrix}$$

In the above augmented cost matrix the left submatrix represents the classification accuracy of the IRIS data. The right submatrix represents the classification accuracy of the WINE data. The cost matrix of BUPA and PIMA are given in the following matrix. In this matrix the left submatrix represents the classification accuracy of PIMA and the right one represents the classification accuracy of BUPA.

$$\begin{pmatrix} 134 & 134 & | & 85 & 60 \\ 44 & 456 & | & 30 & 170 \end{pmatrix}$$

Fig. 6.8 illustrates the classification accuracy and the architectural complexity obtained with the proposed method. The classification accuracy is obtained, averaged over 10 independent runs of the algorithm.

Fig. 6.9 illustrates the Pareto front obtained from the EX_GSN using the training patterns of the IRIS and WINE datasets, respectively. Moreover, after

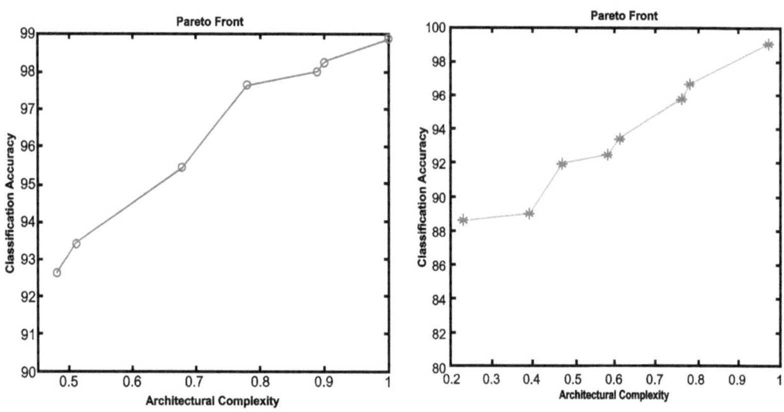

Fig. 6.9. Pareto front of the IRIS and WINE datasets, obtained by the proposed method

Table 6.6. Classification accuracy and architectural complexity obtained by PIMA and BUPA

PIMA		BUPA	
Arch. Complexity	Class. Accuracy	Arch. Complexity	Class. Accuracy
0.23	66.91	0.19	62.80
0.39	67.00	0.49	65.36
0.63	72.00	0.62	69.55
0.73	74.00	0.88	72.42
0.89	75.23	0.94	73.51

the training process is finished we run once again the k-medoid algorithm in EX_GSN to keep a few non-dominated solutions. In the case of IRIS, we kept 7, whereas in the case of WINE we kept 8.

Table 6.6 shows the classification accuracy and the architectural complexity obtained by PIMA and BUPA with a value of $k = 5$ applied in EX_GSN.

6.7 Conclusions and Future Research Directions

In this work, we have proposed a discrete Pareto-based particle swarm optimization method for the simultaneous optimization of architectural complexity and predictive accuracy of a PNN. The proposed method generates PDs for a single hidden layer. Then, we add the original set of input features in the hidden layer. Our proposed approach selects the optimal set of PDs and features from the hidden layer. Further, the method implicitly optimizes the weight vectors between the hidden layer and the output layer. The set of non-dominated solutions is maintained in an external archive called EX_GSN, which is controlled by the k-medoid algorithm. The k-medoid algorithm not only helps to restrict the size of the archive but also maintains diversity in the Pareto front. The experimental studies undertaken showed that our method gives more flexibility to the user for choosing the final solution according to his/her preferences.

Some possible directions for future research include: an empirical comparison of MOPSO methods developed so far in classification tasks, the development of a theoretical foundation for discrete multi-objective particle swarm optimization, and the application of a continuous MOPSO to data mining problems (e.g., rule mining, clustering, feature selection, instance selection, noise removal, etc.), among others.

References

[1] Abbass, H.: An evolutionary artificial neural networks approach to breast cancer diagnosis. Artificial Intelligence in Medicine 25(3), 265–281 (2002)

[2] Abido, M.A.: Two-level of non-dominated solutions approach to multi-objective particle swarm optimization. In: 2007 Genetic and Evolutionary Computation Conference (GECCO 2007), vol. 1, pp. 726–733. ACM Press, New York (2007)

[3] Aksyonova, T.I., Volkovich, V.V., Tetko, I.V.: Robust polynomial neural networks in quantitative-structure activity relationship studies. SAMS 43(10), 1331–1339 (2003)

[4] Banks, A.: A review of particle swarm optimization. Part II: hybridization, combinatorial, multi-criteria and constrained optimization, and indicative applications. Natural Computing 7, 109–124 (2008)

[5] Blake, C.L., Merz, C.J.: UCI repository of machine learning databases, `http://www.ics.uci.edu/~mlearn/MLRepository`

[6] Bozdogan, H.: Statistical data mining and knowledge discovery. Chapman and Hall/CRC Press (2003)

[7] Branke, J., Mostaghim, S.: About selecting the personal best in multi-objective particle swarm optimization. In: Runarsson, T.P., Beyer, H.-G., Burke, E.K., Merelo-Guervós, J.J., Whitley, L.D., Yao, X. (eds.) PPSN 2006. LNCS, vol. 4193, pp. 523–532. Springer, Heidelberg (2006)
[8] Carvalho, A.B.D., Pozo, A.: Non-ordered data mining rules through multi-objective particle swarm optimization: dealing with numeric and discrete attributes. In: Proceedings of the 8th International Conference on Hybrid Intelligent Systems, pp. 495–500. IEEE Press, Los Alamitos (2008)
[9] Clerc, M., Kennedy, J.: The particle swarm–explosion, stability, and convergence in a multidimensional complex space. IEEE Transactions on Evolutionary Computation 6(1), 58–73 (2002)
[10] Coello, C.A.C., Salazar Lechuga, M.: MOPSO: A proposal for multiple objective particle swarm optimization. Proceedings of IEEE Congress on Evolutionary Computation 2, 1051–1056 (2002)
[11] Coello Coello, C.A., Lamont, G.B., Van Veldhuizen, D.A.: Evolutionary algorithms for solving multi-objective problems, 2nd edn. Springer, New York (2007)
[12] Correa, E.S., Freitas, A.A., Johnson, C.G.: A new discrete particle swarm algorithm applied to attribute selection in a bioinformatics data set. In: Proceedings of the 2006 Genetic and Evolutionary Computation Conference (GECCO 2006), Seattle, Washington, USA, pp. 35–42 (2006)
[13] Deb, K.: Multi-objective optimization using evolutionary algorithms. Wiley Interscience Series in Systems and Optimization. John Wiley and Sons, Chichester (2001)
[14] Deb, K., Goldberg, D.E.: An investigation of niche and species formation in genetic function optimization. In: Proceedings of the 3rd International Conference on Genetic Algorithms, pp. 42–50 (1989)
[15] Deb, K., Pratap, A., Agarwal, S., Meyarivan, T.: A fast and elitist multiobjective genetic algorithm: NSGA II. IEEE Transactions on Evolutionary Computation 6(2), 182–197 (2002)
[16] Eberhart, R., Kennedy, J.: A new optimizer using particles swarm theory. In: Proceedings of the Sixth International Symposium on Micro Machine and Human Science, pp. 39–43. IEEE Service Center, Piscataway (1995)
[17] Engelbrecht, A.P.: Fundamentals of computational swarm intelligence. Wiley, Chichester (2005)
[18] Everson, R.M., Fieldsend, J.E., Singh, S.: Full elite sets for multi-objective optimization. In: Parmee, I.C. (ed.) Adaptive Computing in Design and Manufacture V, pp. 343–354 (2002)
[19] Fayyad, U., Piatetsky-Shapiro, G., Smyth, P.: From data mining to knowledge discovery in databases. AI Magazine 17(3), 37–54 (1996)
[20] Feng, L., Liu, Z., Ma, C.: Outlier identification and justification using multi-objective PSO based clustering algorithm in power system. In: Proceedings of 5th IEEE International Conference on Industrial Informatics, vol. 1, pp. 365–369 (2007)
[21] Fieldsend, J.E., Everson, R.M., Singh, S.: Using unconstrained elite archives for multi-objective optimization. IEEE Transactions on Evolutionary Computation 7(3), 305–323 (2003)
[22] Fieldsend, J.E., Singh, S.: A multi-objective algorithm based upon particle swarm optimization, an efficient data structure and turbulence. In: Proceedings of the Workshop on Computational Intelligence, Birmingham, UK, pp. 37–44 (2002)

[23] Fogel, D.B.: Evolutionary computation: Toward a new philosophy of machine intelligence. IEEE Press, Piscataway (2006)
[24] Goldberg, D.E.: Genetic algorithms in search, optimization, and machine learning. Addison-Wesley Publishing Company, Reading (1989)
[25] Goldberg, D.E., Richardson, J.: Genetic algorithms with sharing for multi-modal function optimization. In: Proceedings of 2nd International Conference on Genetic Algorithms, pp. 41–49 (1987)
[26] Haykin, S.: Neural Networks. Macmillan, New York (1994)
[27] Hilderman, R.J.: Knowledge discovery and measures of interest, 1st edn. Springer, Heidelberg (2001)
[28] Holland, J.H.: Adaptation in natural and artificial systems. University of Michigan Press, Ann Arbor (1975)
[29] Hu, X., Eberhart, R.: Multi-objective optimization using dynamic neighborhood particle swarm optimization. In: Proceedings of the 2002 Congress on Evolutionary Computation, vol. 2, pp. 1677–1681. IEEE Press, Los Alamitos (2002)
[30] Ishida, C.Y., de Carvalho, A.B., Pozo, A.T.R., Goldbarg, E.F.G., Goldbarg, M.C.: Exploring multi-objective PSO and GRASP-PR for rule induction. In: van Hemert, J., Cotta, C. (eds.) EvoCOP 2008. LNCS, vol. 4972, pp. 73–84. Springer, Heidelberg (2008)
[31] Ivakhnenko, A.G., Ivakhnenko, G.A.: The review of problems solvable by algorithms of the group method of data handling (GMDH). Pattern Recognition and Image Analysis 5(4), 527–535 (1995)
[32] Ivakhnenko, A.G.: The group method of data handling-a rival of the method of stochastic appriximation. Soviet Automatic Control c/c of Avtomatika 1(3), 43–55 (1968)
[33] Ivakhnenko, A.G.: Polynomial theory of complex systems. IEEE Transactions on Systems, Man, and Cybernetics SMC-1(4), 1–13 (1971)
[34] Jiang, M.: Stochastic convergence analysis and parameter selection of the standard particle swarm optimization algorithm. Information processing Letters 102, 8–16 (2007)
[35] Kennedy, J., Eberhart, R.C.: Swarm intelligence. Morgan Kaufmann Publishers, Inc, San Francisco (2001)
[36] Kennedy, J., Eberhart, R.: Particle swarm optimization. In: Proceedings of the 1995 IEEE International Conference on Neural Networks, pp. 1942–1948. IEEE Press, Los Alamitos (1995)
[37] Kennedy, J., Eberhart, R.C.: A discrete binary version of the particle swarm algorithm. In: Proceedings of the 1997 Conference on Systems, Man, and Cybernetics, pp. 4104–4109 (1997)
[38] Khaneswar, M.A., Teshnehlab, M., Shoorehdeli, M.A.: A novel binary particle swarm optimization. In: Proceedings of 2007 Mediterranean conference on Control and Automation, Athens-Greece, pp. 1–6. IEEE Press, Los Alamitos (2007)
[39] Knowles, J.D., Corne, D.W.: Approximating the non-dominated front using the Pareto archived evolution strategy. Evolutionary Computation 8(2), 149–172 (2000)
[40] Kriegel, H.P., Borgwardt, K.M., Kröger, P., Pryakhin, A., Schubert, M., Zimek, A.: Future trends in data mining. Data and Knowledge Discovery 15(1), 87–97 (2007)
[41] Laumanns, M., Thiele, L., Deb, K., Zitzler, E.: Combining convergence and diversity in evolutionary multi-objective optimization. Evolutionary Computation 10(3), 263–282 (2002)

[42] Lee, Y., Van Roy, B., Reed, C.D., Lippmann, R.P., Kennedy, R.L.: Solving data mining problems through pattern recognition. Prentice-Hall, Englewood Cliffs (1997)
[43] Li, S.T., Chen, C.C., Li, J.W.: A multi-objective particle swarm optimization algorithm for rule discovery. In: Proceedings of 3rd International Conference on International Information Hiding and Multi-media Signal Processing (IIH-MSP 2007), pp. 597–600. IEEE Press, Los Alamitos (2007)
[44] Michalewicz, Z.: Genetic Algorithms + Data Structures = Evolution Programs. Springer, Heidelberg (1999)
[45] Miettinen, K.: Nonlinear Multiobjective Optimization. Kluwer Academic Publishers, Boston (1999)
[46] Misra, B.B., Dehuri, S., Dash, P.K., Panda, G.: A reduced and comprehensible polynomial neural network for classification. Pattern Recognition Letters 29(12), 1705–1712 (2008)
[47] Mostaghim, S., Teich, J.: Strategies for finding good local guides in multi-objective particle swarm optimization (MOPSO). In: Proceedings of the 2003 IEEE Swarm Intelligence Symposium, pp. 26–33 (2003)
[48] Mostaghim, S., Teich, J.: The role of ε-dominance in multi-objective particle swarm optimization methods. In: Proceedings of the 2003 Congress on Evolutionary Computation, vol. 3, pp. 1764–1771. IEEE Press, Los Alamitos (2003)
[49] Noghin, V.D.: A logical justification of the Edgeworth-Pareto principle. Comp. Mathematics and Math. Physics 42, 915–920 (2002)
[50] Noghin, V.D.: An axiomatization of the generalized Edgeworth-Pareto principle in terms of choice functions. Mathematical Social Sciences 52(2), 210–216 (2006)
[51] Oh, S.K., Pedrycz, W., Park, B.J.: Polynomial neural networks architecture: analysis and design. Computers and Electrical Engineering 29, 703–725 (2003)
[52] Piatetsky-Shapiro, G.: Data mining and knowledge discovery 1996 to 2005: overcoming the hype and moving from university to business and analytics. Data Mining and Knowledge Discovery 15(1), 99–105 (2007)
[53] Reyes-Sierra, M., Coello, C.A.C.: Multi-objective particle swarm optimizers: A Survey of the state-of-the-art. International Journal of Computational Intelligence Research 2(3), 287–308 (2006)
[54] Shi, Y., Eberhart, R.: Parameter selection in particle swarm optimization. In: Porto, V.W., Waagen, D. (eds.) EP 1998. LNCS, vol. 1447, pp. 591–600. Springer, Heidelberg (1998)
[55] Shi, Y., Eberhart, R.: Empirical study of particle swarm optimization. In: Proceedings of the Congress on Evolutionary Computation (CEC 1999), pp. 1945–1950. IEEE Press, Los Alamitos (1999)
[56] Stacey, A., Jancic, M., Grundy, I.: Particle swarm optimization with mutation. In: Proceedings of the Congress on Evolutionary Computation, pp. 1425–1430. IEEE Press, Los Alamitos (2003)
[57] Tetko, I.V., Aksenova, T.I., Volkovich, V.V., Kashena, T.N., Filipov, D.V., Welsh, W.J., Livingstone, D.J., Villa, A.E.P.: Polynomial neural network for linear and non-linear model selection in quantitative structure activity relationship studies on the interent. SAR QSAR Environ Res. 11, 263–280 (2000)
[58] de Almeida Prado G. Torácio, A., Pozo, A.T.R.: Multiple Objective Particle Swarm for Classification-Rule Discovery. In: Proceedings of IEEE Congress on Evolutionary Computation (CEC 2007), pp. 684–691 (2007)
[59] Trelea, I.C.: The particle swarm optimization algorithm: convergence analysis and parameter selection. Information Processing Letters 85, 317–325 (2003)

[60] Tsai, C.Y., Yeh, S.W.: A multiple objective particle swarm optimization approach for inventory classification. International Journal Production Economics 114, 656–666 (2008)
[61] Witten, I.H., Frank, E.: Data mining: practical machine learning tools and techniques, 2nd edn. Morgan Kaufmann, San Francisco (2005)
[62] Yao, X.: Evolving artificial neural networks. Proceedings of the IEEE 87, 1423–1447 (1999)
[63] Zhang, G.P.: Neural networks for classification: A survey. IEEE Transactions on Systems, Man and Cybernetics, Part C: Applications and Reviews 30(4), 451–462 (2000)
[64] Zhen, Y.L., Ma, L.H., Qian, J.X.: On the convergence analysis and parameter selection in particle swarm optimization. In: Proceedings of the 2nd International Conference on Machine Learning and Cybernetics, pp. 1802–1807. IEEE Computer Society Press, Los Alamitos (2003)
[65] Zitzler, E., Thiele, L.: Multi-objective evolutionary algorithms: acomparative case study and the strength Pareto approach. IEEE Transactions on Evolutionary Computation 3(4), 257–271 (1999)
[66] Zribi, M., Ghorbel, F.: An unsupervised and non-parametric Bayesian classifier. Pattern Recognition Letters 24(1-3), 97–112 (2003)

7

Rigorous Runtime Analysis of Swarm Intelligence Algorithms – An Overview

Carsten Witt

DTU Informatics
Technical University of Denmark
2800 Kgs. Lyngby, Denmark
cfw@imm.dtu.dk

Summary. The theoretical runtime analysis of randomized search heuristics is an emerging research area where many results have been obtained in recent years. Most of these studies deal with evolutionary algorithms rather than swarm intelligence approaches such as ant colony optimization and particle swarm optimization. Despite the overwhelming practical success of swarm intelligence, the first runtime analyses of such approaches date only from 2006. Since then, however, significant progress has been made in analyzing the runtime of simple ACO and PSO variants. The aim of this chapter is to give an overview on existing studies in this area. Moreover, it is elaborated what direction the theory of swarm intelligence is likely to take, and the vision of a unified theory of randomized search heuristics is discussed.

7.1 Introduction

Heuristics are the method of choice to solve optimization tasks in engineering if no tailored method is available. In particular, bio-inspired, randomized search heuristics such as evolutionary algorithms (EA), ant colony optimization (ACO) and particle swarm optimization (PSO) can perform extraordinarily well and exhibit a great generality and flexibility. In recent years, the last two approaches, which are commonly subsumed under the term *swarm intelligence* (SI), have received increasing attention. From a practical perspective, the invention of SI algorithms is an undeniable success story.

This paper considers swarm intelligence from a theoretical point of view. We claim that developing a solid theory for SI algorithms is more than an academic challenge. Rigorously proven results on the efficiency of such approaches will serve as a valuable basis for their design and application. Moreover, proven results will provide long-lasting (actually everlasting) statements on the usefulness and limitations of the approaches. Finally, a solid theory prepares a scientifically founded basis in order to transfer expertise in swarm intelligence to different disciplines, e. g., to theoretical computer science.

There is, however, not *the* single theory of randomized search heuristics. In the domain of evolutionary computation, different strands of theory have emerged, e. g., fitness landscape analysis, convergence theory, infinite population models etc. We focus on an aspect that we consider to be most closely linked to the

C.A. Coello Coello et al. (Eds.): Swarm Intel. for Multi-objective Prob., SCI 242, pp. 157–177.
springerlink.com

efficiency of algorithms, namely the rigorous *theoretical runtime analysis*. Roughly speaking, the key question is how many generations the heuristic takes to find an optimal solution to the problem or a good approximate solution of a prespecified quality. Such an analysis is carried out as in the classical algorithms community and makes use of several strong methods for the analysis of randomized algorithms [1, 2].

Since the early 1990s, the runtime analysis of EAs has evolved rapidly and found its place in theoretical computer science. Starting from toy problems such as OneMax (e.g., [3–5]), impressive results have been obtained regarding the runtime of simple and more complicated EAs in combinatorial optimization (see [6] for a recent survey on this topic). This is different with SI algorithms. First runtime analyses did not appear until 2006, and theory is lagging far more behind practice than in evolutionary computation. Possible reasons are twofold. On the one hand, SI algorithms constitute a much younger approach than EAs. On the other hand, their biological background, the so-called swarm dynamics, seems to make them harder to analyze. Notwithstanding, driven by the successes in the runtime analysis of EAs, significant progress has been made also for SI algorithms in this area. This motivates us to give a first survey covering runtime results for both ACO and PSO algorithms.

This paper is structured as follows. In Sect. 7.2 we provide backgrounds, notions and foundations needed for this survey. Sect. 7.3 summarizes the most important achievements in the runtime analysis of ACO algorithms, while Sect. 7.4 deals with corresponding results for PSO. Finally, we discuss the current status of theory and possible visions and directions for future research in Sect. 7.5.

7.2 Theoretical Runtime Analysis—Backgrounds

The typical domain of application for randomized search heuristics, in particular for the bio-inspired ones, is the so-called *black-box* scenario. This means that the function for which an optimum is sought is completely unknown to the algorithm and information can be gathered solely by evaluating it on certain inputs. In particular, no information on the gradient and other characteristics is available, which excludes classical search methods such as the Newton method from our considerations.

The black-box assumption is met, e.g., in complex systems where the quality of a solution can only be determined as the outcome of a simulation of the system for a given parameter setting. This also motivates our cost measure, i.e., the number of evaluations of the black-box function until the goal of optimization has been reached. So the basic question is: given an objective function $f\colon D^n \to \mathbb{R}$ mapping points from an n-dimensional space to an objective value, how many f-evaluations take place until the objective value has reached a certain quality? We formally capture this by the following definition.

Definition 1. *The runtime $T_{A,f}$ of an algorithm A optimizing f is the number of times that f is evaluated on a search point until an optimum is found.*

If randomized search heuristics are studied, then $T_{A,f}$ is typically a random variable. As usual in the analysis of randomized algorithms, we therefore often focus on the expectation $E(T_{A,f})$, the so-called expected runtime of A on f. However, the expected runtime is not necessarily a reliable measure of the heuristic's efficiency. Even if the expected runtime is exponentially big, the probability of a small polynomial runtime might still be sufficient to allow efficient optimization by using multistart variants of the heuristic. Hence, we are also interested in the distribution of $T_{A,f}$, i. e., we try to estimate $\text{Prob}(T_{A,f}) \leq D$ for a given time bound D.

All our analyses are usually conducted in an asymptotic framework where the number of dimensions n is the parameter that determines the size (or difficulty) of the problem. Similarly to an asymptotic analysis in computer science, where n is the size of the input (e. g., the number of cities in a TSP instance), we would therefore like to represent the required or sufficient runtime as a function of n. As the last sentence suggests, often no exact formula is available but at least upper and lower bounds in O-notation.

As already mentioned, the runtime analysis of bio-inspired search heuristics can be difficult since these heuristics were not designed with their theoretical analysis in mind. Initial theoretical studies therefore mostly considered convergence aspects, i. e., the question whether the expected runtime is finite. Such convergence analyses can be considered as precursors of runtime analyses and, in the domain of SI algorithms, have shown to be very crucial on the way towards first results regarding the runtime.

7.3 Runtime Analysis of ACO

ACO (see [7] for an overview) is a randomized search heuristic inspired by the search of an ant colony for a common source of food. In such artificial ant systems, solutions are constructed by random walks on construction graphs. These random walks are controlled by a probabilistic model induced by so-called pheromone values on edges of the graph. According to the quality of the solutions obtained, the probabilistic model is repeatedly updated in order to reflect good solutions.

Regarding the theory of ACO, only convergence results [8] were known until 2006 and analyzing the runtime of simple ACO algorithms has been pointed out as a challenging task by Dorigo and Blum [9]. Soon after this appeal, first steps into analyzing the runtime of ACO algorithms were made by Gutjahr [10], and, independently, the first runtime analyses of a simple ACO algorithm called 1-ANT were done at the same time by Neumann and Witt [11]. Both studies focus on very simple ACO algorithms using only a single ant. For our purposes, this seems to be the right starting point. At the time where no runtime results are available, one should consider as simple scenarios as possible to minimize the risk of failing to produce such results. Moreover, the choice of the single-ant systems can be motivated by the history of the theory of EAs. There, initial theoretical studies dealt also with single-individual systems such as the well-known

(1+1) EA. Such basic investigations allow for the development of methods for the analysis and build the foundations needed to study more complicated algorithms. In retrospect, numerous successes in the runtime analysis of simple and more complicated EAs justify the choice of the (1+1) EA as starting point. We hope that studies of single-ant systems will initiate a similarly fruitful foundation for the analysis of more complicated ACO algorithms.

In this survey, we consider the runtime behavior of several single-ant ACO algorithms that work by a common principle to construct new solutions. Our focus is pseudo-boolean optimization, i. e., we would like to find a solution $x \in \{0,1\}^n$ that maximizes a given function of the kind $f\colon \{0,1\}^n \to \mathbb{R}$. Such solutions are constructed by the ACO algorithm by performing a random walk on an appropriate directed construction graph $C = (V, E)$, which is equipped with a designated start vertex $s \in V$ and pheromone values $\tau\colon E \to \mathbb{R}$ on the edges. The construction procedure is shown in Fig. 7.1.

Construct(C, τ)
1. $v := s$, mark v as visited.
2. Let N_v be the set of non-visited successors of v in C.
   If $N_v \neq \emptyset$ then
   a) Choose successor $w \in N_v$ with probability $\tau_{(v,w)} / \sum_{(v,u) \mid u \in N_v} \tau_{(v,u)}$.
   b) Mark w as visited, set $v := w$ and go to 2.
3. Return the solution x and the path $P(x)$ constructed by this procedure.

Fig. 7.1. The construction procedure for pseudo-Boolean optimization.

We examine the construction graph given in Fig. 7.2, which is known in the literature as *Chain* [12]. For bit strings of length n, the graph has $3n+1$ vertices and $4n$ edges. The decision whether a bit x_i, $1 \le i \le n$, is set to 1 is made at node $v_{3(i-1)}$. If edge $(v_{3(i-1)}, v_{3(i-1)+1})$ is chosen, x_i is set to 1 in the constructed solution. Otherwise the other edge, $(v_{3(i-1)}, v_{3(i-1)+2})$, is taken, and $x_i = 0$ holds. After this decision has been made, there is only one single edge which can be traversed in the next step. In case that $(v_{3(i-1)}, v_{3(i-1)+1})$ has been chosen, the next edge is $(v_{3(i-1)+1}, v_{3i})$, and otherwise the edge $(v_{3(i-1)+2}, v_{3i})$ will be traversed. Hence, these edges have no influence on the constructed solution and we can assume $\tau_{(v_{3(i-1)}, v_{3(i-1)+1})} = \tau_{(v_{3(i-1)+1}, v_{3i})}$ and $\tau_{(v_{3(i-1)}, v_{3(i-1)+2})} = \tau_{(v_{3(i-1)+2}, v_{3i})}$, $1 \le i \le n$, for their pheromone values. We ensure that $\sum_{(u,\cdot) \in E} \tau_{(u,\cdot)} = 1$ for $u = v_{3i}$, $0 \le i \le n-1$, and $\sum_{(\cdot,v)} \tau_{(\cdot,v)} = 1$ for $v = v_{3i}$, $1 \le i \le n$. Let $p_i = \mathrm{Prob}(x_i = 1)$ be the probability of setting the bit x_i to 1 in the next constructed solution. The probability p_i is often called the *success probability* at a bit. Due to our setting $p_i = \tau_{(3(i-1), 3(i-1)+1)}$ and $1 - p_i = \tau_{(3(i-1), 3(i-1)+2)}$ holds, i. e., the pheromone values correspond directly to the success probabilities at the bits. In addition, following the MAX-MIN ant system by Stützle and Hoos [13], each $\tau_{(u,v)}$ is restricted to the interval $[1/n, 1 - 1/n]$ such that every solution always has a positive probability of being chosen. The choice of these values is inspired by standard mutation operators in evolutionary computation where an incorrect bit has a probability of $1/n$ of being corrected.

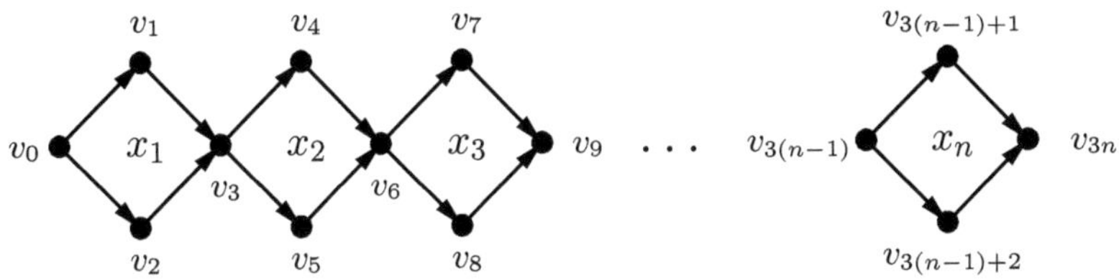

Fig. 7.2. Construction graph for pseudo-Boolean optimization.

Depending on whether the edge (u, v) is contained in the path $P(x)$ of the constructed solution x, the pheromone values are updated to τ' in the update procedure in order to reward the solution x. Controlled by the so-called evaporation factor ρ, an update works as follows:

$$\tau'_{(u,v)} = \begin{cases} \min\{(1-\rho)\cdot\tau_{(u,v)}+\rho,\ 1-1/n\} & \text{if } (u,v)\in P(x) \\ \max\{(1-\rho)\cdot\tau_{(u,v)},\ 1/n\} & \text{otherwise.} \end{cases}$$

So far we have not specified the conditions for pheromone updates to take place. Actually, the update mechanism distinguishes between the two most important types of ACO algorithms which were studied in terms of the runtime behavior. The aim of the following subsections is to describe these two types and the results obtained in greater detail.

7.3.1 1-ANT

The following single-ant ACO algorithm called 1-ANT was introduced by Neumann and Witt [11] and is shown in Fig. 7.3. Inspired by the properties of EAs, pheromone updates take only place if a new good solution is found. More precisely, we accept a solution x as a new best-so-far solution and update pheromones only if the function value is at least as good as for the previous best-so-far solution x^*, i. e., $f(x) \geq f(x^*)$. The attentive reader may notice that in the present survey the pheromone values have been rescaled in comparison to [11].

Having fixed an objective function and the chain construction graph, the choice of the evaporation factor ρ is still open. A simple but far-reaching observation has been stated by Neumann and Witt [11]: if ρ is chosen large enough (a value of $1-1/(n-1)$ is provably sufficient) then each pheromone update results in that either the upper bound $1-1/n$ or the lower bound $1/n$ on the pheromone values is attained. Hence, each bit of the best-so-far solution is reproduced with probability $1-1/n$ and flipped with probability $1/n$. This corresponds to a standard bit-flip mutation as used in the well-known (1+1) EA.

Theorem 1 ([11]). *Choosing $\rho \geq 1-1/(n-1)$, the 1-ANT and the (1+1) EA are the same algorithm.*

In other words, the previous theorem identifies a simple EA as a special case of a simple ACO algorithm. This transfers the variety of results obtained on the

1-ANT
1. Set $\tau_{(u,v)} = 1/2$ for all $(u,v) \in E$.
2. Create x^* using Construct(C,τ).
3. Update pheromones w. r. t. x^*.
4. Repeat
   a) Create x using Construct(C,τ).
   b) If $f(x) \geq f(x^*)$ then
      i) $x^* := x$.
      ii) Update pheromones w. r. t. x^*.

Fig. 7.3. 1-ANT

runtime of the (1+1) EA to the 1-ANT. Moreover, this characterization can be seen as a step towards a unified theory for the analysis of randomized search heuristics.

Test functions

The correspondence of the 1-ANT and the (1+1) EA stimulated a similar line of research as known from the runtime analysis of EAs. Initially, simple example functions were studied as test beds, in particular the well-known function

$$\text{ONEMAX}(x_1,\ldots,x_n) = x_1 + \cdots + x_n,$$

counting the number of ones in a bit string,

$$\text{BINVAL}(x_1,\ldots,x_n) = \sum_{i=1}^{n} 2^{n-i} x_i,$$

which yields the interpretation of the bit string as a binary number, and

$$\text{LEADINGONES}(x_1,\ldots,x_n) = \sum_{i=1}^{n} \prod_{j=1}^{i} x_i,$$

which counts the number of leading ones in the string. Actually, already Dorigo and Blum [9] mention ONEMAX as an appropriate starting point for the runtime analysis of ACO. Again, the aim is to keep the scenario for the theoretical investigations as simple as possible. Also the analysis of EAs started with the toy problem ONEMAX but today comprises many classical problems from combinatorial optimization.

The functions ONEMAX, BINVAL and LEADINGONES were used to study the effect of pheromone updates in the 1-ANT and the impact of ρ in greater detail. Interestingly, the 1-ANT suffers from a sudden phase transition from efficient to totally inefficient behavior on all three functions (see also [14] and [15]). If ρ is above a certain threshold value then the function is optimized in polynomial time, in most cases being $O(n^2)$, with high probability or in expectation. Below

Table 7.1. Runtime behavior of the 1-ANT depending on ρ (taken from [11], [14] and an extended version of [15]); c_1 and c_2 are constants derived in the analysis.

	superpolynomial runtime	polynomial runtime
OneMax	$\rho = o(1/\log n)$	$\rho = 1 - O(n^{-\epsilon})$
LeadingOnes	$\rho \leq c_1/\log n$	$\rho \geq c_2/\log n$
BinVal	$\rho \leq c_1/\log n$	$\rho \geq c_2/\log n$

the threshold, the runtime is no longer polynomial or even exponential. Table 7.1 summarizes the results obtained w. r. t. the 1-ANT on the three example functions.

The observed phase transition suggests that the 1-ANT is very sensitive to the choice of the evaporation factor. On an intuitive level, this can be explained as follows. At each point of time, the 1-ANT maintains a probability distribution over the search space by assigning pheromone values or, equivalently, success probabilities to bits. Since the bits are processed independently, the probability distribution on $\{0,1\}^n$ is just the product of the bits' distributions. In order for the 1-ANT to improve a best-so-far solution x^*, the probability distribution for the next constructed solution must reflect x^* to a sufficient amount. Considering OneMax, if the expected number of ones in the next constructed solution is far below OneMax(x^*) then, according to Chernoff bounds, an improvement will be unlikely. This is exactly what happens when ρ is too small. The best-so-far OneMax-value and the expected next OneMax-value are diverging in the course of optimization, making it finally extremely difficult to obtain further improvements. In other words, ρ is too small for the pheromone updates – which only take place in improving steps – to catch up with the best-so-far function value. Only above the threshold, best-so-far and expected next value stay close to each other, which allows for a polynomial runtime. Similar effects occur with the other two test functions.

From a methodological perspective, the analyses of the 1-ANT seem highly interesting and valuable. The phase transition behavior is proved using inverses of concentration inequalities. Rather than showing by Chernoff bounds etc. that a random variable (the number of ones for OneMax) is closely concentrated around its mean, one has to establish a lower bound on the probability of the random variable overshooting its expectation by a sufficiently large amount. Such bounds can be derived based upon old results by Hoeffding [16] and Gleser [17], which were rarely applied in the analysis of randomized search heuristics before. Also the analysis regarding LeadingOnes and BinVal [15] presents wholly new techniques for the analysis of ACO algorithms. They provide insight into the effect of pheromone updates and characterize the time for a single pheromone value to reach one of its bounds very exactly.

The minimum spanning tree and the shortest path problem

As mentioned above, it is presently hard to analyze the runtime of simple ACO algorithms on problems from combinatorial optimization. Still, Neumann and

Witt [18] as well as Attiratanasunthron and Fakcharoenphol [19] have obtained initial results in this direction. In particular, the first work elaborates on the choice of the construction graph. The chain graph mentioned above is a general one for the optimization of pseudo-Boolean functions and does not take knowledge about the given problem into account. ACO algorithms have the advantage that more knowledge about the structure of a given problem can be incorporated into the construction of solutions. This is done by choosing an appropriate construction graph together with a procedure which allows to obtain feasible solutions. The choice of such a construction graph together with its procedure has been observed experimentally as a crucial point for the success of such an algorithm. Moreover, the use of heuristic information is considered. In many ACO algorithms, each edge e of the construction graph does not only contain a dynamically changing pheromone value $\tau(e)$ but also a static heuristic information $\eta(e)$, which determines the overall attractiveness of an edge. The construction procedure then chooses edge e with a probability proportional to $(\tau(e))^{\alpha} \cdot (\eta(e))^{\beta}$. If $\beta = 0$, no heuristic information is used and, if in addition $\alpha = 1$, we obtain the transition probabilities from Fig. 7.1. If $\alpha = 0$, the construction procedure does not use pheromone information.

Neumann and Witt [18] study the minimum spanning tree (MST) problem. Given an undirected connected graph $G = (V, E)$ with edge costs (weights) $w \colon E \to \mathbb{N}_{\geq 1}$, the goal is to find a spanning tree $E^* \subseteq E$ such that the total cost $\sum_{e \in E^*} w(e)$ becomes minimal. Simple EAs, in particular the (1+1) EA were studied for this problem before [20]. Due to Theorem 1, all these results can be transferred to the 1-ANT with the chain construction graph in case that $\rho \geq 1 - 1/(m-1)$, where $m := |E|$. As a result, the expected optimization time of the 1-ANT is $O(m^2(\log n + \log w_{\max}))$ in this case, where $n := |V|$ and $w_{\max}$ is the largest weight of the input. In addition, a class of instances with polynomial weights has been presented in [20] where the expected time to obtain an optimal solution is $\Theta(m^2 \log n)$.

After having stated the results for the chain graph, the authors of [18] examine construction graphs being more suitable for the problem. The algorithm mentioned in the following is a simplified 1-ANT working with only two different pheromone values, however, with the possibility of adding heuristic information. As a first attempt towards a specialized construction procedure, a random walk on the original graph given by the MST instance itself is considered. It is well known how to choose a spanning tree of a given graph uniformly at random using random-walk algorithms. The construction procedure produces solutions by a variant of Broder's algorithm [21]. Neumann and Witt [18] show a polynomial, but relatively large, upper bound for obtaining a minimum spanning tree by this procedure if no heuristic information influences the random walk. Using only heuristic information for constructing solutions, they show that the 1-ANT together with the Broder-based construction procedure with high probability does not find a minimum spanning tree or even does not present a feasible solution in polynomial time.

After that, the authors consider a more incremental construction procedure that follows a general approach proposed by Dorigo and Stützle [7] to obtain an ACO construction graph. The procedure is called Kruskal-based as in each step an edge that does not create a cycle is chosen to be included into the solution. It turns out that the expected optimization time of the 1-ANT using the Kruskal-based construction procedure is $O(mn(\log n + \log w_{\max}))$. This beats the 1-ANT with the chain construction graph, i. e., in the case that the minimum spanning tree problem is more generally modeled as a pseudo-Boolean function since then the above-mentioned lower bound $\Omega(m^2 \log n)$ of the (1+1) EA carries over. Using the 1-ANT together with the Kruskal-based construction procedure and a large influence of the heuristic information, the algorithm has even a constant expected optimization time.

Finally, Attiratanasunthron and Fakcharoenphol [19] investigate a modified ACO approach for the construction of shortest paths to a single destination in directed acyclic graphs. In particular, their work investigates the behavior of multiple ants in this setting. The idea is to place an ant at each node of the directed graph. Then shortest paths can be built in a way resembling a dynamic programming approach. First, one waits for a closest neighbor of the destination node to be found. Since ants start from all nodes of the graph, this information is maintained and can be used to find shortest paths from further nodes. Similarly as in [18], well-chosen pheromone bounds are used but no heuristic information is incorporated. In the end, an expected number of $O((1/\rho)n^2 m \log n)$ iterations is sufficient in the setting of Attiratanasunthron and Fakcharoenphol [19] in order to find all shortest paths. Altogether, the analyses from [18] and [19] provide first proofs that ACO algorithms for combinatorial optimization can be analyzed rigorously using the toolbox from the analyses of randomized algorithms.

7.3.2 MMAS*

Plain algorithm

At the same time when Neumann and Witt were working on the first runtime analyses of the 1-ANT, Gutjahr [10] investigated the runtime of a single-ant system which we call MMAS* here. It uses the same construction graph and construction procedure as the 1-ANT, however, differs in terms of the pheromone update and selection of the best-so-far solution. Only if a newly constructed solution is strictly better than the previous best-so-far solution, an exchange happens. Moreover, in contrast to the 1-ANT, the best-so-far solution is rewarded in any step, regardless of the result of the construction procedure. The procedure is displayed in Fig. 7.4

The different update rule for pheromone values has a drastic consequence: the phase transition behavior observed for the 1-ANT does not appear, and the runtime on the test functions OneMax and LeadingOnes is bounded by a polynomial in n as long as $1/\rho$ is also a polynomial. These results were shown by Gutjahr [10] and Gutjahr and Sebastiani [22] by applying and generalizing

MMAS*
1. Set $\tau_{(u,v)} = 1/2$ for all $(u, v) \in E$.
2. Create x^* using Construct(C, τ).
3. Update pheromones w. r. t. x^*.
4. Repeat
   a) Create x using Construct(C, τ).
   b) If $f(x) > f(x^*)$ then $x^* := x$.
   c) Update pheromones w. r. t. x^*.

Fig. 7.4. MMAS*

a technique known as the method of fitness-based partitions in the theory of evolutionary algorithms. We describe the general idea.

Suppose that during a run there is a phase where MMAS* never replaces the best-so-far solution x^* in step 4.b) of the algorithm. This implies that the best-so-far solution is reinforced again and again until all pheromone values have reached their upper or lower borders corresponding to the setting of the bits in x^*. We can say that x^* has been "frozen in pheromone". The probability to create a 1 for any bit is now either $1/n$ or $1-1/n$. The distribution of constructed solutions equals the distribution of offspring of the (1+1) EA with x^* as the current search point. We conclude that, as soon as all pheromone values touch their upper or lower borders, MMAS* behaves like the (1+1) EA until a solution with larger fitness is encountered.

We introduce a fitness-based partition of the search space. Let $f_1 < f_2 < \cdots < f_m$ be an enumeration of all fitness values and let A_i, $1 \le i \le m$ contain all search points with fitness f_i. In particular, A_m contains only optimal search points. Now, let s_i, $1 \le i \le m-1$, be a lower bound on the probability of the (1+1) EA creating an offspring in $A_{i+1} \cup \cdots \cup A_m$, provided the current population belongs to A_i. The expected waiting time until such an offspring is created is at most $1/s_i$ and then the set A_i is left for good. As every set A_i has to be left at most once, the expected optimization time for the (1+1) EA and the (1+1) EA* is bounded from above by

$$\sum_{i=1}^{m-1} \frac{1}{s_i}.$$

As MMAS* needs at most $(\ln n)/\rho$ steps to freeze a solution in pheromone and there are at most m fitness levels to leave, the arguments by Gutjahr and Sebastiani [22] lead to the bound

$$\frac{m \ln n}{\rho} + \sum_{i=1}^{m-1} \frac{1}{s_i}.$$

on the expected optimization time of MMAS*. For OneMax, the bound reads as $O((n \log n)/\rho)$, and for LeadingOnes, one obtains $O((n \log n)/\rho + n^2)$.

It seems hard to complement the upper bounds derived by the method of fitness-based partitions by lower bounds. Gutjahr and Sebastiani [22] study lower bounds on the runtime of MMAS* on a different class of functions. It turns out that needle-in-a-haystack functions such as $\text{NEEDLE}(x_1, \ldots, x_n) = \prod_{i=1}^{n} x_i$ are extremely difficult to optimize for MMAS* if ρ is in its usual range. In this case, the expected optimization time is $\Omega(n^n)$, i. e., exponentially worse than blind search, which optimizes NEEDLE in expected time $O(2^n)$. A reason is that MMAS* is unable to explore plateaus. Once MMAS* has created a best-so-far solution of fitness 0, it will only replace this solution if the needle is found.

Neumann, Sudholt and Witt [23] pick up the ideas by Gutjahr and Sebastiani [22] and study the behavior of MMAS* in more detail. In particular, they investigate a variant called MMAS where Step 4.b) of the algorithm is replaced by $f(x) \geq f(x^*)$, i. e., the best-so-far solution is replaced also in case of equal fitness. Interestingly, this small change allows a drastic performance increase on NEEDLE since the bound on the expected runtime drops to $2^{O(n)}$. Moreover, Neumann, Sudholt and Witt [23] prove the first lower bounds for MMAS* on ONEMAX and LEADINGONES. These results show that upper bounds obtained before using the method of fitness-based partitions are almost tight. Still, there is room for improvement. For the LEADINGONES example, they present two improved upper bounds, amongst them $O(n^2 + n/\rho)$, which are tight with the lower bound for almost all values of ρ. This analysis fathoms the limits of the general method. Moreover, the detailed study of MMAS* on LEADINGONES is interesting also from a methodological point of view. The random walk described by pheromone values for "unimportant" bits is investigated in detail and its hitting times for boundary values are bounded using a general study of martingale processes. This technique fosters the understanding of the stochastic process behind random pheromone values and is a further step towards the analysis of ACO on more complicated problems.

Hybridizations with local search

In practice, heuristics are rarely used in their pure form. Often ACO is combined with local search methods [7, 24]. Experimental investigations show that the combination of ACO with a local search procedure improves the performance significantly. On the other hand, there are examples where local search cannot help to improve the search process or even mislead the search process [25]. Neumann, Sudholt and Witt [26] therefore study how the incorporation of local search into ACO algorithms can impact the runtime of the algorithm. Their aim is to provide examples where local search can be highly beneficial and examples where it is detrimental.

The algorithm studied by Neumann, Sudholt and Witt [26] is an MMAS* where each construction procedure is followed by local search. In the following description (see Fig. 7.5), LocalSearch(x) is a procedure that, starting from x, repeatedly replaces the current solution by a Hamming neighbor with strictly larger fitness until a local optimum is found. We do not specify a pivoting rule, hence we implicitly deal with a class of algorithms.

1.) Set $\tau_{(u,v)} = 1/2$ for all $(u, v) \in E$.
2.) Create x using Construct(C, τ).
3.) Set $x^* :=$ LocalSearch(x).
4.) Update pheromones w. r. t. x^*.
5.) Create x using Construct(C, τ).
6.) Set $z :=$ LocalSearch(x).
7.) If $f(z) > f(x^*)$ then $x^* := z$.
8.) Update pheromones w. r. t. x^*.
9.) Go to 5.).

Fig. 7.5. MMAS*-LS

Neumann, Sudholt and Witt [26] define two functions where MMAS*-LS behaves totally different from MMAS*. These functions are based on the following general insight about the effect of local search. We already know that pheromone values induce a sampling distribution over the search space. On a typical fitness landscape, once the best-so-far solution has reached a certain quality, sampling new solutions with a high variance becomes inefficient and the current best-so-far solution x^* is maintained for some time. The above-mentioned studies of MMAS variants have shown that then the pheromones quickly reach the upper and lower bounds corresponding to x^*. This means that the algorithm turns to sampling close to x^*. In other words, MMAS variants typically reach a situation where the "center of gravity" of the sampling distribution follows the current best-so-far solution and the variance of the sampling distribution is low.

When introducing local search into an MMAS algorithm, this may not be true. Local search is able to find local optima that are far away from the current best-so-far solution. In this case the "center of gravity" of the sampling distribution is far away from the best-so-far solution. Assume now there is a path of Hamming neighbors with increasing fitness leading to a local optimum. Assume further that all points close to the path have lower fitness. Then for MMAS* it is likely that the sampling distribution closely follows the path. The path of increasing fitness need not be straight. In fact, it can make large bends through the search space until a local optimum is reached. On the other hand, MMAS*-LS, when starting with the same setting, will reach the local optimum within a single iteration of local search. Then the local optimum becomes the new best-so-far solution x^* while the sampling distribution is still concentrated around the starting point. In the following generations, as long as the best-so-far solution is not exchanged, the pheromone values on all bits synchronously move towards their respective bounds in x^*. This implies for the sampling distribution that the "center of gravity" takes a (sort of) direct route towards the local optimum, irrespective of the bent path taken by local search. An illustration is given in Fig. 7.6.

Consequences are that different parts of the search space are sampled by MMAS* and MMAS*-LS, respectively. Moreover, with MMAS* the variance in the solution construction is always quite low as the sampling distribution is concentrated on certain points on the path. But when the best-so-far solution with local search suddenly moves a long distance, the variance in the solution

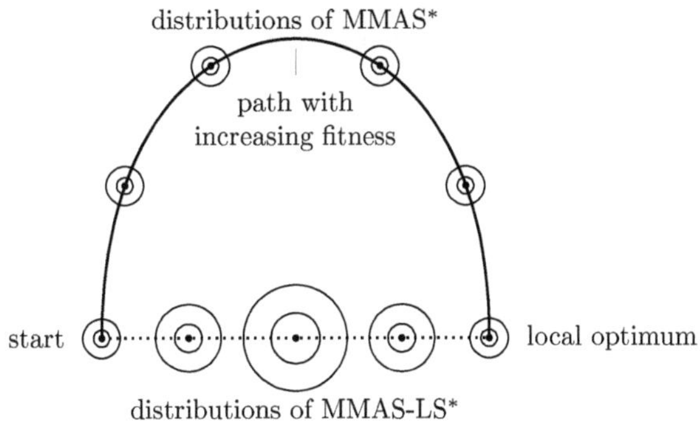

Fig. 7.6. A sketch of the search space showing the behavior of MMAS* and MMAS*-LS. The dots and circles indicate the sampling distributions of MMAS* and MMAS*-LS, respectively, at different points of time. While the distribution of MMAS* tends to follow the fitness-increasing path from left to right, the distribution of MMAS*-LS takes a direct route towards the local optimum.

construction may be very high as the bits differing between the starting point and x^* may have pheromones close to $1/2$. These bits are assigned almost randomly, which strongly resembles a uniform crossover operation well-known in evolutionary computation.

Neumann, Sudholt and Witt [26] create functions where MMAS* and MMAS*-LS have a different runtime behavior based on the above insight. For one function they place a target region with many global optima on the straight line between starting point and local optimum and turn the local optimum into a trap that is hard to overcome. In such a setting, MMAS*-LS drastically outperforms MMAS*; the expected optimization time of MMAS*-LS is polynomial while MMAS* needs an exponential number of steps with overwhelming probability. On the other hand, if the region of global optima is made a region of traps and the global optimum is very close to the local optimum, MMAS* has a clear advantage over MMAS*-LS. Another function following this idea is defined and analyzed. Here MMAS*-LS fails badly and needs an exponential number of steps with overwhelming probability. On the other hand, MMAS* typically has polynomial runtime. These analyses serve as first steps towards the understanding of ACO hybrids on more complicated problems.

7.4 PSO

Particle Swarm Optimization (PSO) was introduced by Kennedy and Eberhart [27] based on a model of social behaviour as observed in bird flocks, fish schools, insect swarms etc. Just like EAs and ACO, PSO is a bio-inspired search heuristic that is implemented to solve complex optimization problems. Comprehensive treatments are given in the text books [28, 29].

The standard formulation of PSO [27] aims at minimizing a function $f\colon \mathbb{R}^n \to \mathbb{R}$. Generally, the algorithm maintains m triples $(\mathbf{x}^{(i)}, \mathbf{x}^{*(i)}, \mathbf{v}^{(i)})$, $1 \le i \le m$, denoted as particles. Each particle i consists of its current position $\mathbf{x}^{(i)} \in \mathbb{R}^n$, its own best-so-far position $\mathbf{x}^{*(i)} \in \mathbb{R}^n$ and its velocity $\mathbf{v}^{(i)} \in \mathbb{R}^n$. The movement for each particle is influenced by the best particle in its neighborhood. In this work, we only use the trivial neighborhood consisting of the whole swarm. This means that all particles are influenced by a single globally best particle, denoted as $\mathbf{x}^*$.

The velocities are updated as follows. The velocity vector is changed towards the particle's own best solution and towards the globally best solution $\mathbf{x}^*$. The impact of these two components is determined by so-called learning factors $c_1, c_2 \in \mathbb{R}_0^+$, which are fixed parameters to be chosen in advance (time-varying extensions are outside the scope of this paper). Additionally, a so-called inertia weight $\omega \in \mathbb{R}_0^+$ is used as a third parameter in order to continuously decrease the impact of previous velocity values.

In an iteration of the PSO algorithm, the particles' velocities and positions are first updated according to

$$\begin{aligned}\mathbf{v}^{(i)} &:= \omega \mathbf{v}^{(i)} + c_1 \mathbf{r} \otimes (\mathbf{x}^{*(i)} - \mathbf{x}^{(i)}) \\ &\qquad + c_2 \mathbf{r}' \otimes (\mathbf{x}^* - \mathbf{x}^{(i)}), \\ \mathbf{x}^{(i)} &:= \mathbf{x}^{(i)} + \mathbf{v}^{(i)}\end{aligned}$$

for $1 \le i \le m$, where $\mathbf{r} \in (U[0,1])^n$ and $\mathbf{r}' \in (U[0,1])^n$ are vectors drawn independently and componentwise from the uniform distribution $U[0,1]$ and $\otimes$ denotes the componentwise multiplication.

Afterwards, the own best solutions are exchanged if the newly constructed solution is strictly better, i. e.,

$$\mathbf{x}^{*(i)} := \begin{cases} \mathbf{x}^{(i)} & \text{if } f(\mathbf{x}^{(i)}) < f(\mathbf{x}^{*(i)}) \\ \mathbf{x}^{*(i)} & \text{otherwise,} \end{cases}$$

and $\mathbf{x}^*$ is set to the best among the $\mathbf{x}^{*(i)}$. The PSO algorithm then proceeds to the next iteration unless some stopping criterion is fulfilled.

As mentioned in Sect. 7.2, often studies of convergence aspects of a heuristic serve as precursors for their rigorous runtime analysis. Such analyses have been conducted for PSO for several years [30–33] but no rigorous runtime analysis of a PSO algorithm was known until 2008.

7.4.1 Discrete Search Spaces

Although PSO is mostly applied in continuous search spaces, some variants were proposed that fit the scenario of pseudo-boolean optimization, i. e., of functions $f\colon \{0,1\}^n \to \mathbb{R}$. Sudholt and Witt [34] study the earliest such variant, which was defined by Kennedy and Eberhart [35] and is called *Binary PSO* hereinafter. As in standard PSO, velocities are used to determine the next position of a particle. However, as each bit of the position may only obtain discrete values 0 and 1, velocities are used in a stochastic solution construction process. More precisely, the velocity value of a bit determines the probability to set this bit to 1 in the next solution construction in the following way: each bit v_j, $1 \le j \le n$, of the velocity vector is mapped to a probability via the sigmoid function $s(v) := 1/(1 + e^{-v})$. Then the new particle position is obtained by setting x_j, $1 \le j \le n$, to 1 with probability $s(v_j)$ independently for each bit. To ensure convergence of the system, also a maximum velocity $v_{\max} := \ln(n - 1)$ and a minimum velocity $-v_{\max}$ is introduced. Apart from that, the algorithm corresponds to the standard PSO for $\omega = 1$.

Sudholt and Witt [34] investigate the Binary PSO on simple functions. To understand how the algorithm makes progress towards optima, they analyze the process that updates the velocity vector in phases where no improvement of the globally best solution happens. For certain settings of the learning factors, similar conclusions can be drawn as for the MMAS* algorithm described in Sect. 7.3.2. While no improvement happens, the velocity vector is drawn towards the globally best solution and will reflect this solution better and better. Finally, the upper or lower bounds $\pm v_{\max}$ on velocity entries are attained, and the probabilistic construction will reproduce each bit of the globally best solution with probability $s(v_{\max}) = 1 - 1/n$. Hence, we are again confronted with the sampling distribution of the (1+1) EA. As a result, the fitness-level method carries over as in the case of MMAS*. The freezing time now corresponds to the time until velocity entries attain their bounds. Using the adapted fitness-level method, the authors get a bound on the expected optimization time of

$$O(mn \log n) + \sum_{i=1}^{m-1} \frac{1}{s_i}.$$

(reusing the notation from Sect. 7.3.2) for the Binary PSO with $c_1 + c_2 = O(1)$. An an application, this bound is concretized for unimodal functions.

Since the fitness-level method relies on pessimistic assumptions, it only yields the bound $O(n^2 \log n)$ w.r.t. the simple function ONEMAX. Sudholt and Witt [34] show in a more detailed analysis that it is not necessary to spend $\Theta(n \log n)$ steps on each fitness level until all bits have been frozen to the velocity bounds of the globally best. In particular, they improve the optimization time bound of the Binary PSO with one particle on ONEMAX to $O(n \log n)$, which is the same asymptotic upper runtime bound as for the (1+1) EA and the 1-ANT. This analysis is crucial to gain an understanding of the probabilistic behavior of velocities and to show that not all of them need to stay at $\pm v_{\max}$ in the course of optimization. Hence, in terms of the velocity values, the stochastic processes behind Binary PSO and the (1+1) EA differ significantly.

7.4.2 Continuous Search Spaces

The work by Sudholt and Witt [34] studies one of the few discrete variants of PSO, which lives in finite search spaces and optimizes each function in expected time $O(n^n)$. Typically, PSO is however applied to real-parameter optimization. There convergence, i. e., the question whether there exists – under a reasonable stochastic convergence mode – a limit for the particles' positions $\mathbf{x}_t^{(i)}$ at time t

as $t \to \infty$, was not obvious. It was not until 2007 that Jiang, Luo and Yang [33] provided a first such convergence analysis for certain choices of the parameter set (c_1, c_2, ω). In particular, $\omega < 1$ is essential for the velocity vectors to shrink and the system to reach a stable state.

Surprisingly, in terms of standard PSO, no statement can be made on the relationship of the limit and the optima of the given function. Actually, the limit is identified as the globally best particle position $\mathbf{x}^*$, which, for an unfortunate (or pathological) swarm configuration, may be arbitrarily far from optimality. Such pathological configurations, which lead to so-called stagnation away from optima, are already stated by van den Bergh [31]. One readily checks that stagnation occurs in the standard PSO when $\mathbf{x}^{(1)} = \mathbf{x}^{*(1)} = \cdots = \mathbf{x}^{(m)} = \mathbf{x}^{*(m)}$ and $\mathbf{v}^{(1)} = \cdots = \mathbf{v}^{(m)} = \mathbf{0}$. Despite looking like a really pathological configuration, such stagnation becomes visible in experiments for swarms of small size even with the SPHERE function [36]. Using larger swarms, stagnation becomes less pronounced since a diversity of particle positions seems to average the swarm around optima. Still, a careful parametrization is needed to find good enough solutions. If this is achieved, the PSO might be called efficient and attractive from a practical point of view.

From our theoretical perspective, the situation is however unsatisfactory since, strictly speaking, PSO is not an optimizer in this case. This also holds for many PSO variants which introduce more stochasticity (e. g., the well-known Bare Bones PSO [37]) but still stagnate when the whole swarm has collapsed to a single, possibly non-optimal point. Regarding PSO as a randomized optimization technique, it should however be able to converge to local optima from any swarm configuration (and even with a swarm of size 1). Different concepts have therefore been proposed to ensure convergence to local optima, most notably the introduction of a *mutation* operator similar to EAs (e. g., [38, 39]). In a simplified form, this means that a tentative solution is sampled around the globally best solution according to

$$\tilde{\mathbf{x}} := \mathbf{x}^* + \mathbf{p},$$

where the mutation vector $\mathbf{p}$ is a perturbation with each component drawn independently according to a probability distribution with zero mean. Under reasonable assumptions on the probability distribution and the function to be optimized, such mutations in conjunction with a memory for the globally best solution provide enough stochasticity to make the PSO variant converge at least to local optima.

The most commonly named PSO variant with proven convergence to optima is the Guaranteed Convergence PSO (GCPSO) by van den Bergh and Engelbrecht [40]. It brings in a "cube mutation" by drawing $\mathbf{p} \in (U[-\ell, \ell])^n$, i. e., uniformly from a hypercube with side lengths 2ℓ. Morover, it introduces step-length control, i. e., ℓ is adapted in the course of optimization. Witt [41] has conducted the first rigorous runtime analysis for a specific instantiation of GCPSO called GCPSO_1. He starts with $m = 1$, i. e., a swarm of size 1 since only the globally best solution is subject to mutation while the remaining particles might still suffer from stagnation. Moreover, the famous 1/5-rule is used to adapt ℓ: after an observation phase consisting of n steps has elapsed, ℓ is doubled if the total number of successes was at least $n/5$ in the phase and halved otherwise; then a new phase is started.

Witt's [41] runtime analysis is carried out for the "cousin" of the ONEMAX function in continuous search spaces, namely the SPHERE function defined by

$$\text{SPHERE}(\mathbf{x}) := \|\mathbf{x}\|^2 = x_1^2 + \cdots + x_n^2$$

with its global optimum in $\mathbf{0}$. This function is a well-accepted test bed for the investigation of simple evolutionary algorithms in this scenario, see, e. g., [42]. Actually one of the simplest such EAs, the famous (1+1) ES, shares remarkable similarities with the GCPSO_1. Only the mutation operator is different. While the (1+1) ES employs a normal distribution with zero mean to draw the components of $\mathbf{p}$, GCPSO_1 uses the above-mentioned cube mutation. The first concept has the nice property of being isotropic, i. e., each direction of the mutation vector has the same probability density. This does not hold for the cube mutation, which favors the "corners" of the hypercube. One could therefore conjecture that the progress of the GCPSO_1 to the optimum cannot be bounded well enough for all possible orientations of the coordinate system.

Surprisingly, it turns out that GCPSO_1 has the same asymptotic runtime behavior as the (1+1) ES on SPHERE irrespectively of rotations of the coordinate system. In order to halve the distance from the globally best particle to the optimum, a typical number of $\Theta(n)$ steps is needed. This bound is in the same order as for the (1+1) ES since the mutation operator of GCPSO is still "close" to being isotropic. One insight of Witt's [41] analysis is that ℓ should be chosen so small that basically, the portion of the hypercube corresponding to a success is almost half the hypercube itself. If ℓ is too big, the curse of dimensionality with high probability prevents the mutation from hitting the "right" region of the hypercube. The same intuition holds in the analysis of the (1+1) ES with the isotropic mutation operator [42]. However, the geometric objects appearing in the latter analysis, namely intersections of hyperspherical volumes, are much more difficult to handle than the intersection of a hypersphere and a hypercube that is implicit in the analysis of the GCPSO_1. In this respect, GCPSO_1 combines asymptotical optimal progress with a considerably simplified analysis. This result might come as a surprise also to those who are familiar with the theory of evolution strategies.

7.5 Outlook

7.5.1 Towards a Unified Theory

The theory of SI algorithms is still in its early infancy. Still, it seems possible to identify promising future lines of research and to discuss possibilities for a unified theory of SI algorithms. A first step into this direction was made by the analyses of single-individual ant systems such as 1-ANT and MMAS on the one hand and the simple Binary PSO on the other hand. As mentioned above, both build a probabilistic model for good solutions in the course of optimization. Moreover, in these simple cases, this distribution is a product of independent distributions for the single bits. This scenario was studied with similar techniques both for the simple ACO and Binary PSO algorithms. In particular, the method of fitness-based partitions known from the analysis of evolutionary algorithms could be carried over to both types of SI algorithms. From a more general perspective, these are nothing else than probabilistic model-building algorithms, also known as estimation-of-distribution algorithms (EDAs). The simple SI algorithms studied so far follow a common principle and can be analyzed as specific instantiations of EDAs with different update mechanism for the probability distribution. We think that the methods for the analysis of simple ACO and PSO algorithms can cross-pollinate the theory of EDAs and contribute to a classification of useful update mechanisms in such algorithms.

7.5.2 Future Research

Another outstanding open problem in ACO is the rigorous runtime analysis of multi-ant systems for non-pathological situations. Here a variety of starting points is conceivable. ACO algorithms typically allow for different strategies for the selection of the solution with respect to which a pheromone update takes place: either a memory over the whole run is maintained, corresponding to the best-so-far strategy in the 1-ANT and MMAS, or the memory is limited, and in each generation a new solution for the pheromone update is selected by multiple ants. The latter strategy is called iteration-best and does not make sense in single-ant systems since the process will collapse with pure random search then. This is reminiscent of (1,λ)-strategies in evolutionary computation, which update their current solution based solely on the best of λ offspring. Results on multiple-ant systems with iteration-best update might therefore parallel the analysis of (1,λ)-EAs and allow for the transfer of an existing toolbox for the latters' analysis. Regarding PSO algorithms with a non-trivial swarm, the impact of swarm size and the interaction of particles on the runtime is widely misunderstood so far. At the moment, it is unclear how existing methods for the analysis of randomized search heuristics allow for the analysis of standard PSO algorithms. Therefore, a first result on the time for the standard PSO in continuous search spaces to converge to its swarm leader, assuming a non-trivial parametrization of the learning factors, is a challenging open problem.

Finally, a widely open field is the runtime analysis of multi-objective swarm algorithms. Also here the hope is to establish such results based on examples from the domain of evolutionary computation. In the last years, the rigorous runtime analysis of multi-objective evolutionary algorithms has developed at rapid pace. Starting from simple algorithms such as SEMO on toy problems [43], a rich theory regarding different types of multi-objective evolutionary optimizers has emerged in recent years (e. g., [44, 45]). As we have pointed out in this survey, a quite general framework for the analysis of evolutionary algorithms on the one hand and SI algorithms on the other hand has been build up in the single-objective case. The time is ripe for an attempt to analyze also multi-objective swarm algorithms in this framework.

Acknowledgement. Carsten Witt was partly supported by the Deutsche Forschungsgemeinschaft (DFG) as a part of the Collaborative Research Center "Computational Intelligence" (SFB 531).

References

[1] Motwani, R., Raghavan, P.: Randomized Algorithms. Cambridge University Press, Cambridge (1995)
[2] Mitzenmacher, M., Upfal, E.: Probability and Computing – Randomized Algorithms and Probabilistic Analysis. Cambridge University Press, Cambridge (2005)
[3] Mühlenbein, H.: How genetic algorithms really work: Mutation and hillclimbing. In: Proc. of Parallel Problem Solving from Nature II (PPSN 1992), pp. 15–26. Elsevier, Amsterdam (1992)
[4] Droste, S., Jansen, T., Wegener, I.: On the analysis of the (1+1) evolutionary algorithm. Theoretical Computer Science 276, 51–81 (2002)
[5] He, J., Yao, X.: A study of drift analysis for estimating computation time of evolutionary algorithms. Natural Computing 3, 21–35 (2004)
[6] Oliveto, P.S., He, J., Yao, X.: Time complexity of evolutionary algorithms for combinatorial optimization: A decade of results. International Journal of Automation and Computing 4, 281–293 (2007)
[7] Dorigo, M., Stützle, T.: Ant Colony Optimization. MIT Press, Cambridge (2004)
[8] Gutjahr, W.J.: ACO algorithms with guaranteed convergence to the optimal solution. Information Processing Letters, 145–153 (2002)
[9] Dorigo, M., Blum, C.: Ant colony optimization theory: A survey. Theoretical Computer Science 344, 243–278 (2005)
[10] Gutjahr, W.J.: First steps to the runtime complexity analysis of Ant Colony Optimization. Computers and Operations Research 35, 2711–2727 (2008)
[11] Neumann, F., Witt, C.: Runtime analysis of a simple ant colony optimization algorithm. In: Asano, T. (ed.) ISAAC 2006. LNCS, vol. 4288, pp. 618–627. Springer, Heidelberg (2006)
[12] Gutjahr, W.J.: On the finite-time dynamics of ant colony optimization. Methodology and Computing in Applied Probability, 105–133 (2006)
[13] Stützle, T., Hoos, H.H.: MAX-MIN ant system. Journal of Future Generations Computer Systems 16, 889–914 (2000)
[14] Doerr, B., Johannnsen, D.: Refined runtime analysis of a basic ant colony optimization algorithm. In: Proc. of the Congress on Evolutionary Computation (CEC 2007), pp. 501–507. IEEE Press, Los Alamitos (2007)
[15] Doerr, B., Neumann, F., Sudholt, D., Witt, C.: On the runtime analysis of the 1-ANT ACO algorithm. In: Proc. of the Genetic and Evolutionary Computation Conference (GECCO 2007), pp. 33–40. ACM Press, New York (2007)
[16] Hoeffding, W.: On the distribution of the number of successes in independent trials. Annals of Mathematical Statistics 27, 713–721 (1956)
[17] Gleser, L.J.: On the distribution of the number of successes in independent trials. The Annals of Probability 3, 182–188 (1975)

[18] Neumann, F., Witt, C.: Ant colony optimization and the minimum spanning tree problem. In: Maniezzo, V., Battiti, R., Watson, J.-P. (eds.) LION 2007 II. LNCS, vol. 5313, pp. 153–166. Springer, Heidelberg (2008)
[19] Attiratanasunthron, N., Fakcharoenphol, J.: A running time analysis of an ant colony optimization algorithm for shortest paths in directed acyclic graphs. Information Processing Letters 105, 88–92 (2008)
[20] Neumann, F., Wegener, I.: Randomized local search, evolutionary algorithms, and the minimum spanning tree problem. Theoretical Computer Science 378, 32–40 (2007)
[21] Broder, A.: Generating random spanning trees. In: Proc. of the 30th Annual Symposium on Foundations of Computer Science (FOCS 1989), pp. 442–447. IEEE Press, Los Alamitos (1989)
[22] Gutjahr, W.J., Sebastiani, G.: Runtime analysis of ant colony optimization with best-so-far reinforcement. Methodology and Computing in Applied Probability 10, 409–433 (2008)
[23] Neumann, F., Sudholt, D., Witt, C.: Analysis of Different MMAS ACO Algorithms on Unimodal Functions and Plateaus. Swarm Intelligence 3, 35–68 (2009)
[24] Hoos, H.H., Stützle, T.: Stochastic Local Search: Foundations & Applications. Elsevier/Morgan Kaufmann, San Francisco (2004)
[25] Balaprakash, P., Birattari, M., Stützle, T., Dorigo, M.: Incremental local search in ant colony optimization: Why it fails for the quadratic assignment problem. In: Dorigo, M., Gambardella, L.M., Birattari, M., Martinoli, A., Poli, R., Stützle, T. (eds.) ANTS 2006. LNCS, vol. 4150, pp. 156–166. Springer, Heidelberg (2006)
[26] Neumann, F., Sudholt, D., Witt, C.: Rigorous analyses for the combination of ant colony optimization and local search. In: Dorigo, M., Birattari, M., Blum, C., Clerc, M., Stützle, T., Winfield, A.F.T. (eds.) ANTS 2008. LNCS, vol. 5217, pp. 132–143. Springer, Heidelberg (2008)
[27] Kennedy, J., Eberhart, R.C.: Particle swarm optimization. In: Proc. of the IEEE International Conference on Neural Networks, pp. 1942–1948. IEEE Press, Los Alamitos (1995)
[28] Clerc, M.: Particle Swarm Optimization. ISTE (2006)
[29] Kennedy, J., Eberhart, R.C., Shi, Y.: Swarm Intelligence. Morgan Kaufmann, San Francisco (2001)
[30] Clerc, M., Kennedy, J.: The particle swarm - explosion, stability, and convergence in a multidimensional complex space. IEEE Transactions on Evolutionary Computation 6, 58–73 (2002)
[31] van den Bergh, F.: An Analysis of Particle Swarm Optimizers. PhD thesis, Department of Computer Science, University of Pretoria, South Africa (2002)
[32] Trelea, I.C.: The particle swarm optimization algorithm: convergence analysis and parameter selection. Information Processing Letters 85, 317–325 (2003)
[33] Jiang, M., Luo, Y.P., Yang, S.Y.: Stochastic convergence analysis and parameter selection of the standard particle swarm optimization algorithm. Information Processing Letters 102, 8–16 (2007)
[34] Sudholt, D., Witt, C.: Runtime analysis of binary PSO. In: Proc. of the Genetic and Evolutionary Computation Conference (GECCO 2008), pp. 135–142. ACM Press, New York (2008); extended version to appear in Theoretical Computer Science
[35] Kennedy, J., Eberhart, R.C.: A discrete binary version of the particle swarm algorithm. In: Proc. of the World Multiconference on Systemics, Cybernetics and Informatics (WMSCI), pp. 4104–4109 (1997)

[36] Angeline, P.J.: Evolutionary optimization versus particle swarm optimization: Philosophy and performance differences. In: Porto, V.W., Waagen, D. (eds.) EP 1998. LNCS, vol. 1447, pp. 601–610. Springer, Heidelberg (1998)

[37] Kennedy, J.: Bare bones particle swarms. In: Proc. of the IEEE Swarm Intelligence Symposium, pp. 80–87 (2003)

[38] Ratnaweera, A., Halgamuge, S., Watson, H.C.: Self-organizing hierarchical particle swarm optimizer with time-varying acceleration coefficients. IEEE Transactions on Evolutionary Computation 8, 240–255 (2004)

[39] Li, C., Liu, Y., Zhou, A., Kang, L., Wang, H.: A fast particle swarm optimization algorithm with Cauchy mutation and natural selection strategy. In: Kang, L., Liu, Y., Zeng, S. (eds.) ISICA 2007. LNCS, vol. 4683, pp. 334–343. Springer, Heidelberg (2007)

[40] van den Bergh, F., Engelbrecht, A.P.: A new locally convergent particle swarm optimiser. In: Proc. of the IEEE International Conference on Systems, Man and Cybernetics (2002)

[41] Witt, C.: Why standard particle swarm optimisers elude a theoretical runtime analysis. In: Proc. of Foundations of Genetic Algorithms 10, FOGA 2009 (to appear, 2009)

[42] Jägersküpper, J.: Analysis of a simple evolutionary algorithm for minimization in Euclidean spaces. Theoretical Computer Science 379, 329–347 (2007)

[43] Laumanns, M., Thiele, L., Zitzler, E.: Running time analysis of multiobjective evolutionary algorithms on pseudo-boolean functions. IEEE Transactions on Evolutionary Computation 8, 170–182 (2004)

[44] Brockhoff, D., Friedrich, T., Neumann, F.: Analyzing hypervolume indicator based algorithms. In: Rudolph, G., Jansen, T., Lucas, S., Poloni, C., Beume, N. (eds.) PPSN 2008. LNCS, vol. 5199, pp. 651–660. Springer, Heidelberg (2008)

[45] Neumann, F., Wegener, I.: Can single-objective optimization profit from multi-objective optimization? In: Knowles, Corne, Deb (eds.) Multiobjective Problem Solving from Nature – From Concepts to Applications, pp. 115–130. Springer, Heidelberg (2007)

8

Mining Rules: A Parallel Multiobjective Particle Swarm Optimization Approach

André B. de Carvalho and Aurora Pozo

Computer Sciences Department
Federal University of Paraná
Curitiba PR CP: 19081, Brazil
{andrebc,aurora}@inf.ufpr.br

Summary. Data mining is the overall process of extracting knowledge from data. In the study of how to represent knowledge in data mining context, rules are one of the most used representation form. However, the first issue in data mining is the computational complexity of the rule discovery process due to the huge amount of data. In this sense, this chapter proposes a novel approach based on a previous work that explores Multi-Objective Particle Swarm Optimization (MOPSO) in a rule learning context, called MOPSO-N. MOPSO-N applies MOPSO to search for rules with specific properties exploring Pareto dominance concepts. Besides, these rules can be used as an unordered classifier, so the rules are more intuitive and easier to understand because they can be interpreted independently one of the other. In this chapter, first some extensions to MOPSO-N are presented. These extensions are enhancements to the original algorithm to increase its performance, and to validate them, a wide set of experiments is conducted. Second, the main goal of this chapter, the parallel approach of MOPSO-N, called MOPSO-P, is described. MOPSO-P allows the algorithm to be applied to large datasets. The proposed MOPSO-P is evaluated, and the results showed that MOPSO-P is efficient for mining rules from large datasets.

8.1 Introduction

Data mining is the overall process of extracting knowledge from data. In this area, rules are one of the most used form to represent the extracted knowledge. This is because of their simplicity, intuitive aspect, modularity, and because they can be obtained directly from a dataset [11]. Therefore, rules induction has been established as a fundamental component of many data mining systems. Furthermore, it was the first machine learning technique to become part of successful commercial data mining applications [6].

Although some techniques have been proposed and successfully implemented, like GAssist [2] and XCS [26], few of them are able to deal with large datasets. Considering this fact, this chapter describes an algorithm, named MOPSO-P (Multi-Objective Particle Swarm Optimization-P) that takes advantage of the inherent benefits of the Particle Swarm Optimization (PSO) technique and puts them in a data mining context to obtain comprehensible, accurate classifiers in the form of simple if-then rules.

C.A. Coello Coello et al. (Eds.): Swarm Intel. for Multi-objective Prob., SCI 242, pp. 179–198.
springerlink.com

MOPSO-P is based on a previous work [8] that explores the Multi-objective Particle Swarm Optimization (MOPSO) in a rule learning context, called MOPSO-N. MOPSO-N applies MOPSO to search for rules with specific properties exploring Pareto dominance concepts. Besides, these rules can be used as an unordered classifier, so the rules are more intuitive and easier to understand because they can be interpreted independently one of the other. MOPSO-N has the goal to produce a good set of rules from numerical or discrete data. For each numerical attribute, the rule learning algorithm tries to discover the best range of values for certain class. Furthermore, MOPSO-N has the challenge to produce a classifier with good performance in terms of the Area Under the Receiver Operating Characteristics curve (AUC) [12]. To tackle this purpose, two objectives were chosen, the sensitivity and specificity criteria which are directly related with the Receiver Operating Characteristics curve (ROC). The sensitivity is equivalent to the Y axis of ROC graph and the specificity is the complement of the X axis.

In [8], two studies about the behavior of the algorithm were made: the first study analyzed the AUC, and the second study examined the Pareto Front coverage applying a multi-objective methodology. In the AUC comparison, MOPSO-N was confronted to four benchmark algorithms usually found in the literature, using eight datasets from the UCI machine learning repository [1]. In this experiment, MOPSO-N obtained a very good result, equivalent to the chosen algorithms. The second study was the analysis of the Pareto Front for the proposed algorithm. This experiment compared the approximation sets generated from MOPSO-N to the real Pareto Front obtained from an exhaustive algorithm. To obtain the Pareto Front for the exhaustive algorithm, each chosen dataset had to be discretized, so all possible attributes-values combinations could be determined. This analysis measures which algorithm generates the best set of rules, according to the chosen objectives. The results showed that MOPSO-N obtained a better rule set.

This works proposes some extensions of MOPSO-N that tackles some problems presented in [8]. First, it is proposed a new procedure that reduces the number of rules without losing quality in classification. This procedure removes from the approximation of the Pareto Front all more specific rules that cover the same set of instances than a more generic rule. The parameters of the MOPSO approach are adjusted and a procedure that allows the algorithm to run with incomplete datasets is implemented. Furthermore, a large number of experiments are performed to validate the initial results presented in [8].

Despite this extension of MOPSO-N, the main goal of this chapter is to introduce a parallel approach of MOPSO-N, that allows the algorithm to deal with large datasets. This extension, MOPSO-P, is evaluated comparing its results with the execution of the MOPSO-N algorithm using the AUC measure. Furthermore, the time of execution of both algorithms is confronted.

The rest of this chapter is organized as follows. Sect. 8.2 reviews rule learning concepts and some classification measures. Sect. 8.3 presents the main multiple objective particle swarm aspects. Sect. 8.4 describes the MOPSO-N algorithm and details of the extensions. Sect. 8.5 explains the experiments performed with MOPSO-N. The parallel approach and its evaluation are described on Sect. 8.6. Finally, Sect. 8.7 concludes the paper and discusses future works.

8.2 Rule Learning Concepts

The algorithm described in this chapter is applied to the learning rules problem. A rule is a pair <antecedent, consequent> or if *antecedent* then *consequent* where both the antecedent and the consequent are constructed from the set of attributes.

Let Q be a finite set of attributes, which in practice correspond to the fields in the database. Each $q \in Q$ has an associated domain, $Dom(q)$. An attribute test, b, consists of an attribute, $at(b) \in Q$, and a value set, $Val(b) \in Dom(at(b))$, and may be written $at(b) \in Val(b)$. A record satisfies this test if its value for attribute $at(b)$ belongs in the set $Val(b)$. An algorithm may allow only certain types of value sets $Val(b)$. Types of categorical attribute tests are as follows:

- Value: $Val(b) = \{v(b)\}$, where $v(b) \in Dom(at(b))$. This may be written $at(b) = v(b)$.
- Inequality: $Val(b) = \{x \in Dom(at(b)) : x \neq v(b)\}$, where $v(b) \in Dom(at(b))$. This may be written $at(b) \neq v(b)$.
- Subset: Val(b) unrestricted, i.e. any subset of Dom(at(b)).

Types of numerical attribute tests are as follows:

- Binary partition: $Val(b) = \{x \in Dom(at(b)) : x \leq v(b)\}$ or $Val(b) = \{x \in Dom(at(b)) : x \geq v(b)\}$, where $v(b) \in Dom(at(b))$. In this case, an attribute test may be written $at(b) \leq v(b)$ or $at(b) \geq v(b)$, respectively.
- Range: $Val(b) = \{x \in Dom(at(b)) : l(b) \leq x \leq u(b)\}$, where $l(b), u(b) \in Dom(at(b))$. Here, the AT is written $l(b) \leq at(b) \leq u(b)$.

Each rule created has the same consequent, consisting of just one AT (c_k). This defines the class of interest. We say that the rule r_k covers the example e_i if the example satisfies the antecedent of r_k. And, if the rule r_k covers e_i and $c_k = c_i$ then the rule correctly classifies the example e_i, otherwise, there is an incorrect classification.

The number of examples correctly and incorrectly classified are used to calculate the measures for rules evaluation, and are summarized in the contingency table. Table 8.1 shows the contingency table, where B denotes the set of instances for which the body of the rule is true, i.e., the set of examples covered by the rule r_k, and $\overline{B}$ denotes its complement (the set of instances not covered by the rule r_k). H is the set of instances where the consequent is true ($c_i = c_k$) and $\overline{H}$ is its complement. HB then denotes $H \cap B$, $\overline{H}B$ denotes $\overline{H} \cap B$, and so on.

Where $n(X)$ denotes the cardinality of the set X, e.g., $n(\overline{H}B)$ is the number of instances for which H is false and B is true (i.e., the number of instances erroneously covered by the rule). N denotes the total number of instances in the dataset.

Table 8.1. A Contingency Table

	B	$\overline{B}$	
H	$n(HB)$	$n(H\overline{B})$	$n(H)$
$\overline{H}$	$n(\overline{H}B)$	$n(\overline{H}\overline{B})$	$n(\overline{H})$
	$n(B)$	$n(\overline{B})$	N

From the contingency matrix it is possible to calculate measures such as: True Positive rate (*TP rate*), True Negative rate (*TN rate* or specificity specificity), False Positive rate (*FP rate*) and False Negative rate (*FN rate*). *TP rate*, also called sensitivity sensitivity, is the precision between the positive examples (equation 8.1). Its complement is the *FN rate* (i.e. *FN rate = 1 - FP rate*). Specificity is the precision between the negative examples (equation 8.2). Its complement is the *FP rate*.

$$sensitivity = \frac{n(HB)}{n(H)} \tag{8.1}$$

$$specificity = \frac{n(\overline{H}\overline{B})}{n(\overline{H})} \tag{8.2}$$

8.2.1 Non-ordered Rules

Rules are usually aggregated into a rule set to build a classifier. After, the classifier can be used to classify "unseen" instances. The classification of a new instance is performed by a voting process. In the voting process, all rules vote the class of the instances.

A scheme of the voting process is presented at Fig. 8.1. The process contains the following steps:

- For each class, all rules that cover the input instance are identified.
- The identified rules are sorted according to some ordering criteria. The ordering can be defined according to many different mechanisms, including: (1) combinations of confidence, support and size of antecedent, with confidence being the most significant factor; (2) the size of the rule antecedent; (3) Laplace Accuracy and other rule measures. These mechanisms are generic ordering schemas and are discussed in [28]. This work uses the third schema in AUC calculus.
- The ordering process is applied to allow the selection of only the best k rules, according to the ordering criteria, to vote the class of the instance. This number can vary from one rule to all rules.
- The next step counts the votes of each class. Usually each rule votes with a weight and all weights are added. Finally, the class that scores higher is declared as the class of the example.

More details of this voting process can be found in [28]. There, different kinds of sort and weights are proposed, and all methods are evaluated. Based on these

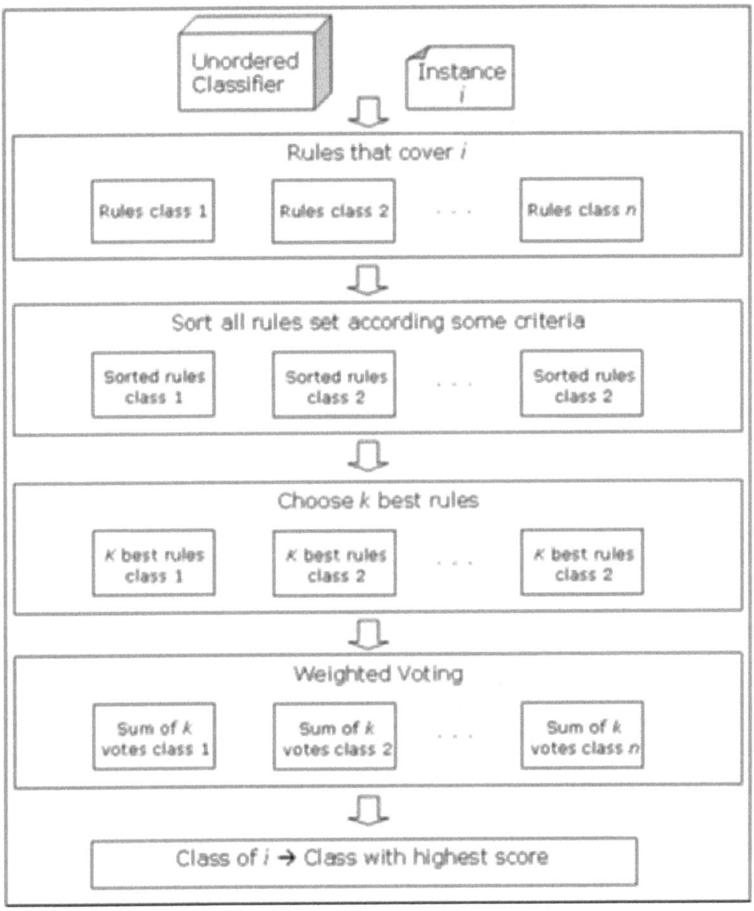

Fig. 8.1. Rules voting process.

methods, the chosen voting process for MOPSO-N and MOPSO-P sort the rules through the Laplace confidence (equation (8.4)) of each rule, k is the number of class present at the dataset, and the defined weight is the positive confidence (equation (8.3)) of the rule.

$$Positive\ Confidence = \frac{n(HB)}{n(B)} \tag{8.3}$$

$$Laplace\ Confidence = \frac{n(HB+1)}{n(B) + NumberOfClasses} \tag{8.4}$$

AUC

For several years, the most used performance measure for classifiers was the accuracy [3]. The accuracy is the fraction of examples correctly classified

(True Positive (TP) and True Negative (TN)), shown on equation (8.5). In spite of its use, accuracy maximization is not an appropriate goal for many real-world tasks [21]. A tacit assumption in the use of classification accuracy as an evaluation metric is that the class distribution among examples is constant and relatively balanced. In real world applications, this case is however, rare, since the cost associated with the incorrect classification of each class can be different because some classifications can lead to actions which could have serious consequences [22].

$$accuracy = \frac{TP + TN}{N} \tag{8.5}$$

$$TPrate = \frac{TP}{P} \tag{8.6}$$

$$FPrate = \frac{FP}{N} \tag{8.7}$$

Based on the explanation above, many people have preferred the Receiver Operating Characteristic (ROC) analysis. The ROC graph has been used in signal detection theory to depict tradeoffs between TP rate (equation (8.6)) and FP rate (equation (8.7)) [10]. However, it was only introduced in the Machine Learning context at the end of the 1990s [22]. The ROC graph has two dimensions: the TP rate at the y-axis and the FP rate at the x-axis.

The best classifiers have high TP rate and low FP rate. The most northwest the point, the best will be the associated classifier. In Fig. 8.2, the classifier associated to Point B is better than the classifier of Point C. We can say that Classifier B dominates Classifier C. Otherwise, we can not say the same about Classifiers A and C, because Point A has a lower value for the FP rate, however it has also a lower value for TP rate. For this case, Point A is not dominated by Point C, neither C is dominated by A.

If one classifier is not dominated by any other, this classifier is called as a non-dominated one. That is, there is no other classifier with a higher TP rate and a lower FP rate. The set of all non-dominated classifiers, when considered in objective function space, is known as the Pareto Front. The continuous line of Fig. 8.2 shows the Pareto Front.

The ROC curve gives a good visualization of the performance of a classifier [4]. However, frequently, for comparison purposes of learning algorithms, a single measure for the classifier performance is needed. In this sense, the Area Under the ROC Curve (AUC) was proposed [13, 18, 24]. The AUC measure has an important statistical property: it is equivalent to the probability that a randomly chosen positive instance will be rated higher than a negative instance and thereby is also estimated by the Wilcoxon test of ranks [14]. The AUC values for non-ordered set of rules are calculated using a weighted voted classification process [11].

Given a rule set containing rules for each class, we use the best k rules of each class for prediction, with the following procedure: (1) select all the rules

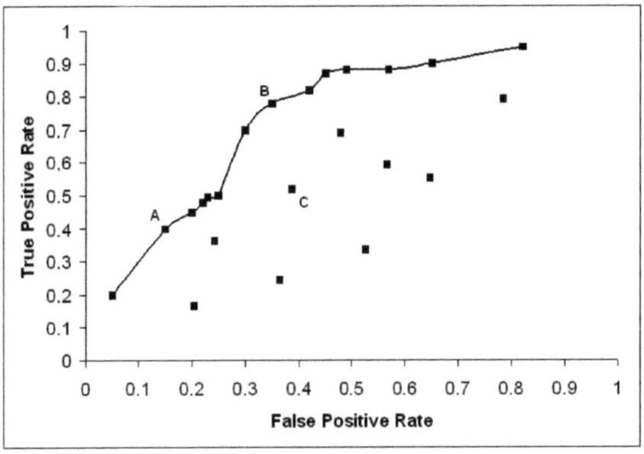

Fig. 8.2. A ROC Graph plotted with a set of classifiers. The continuous line shows the Pareto Front

whose bodies are satisfied by the example; (2) from the rules selected in Step 1, select the best k rules for each class; and (3) compare the expected average of the positive confidence of the best k rules of each class and, choose the class with the highest expected positive confidence as the predicted class. After the vote of all rules, each example has an associated value that is the sum of confidence of positive rules less the sum of confidence of negative rules. Thus, for each instance of the database, we obtain a numerical rank with the associated values. This rank can be used as a threshold to produce a binary classifier. If the rank of the instance goes beyond the threshold, the classifier produces a "yes", otherwise, a "no". Each threshold value generates a different point in the ROC plane. So, varying the threshold from $-\infty$ to $+\infty$ one produces a curve on the ROC plane and we are able to calculate the area under the curve (AUC) [20]. This value gives a good visualization of the classifier performance [4]. For a perfect classification, the positive examples have the greatest values grouped at the top of the ranking; and AUC is equal to 1 (its greatest value).

For maximizing the AUC value, it is preferable that positive examples receive more votes from positive rules than from negative ones, i.e., it is expected: high precision between the positive examples (sensitivity) and high precision between the negatives (specificity). For this reason, it is expected that the Pareto Front with sensitivity and specificity criteria maximizes AUC.

8.3 Multi-Objective Particle Swarm Optimization

Optimization problems with two or more objective functions are called Multi-objective. In such problems, the objectives to be optimized are usually in conflict, which means that there is not a single solution for these problems. In this way, the goal is to find a good "trade-off" of solutions that represent the best possible compromise among the objectives. The general multi-objective maximization problem (with no restrictions) can be stated as to maximize equation (8.8).

$$\overrightarrow{f}(\overrightarrow{x}) = (f_1(\overrightarrow{x}), ..., f_Q(\overrightarrow{x})) \tag{8.8}$$

subjected to $\overrightarrow{x} \in \Pi$, where: $\overrightarrow{x}$ is a vector of decision variables and Π is a finite set of feasible solutions.

Let $\overrightarrow{x} \in \Pi$ and $\overrightarrow{y} \in \Pi$ be two solutions. For a maximization problem, the solution $\overrightarrow{x}$ dominates $\overrightarrow{y}$ if:

$$\forall f_i \in \overrightarrow{f}, i = 1...Q, f_i(\overrightarrow{x}) \geq f_i(\overrightarrow{y}), and \;\; \exists f_i \in \overrightarrow{f}, f_i(\overrightarrow{x}) > f_i(\overrightarrow{y})$$

$\overrightarrow{x}$ is a non-dominated solution if there is no solution $\overrightarrow{y}$ that dominates $\overrightarrow{x}$.

The goal is to discover solutions that are not dominated by any other in objective function space. A set of non-dominated objective vectors is called Pareto optimal and the set of all non-dominated vectors is called Pareto Front. The Pareto optimal set is helpful for real problems, e.g., engineering problems, provides valuable information about the underlying problem [16]. In most applications, the search for the Pareto optimal is NP-hard [16], then the optimization problem focuses on finding an approximation set, as close as possible to the Pareto optimal.

Particle Swarm Optimization (PSO) works with a population-based heuristic inspired by the social behavior of bird flocking aiming to find food. It was introduced by Kennedy and Eberhart [15]. In Particle Swarm Optimization the system initializes with a set of solutions, possibly random, and search for optima by updating generations. The set of possible solutions is a set of particles, called swarm, which moves in the search space, in a cooperative search procedure. These moves are performed by an operator that is guided by a local and a social component [15]. This operator is called velocity of a particle and moves it through an n-dimensional space based on the best positions of their neighbors (social component), of the leader, and on their own best position (local component).

The best particles are found based on the fitness function, which is the problem objective function. There are many fitness functions in Multi-objective Particle Swarm Optimization (MOPSO). By exploring Pareto dominance concepts, it is possible to obtain results with specific properties. Based on this concept each particle of the swarm could have different leaders, but only one may be selected to update the velocity. This set of leaders is stored in a repository, which contains the best non-dominated solutions found. The MOPSO components are defined as follows.

Each particle p_i, at a time step t, has a position $x(t) \in R^n$, that represents a possible solution. The position of the particle, at time $t+1$, is obtained by adding its velocity, $v(t) \in R^n$, to $x(t)$:

$$\overrightarrow{x}(t+1) = \overrightarrow{x}(t) + \overrightarrow{v}(t+1) \tag{8.9}$$

The velocity of a particle p_i is based on the best position already fetched by the particle, $\overrightarrow{p}_{best}(t)$, and the best position already fetched by the set of neighbors of p_i, $\overrightarrow{R_h}(t)$, that is a leader from the repository. The velocity update function, in time step $t+1$ is defined as follows:

$$\overrightarrow{v}(t+1) = \varpi * \overrightarrow{v}(t) + (c_1 * \phi_1) * (\overrightarrow{p}_{best}(t) - \overrightarrow{x}(t)) + (c_2 * \phi_2) * (\overrightarrow{R_h}(t) - \overrightarrow{x}(t)) \quad (8.10)$$

The variables ϕ_1 and ϕ_2, in equation (8.10), are coefficients that determine the influence of the particle's best position, $\overrightarrow{p_{best}}(t)$, and the particle's global best position, $\overrightarrow{R_h}(t)$. The constants c_1 and c_2 indicate how much each component influences on the velocity. The coefficient ϖ is the particle inertia, and controls how much the previous velocity affects the current one.

$\overrightarrow{R}_h$ is a particle from the repository, chosen as a guide of p_i. There are many ways to make this choice, as described in [25]. At the end of the algorithm, the solutions in the repository are the final output. One possible way to make the leader choice is called the sigma distance [19]. This method, according to the results presented in [25], is very adequate for the MOPSO technique. It tries to improve the convergence and the diversity of the MOPSO approach, so it tries to produce good and different solutions in a smaller number of iterations. However, this technique can cause premature convergence in some cases. Furthermore, this method produces good results and it motivated the use of this technique in this work.

In this method for each particle, represented as a point in the objective space, is assigned a value σ_i. For a two-objective problem, the sigma value is defined in the following way:

$$\sigma = \frac{f_1(x)^2 - f_2(x)^2}{f_1(x)^2 + f_2(x)^2} \quad (8.11)$$

For problems with more than two objectives, the sigma value is a vector with $\binom{n}{2}$ elements, where n is the number of objectives of the problem. Each element of this vector represents a combination of two elements applied to equation (8.11). The leader for a particle of the swarm, R_h, is the particle of the repository which has the smallest Euclidian distance between its sigma vector and the sigma vector of the swarm particle.

8.4 MOPSO-N

MOPSO-N was first introduced in [8], and is able to handle both numerical and discrete attributes. In this way, it can be used in different domains, mainly those ones with continuous attributes, such as fault-prediction [9]. This section describes the main aspects of MOPSO-N , and after, the proposed extensions.

8.4.1 Rule Learning Algorithm

The algorithm uses the Michigan approach where the position of each particle represents a single solution, i.e., a rule. A position is an n-dimensional vector of real numbers and contains the attributes restrictions. This vector contains all the attributes restrictions of the learned rule. One real number represents

```
Initialization procedure.
Particles evaluation using fitness functions.
Find non-dominated solutions and divide the objective space
Evolutionary loop
   Calculate the new velocity and new position of each particle.
   Evaluate all particles.
   Update the repository with non-dominated particles and divide the search space.
Return repository
```

Fig. 8.3. MOPSO-N, Rule learning algorithm.

the value for each discrete attribute and two real numbers represent an interval for numerical attributes. The interval is defined by its lower and upper values. Each attribute can accept the value '?', which means, for that rule, that the attribute does not matter for the classification. To represent the particle as a possible solution, the attributes values must be encoded into real numbers. The encoding of the discrete attributes is conceived by integer numbers related to each attribute value of the database. The numerical attributes do not need to be encoded, because their values can be treated directly by the algorithm. When a numerical attribute has a void value, their cells in the particle representation receive the lower bound value available on the database. For a discrete attribute, a void value is represented with -1. In the proposed approach the class value is set at the beginning of the execution.

The rule learning algorithm works as follows (see algorithm in Fig. 8.3). First, an initial procedure is executed where the position of each particle is initialized and all particles are spread randomly in the search space. In this procedure all particles components are initialized too. The next step is the evaluation of each particle for all objectives. In this approach the chosen objectives are sensitivity (equation (8.1)) and specificity (equation (8.2)). The sensitivity and specificity criteria are directly related to the ROC curve. Sensitivity measures how much the classifier predicts correctly the positive examples. Specificity is the same in relation to the negative examples.

Following the evaluation of each particle, the repository is initialized with the non-dominated solutions. Then, the objective space is divided among the particles in the repository, accordingly to the sigma distance [19]. Once the initial configuration is performed, the evolutionary loop is executed performing the moves of all particles in the search space. This loop is executed until a stop criterion is reached. In this work this criterion is the maximum number of generations. In each iteration, initially, the operations discussed in the previous section are implemented. The velocity of the particle is updated and then, the new positions are calculated using equations (8.10) and (8.9), respectively. After all particles have been moved through the search space, they are evaluated using the objectives and once again the best particles are loaded in the repository and the global leaders are redefined.

Table 8.2. Number of rules and AUC, Removing Specific Rules Procedure

Datasets	with specific rules		without specific rules	
	AUC	# Rules	AUC	# Rules
breast	98.97 (1.33)	334	98.94 (1.32)	91
ecoli	78.62 (31.59)	182	93.92 (6.68)	52
glass	75.23 (16.36)	83	71.91 (16.05)	71
haberman	69.63 (10.72)	412	70.18 (8.64)	155

8.4.2 MOPSO-N Extensions

The proposed extensions of MOPSO-N have the following objectives: first, to reduce the total number of rules without losing information of the extracted knowledge, and without losing quality in the classification. Second, tuning the MOPSO parameters to obtain better results. Finally, handling missing attributes to allow the application of MOPSO-N to a large number of datasets.

The first objective is solved by a procedure that leaves in the repository of the non-dominated solutions only the most generic solutions. One rule r_1 is said to be more generic than a rule r_2 if r_1 has less attributes restrictions than r_2 and all the restrictions of r_1 are equal to the restrictions of r_2. Besides, the two rules have to cover the same set of instances, so, their contingency matrices have to be the same. The opposite relation, that is, a rule r_2 has the same attributes restrictions than a rule r_1, and more other attributes restrictions, is called more *specific* than a rule r_1. The procedure, called Removing Specific Rules, works as follows: for each rule that is going to be added to the repository look for a more generic rule or a more specific rule in the repository. Only the more generic rule will remains in the repository. This procedure is executed at each generation, at the end of the selection of the best solution (step 7 in algorithm at Fig. 8.3). Table 8.2 presents the average number of rules and AUC obtained with the execution of MOPSO-N with and without this procedure, for the datasets *breast*, *ecoli*, *glass* and *haberman* of UCI (see Table 8.4). The methodology adopted for this experiment is described in Sect. 8.5. Observing Table 8.2, there is a significant reduction in the number of rules, in all the datasets analyzed. Furthermore, the algorithm does not degrade its performance in the classification.

The second extension is related to the tuning of the MOPSO parameters. In the work presented in [8] the parameters were defined empirically, based on initial experiments. Here, the adjustment of the parameters was defined through the analysis of some related works [25] [5]. The inertia weight, ϖ, was left dynamically, varying randomly in the range of $[0, 0.8]$, at each generation. An alternative velocity update formula [5], using a constriction factor instead of the inertia weight was experimented, but did not present good results. The personal learning coefficients were set to $c_1 = c_2 = 2.05$ and ϕ_1 and ϕ_2 are dynamic, varying within the range [0,1], at each generation. This configuration presented the best MOPSO-N results and was applied in the experiments of Sect. 8.5. Table 8.3 presents the difference between the old and the new parameter configuration.

Table 8.3. Difference between the parameters configuration

Parameter	Old configuration	New configuration
ϖ	dynamic [0, 0.8]	dynamic [0, 0.8]
c_1 and c_2	1	2.05
ϕ_1 and ϕ_2	dynamic [0, 0.4]	dynamic [0, 1]

Finally, the last extension developed is related to incomplete datasets. Usually, datasets have some instances with missing values. This can occur in real domains, since sometimes all the information could not be obtained. So, to handle this problem, MOPSO-N identifies missing values in the input instances and considers these values as void, i.e, for all rules that have an attribute restriction associated to this value, the algorithm assumes that the rule matches this restriction. This procedure has been developed to allow the execution of MOPSO-N in different datasets. The following section presents a set of experiments to evaluate the algorithm.

8.5 MOPSO-N Evaluation

This section presents a set of experiments to validate the performance of MOPSO-N, through an AUC analysis, using a large number of datasets. Here, the results of the new version of MOPSO-N are compared with the same set of literature algorithms used in [8]: C4.5 [23], C4.5 with No Pruning, RIPPER [7] and NNge [17]. The classification study considers a set of experiments with 18 databases from the UCI [1] machine learning repository. The datasets are presented on Table 8.4. All the chosen algorithms have being applied using the tool Weka [27]. The experiments were executed using 10-fold-stratified cross validation and for all algorithms were given the same training and test files. The database with more than two classes were reduced to two-class problems, positive and negative, selecting the class with the lower frequency as positive and joining the remaining examples as negative. All the datasets were used with their original values and no pre-processing step was applied.

MOPSO-N was executed during 100 generations using 500 particles for each class, and it was executed 10 runs. The parameters ω, ϕ_1, ϕ_2, c_1 and c_2 were defined through the adjusting procedure discussed in Sect. 8.4.2. ω varies in the interval $[0, 0.8]$, and both ϕ_1 and ϕ_2 vary in $[0, 1]$. c_1 and c_2 have the same value equal to 2.05. The AUC values for the MOPSO-N algorithm were calculated by using a confidence voted classification process as explained in Sect. 8.2.1.

The AUC values are shown in Table 8.5 and the number between brackets indicates the standard deviation. The non-parametric Wilcoxon test, U-Test, with a 5% confidence level, was executed to define which algorithm is better. In this test, all the ten AUC fold values of all ten executions of MOPSO (in a total of a hundred AUC values) were compared with the ten AUC fold values of the other algorithms. Because the algorithms chosen from the literature are deterministic,

Table 8.4. Description of the experimental datasets

#	Data set	Attributes	Examples	% of Majority Class
1	adult	15	48842	76.07
2	breast	10	683	65.00
3	bupa	7	345	57.97
4	digits	17	9892	90.40
5	ecoli	8	336	89.58
6	flag	29	174	91.23
7	german	21	1000	70.00
8	glass	10	214	92.05
9	haberman	4	306	73.52
10	heart	14	270	55.00
11	ionosphere	34	351	64.10
12	kr-vs-kp	37	3196	52.00
13	lettera	17	19999	96.05
14	new-thyroid	6	215	83.72
15	nursery	9	12960	97.45
16	pima	9	768	65.51
17	satimage	36	6435	90.27
18	vehicle	19	846	76.47

Table 8.5. Experiments Results: Mean AUC

Datasets	MOPSO-N	C4.5	C4.5 NP	RIPPER	NNge
1	87.79 (0.61)	**89.08 (0.63)**	86.07 (0.95)	75.11 (0.90)	72.66 (0.55)
2	**98.94 (1.32)**	95.84 (3.04)	95.33 (3.03)	96.98 (2.30)	96.71 (2.20)
3	**70.73 (9.28)**	63.60 (7.40)	64.2 (6.35)	66.51 (6.44)	60.32 (8.11)
4	94.88 (1.84)	**99.27 (0.45)**	**99.30 (0.44)**	98.99 (0.54)	**99.54 (0.40)**
5	**93.92 (6.68)**	78.90 (19.85)	79.12 (19.65)	79.90 (15.30)	75.17 (12.20)
6	**79.82 (20.47)**	50.00 (0.0)	54.07 (24.31)	55.26 (15.47)	64.44 (20.81)
7	**73.13 (6.04)**	69.30 (6.49)	61.05 (6.28)	63.18 (5.68)	61.57 (5.36)
8	71.91 (16.05)	**75.07 (21.06)**	**75.82 (21.48)**	56.00 (16.08)	49.5 (1.05)
9	**70.18 (8.64)**	59.59 (9.30)	62.28 (9.66)	61.41 (8.49)	56.14 (7.24)
10	**87.47 (7.44)**	76.55 (7.49)	75.30 (7.25)	77.75 (10.15)	76.33 (8.00)
11	87.58 (7.34)	86.60 (6.70)	88.22 (6.28)	**90.97 (4.94)**	**90.47 (4.07)**
12	96.60 (2.53)	**99.85 (0.24)**	**99.86 (0.18)**	99.50 (0.36)	98.50 (0.81)
13	95.21 (1.23)	**98.30 (1.04)**	**98.48 (1.26)**	97.68 (1.15)	95.68 (1.19)
14	**96.18 (9.71)**	91.80 (12.06)	91.89 (12.11)	92.77 (12.17)	92.50 (12.23)
15	85.23 (4.04)	**98.93 (1.27)**	98.79 (1.98)	82.70 (5.27)	90.91 (6.44)
16	**80.90 (4.50)**	75.57 (3.77)	75.46 (4.54)	71.30 (4.07)	68.51 (7.17)
17	**91.57 (1.29)**	73.03 (6.85)	75.95 (7.32)	74.13 (3.38)	72.22 (4.07)
18	**95.59 (1.97)**	93.49 (3.33)	93.34 (3.31)	92.61 (4.83)	85.82 (7.85)
Avg	**86.53**	81.93	81.91	79.59	78.16

in order to perform the Wilcoxon test, the AUC values were repeated thirty times. In Table 8.5, the cells highlighted indicate which algorithm obtained the best result.

Table 8.5 shows that MOPSO-N provided very good AUC results, when compared to the benchmark algorithms. In Datasets 2, 3, 5, 6, 7, 9, 10, 14, 16, 17 and 18, MOPSO-N presents the best result, according to the U-Test. Furthermore, in some datasets, like Datasets 5, 6 and 17, the proposed algorithm achieved a very good classification result, when compared with the chosen algorithms. For the others datasets, C4.5 presents the best results for Dataset 1, 4, 8, 12, 13 and 15. C4.5 NP presents the best results in Datasets 4, 8, 12, 13. NNge achieved the best in results in Datasets 4 and 11, while RIPPER and NNge have the best result only in Dataset 11. MOPSO-N obtained the best average value considering all datasets. The results confirm the initial hypothesis presented in [8], that MOPSO-N has good classification results when dealing with numerical datasets and that a good classifier can be generated using as objectives the sensitivity and specificity criterion.

8.6 A Parallel Multi-Objective Particle Swarm Algorithm

This section presents the proposed Parallel Multi-Objective Particle Swarm Algorithm, MOPSO-P. As said before, MOPSO-P allows the algorithm to be applied to large datasets, taking advantage of the inherent benefits of the Particle Swarm Optimization (PSO) technique.

The algorithm follows the same methodology presented in Sect. 8.3. It is based on the MOPSO-N algorithm, but it splits the training set into p sub-sets, instead of using only one. p is the number of distributed nodes available for a task. Like MOPSO-N, MOPSO-P tries to generate a good set of rules using numerical or discrete data and tries to build a good classifier in terms of AUC. But, the main goal of the proposed algorithm is to reduce the execution time by distributing the process in many nodes, mainly when dealing with large datasets. In the proposed approach, the algorithm was executed in a sequential fashion, i.e., it was not applied in a parallel environment. So, execution problems like synchronization and load balancing were not addressed at this time.

Next, the main aspects of MOPSO-P are described in subsection 8.6.1. After that, in subsection 8.6.2, MOPSO-P is evaluated through an experiment in which the results of MOPSO-N and MOPSO-P are compared using the AUC measure. Furthermore, the execution time of both algorithms is also confronted.

8.6.1 MOPSO-P Algorithm

MOPSO-P has the major steps presented in Sect. 8.4.1. It executes all the same steps presented at Fig. 8.3. The main goal of the algorithm is to allow multiple parallel runs of the MOPSO technique. To do this, two new procedures were designed, one to split the original data and execute multiple runs of MOPSO-N and the second to join the rules generated in each node obtaining the classifier. After the classifier has been built, it can be used to classify new instances (the test step). Fig. 8.4 presents the main aspects of the MOPSO-P algorithm. The following describes each step of the algorithm.

First, MOPSO-P receives as input the training and test sets and the number p of parallel nodes. Then, it executes one procedure that splits the training set into p partitions, each partition has almost the same number of training instances. These partitions are distributed to the nodes. Once each node has received its

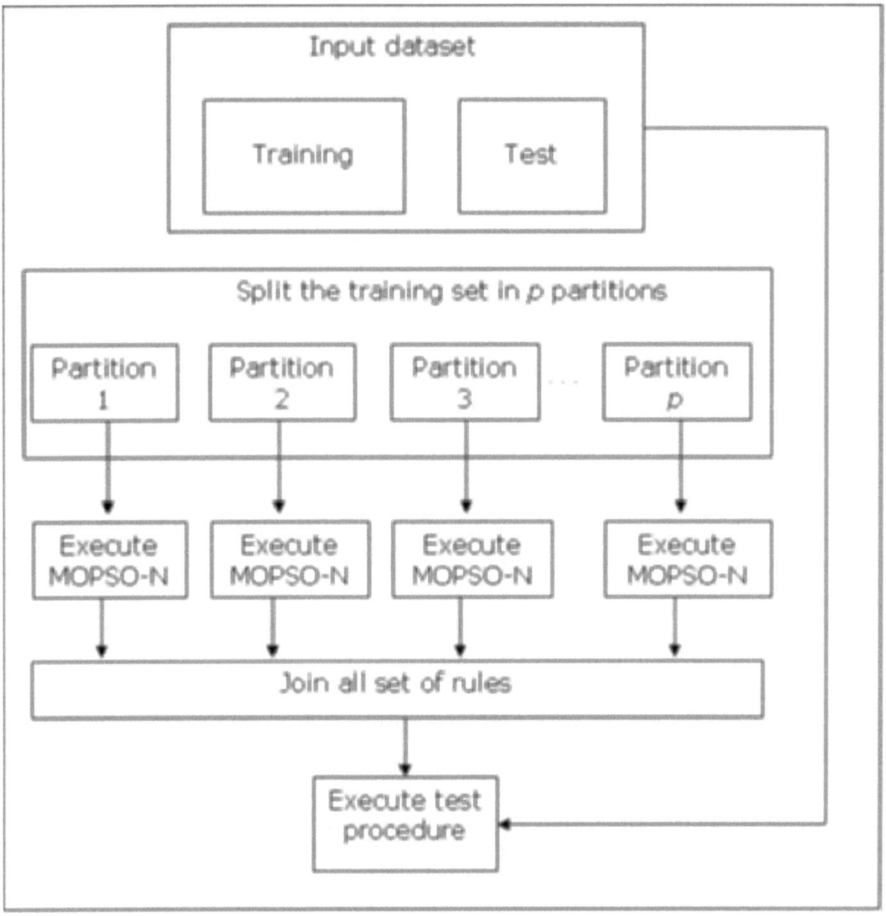

Fig. 8.4. MOPSO-P, Parallel Approach

training set, the nodes start the procedure presented in Fig. 8.3. After all nodes have finished the execution of MOPSO-N, there are p sets of rules. All these rules will form the unordered classifier. So, a simple step that joins all sets of rules is performed. In this step, different decision can be made to avoid rule redundancy. For example, it's possible to perform a selection procedure to identify best rules and remove the equivalent rules. In this work, no selection is applied, so there is no procedure to avoid rule redundancy, and all the rules are used to compose the final classifier. However, future works will investigate different methods of joining the rules. Finally, after this procedure, the classifier is ready to classify new instances. For evaluation purposes, at the last step, a test is accomplished. In the test step, the classifiers are tested in the original input test dataset, without any split procedure. The next subsection presents and discusses a set of experiments to validate the classification results of MOPSO-P.

Table 8.6. Mean AUC, MOPSO-P and MOPSO-N algorithms

Datasets	MOPSO-P		MOPSO-N	
	AUC	# Rules	AUC	# Rules
adult	88.13 (0.17)	785.9	87.82 (0.33)	203.7
digits	97.55 (0.71)	223.4	97.27 (0.65)	64.3
lettera	96.47 (0.26)	302	96.67 (0.19)	94.1
nursery	95.74 (0.20)	64.1	87.20 (0.18)	19.9

8.6.2 MOPSO-P Evaluation

A set of experiments will be discussed to evaluate the classification results of MOPSO-P. In this investigation, the main concern is to validate the MOPSO-P approach. It is worth noting, however, that the experiments are not executed in a parallel domain, but instead, each partition is executed sequentially. First, a study is made that compares the AUC results of MOPSO-P with the MOPSO-N algorithm. In this comparison, four UCI datasets presented in Table 8.2 are analyzed. After this, the running time of MOPSO-P is confronted to the running time of MOPSO-N.

The experiments assume 4 nodes. Both MOPSO algorithms were executed with the same set of parameters presented in the previous experiment. Both algorithms were executed 10 times. For all datasets, the training dataset is composed by 2/3 of the instances and the remaining 1/3 composes the test dataset. The AUC values are shown in Table 8.6 and the figures between brackets indicate the standard deviation. Also, the rules mean are presented.

It can be observed that MOPSO-P obtained very good values of AUC, and that, in almost all datasets, MOPSO-P obtained better results than MOPSO-N. One explanation for these better results can be the larger number of rules generated by MOPSO-P. As said before, MOPSO-N generates an approximation of the Pareto Front considering sensitivity and specificity as objectives. Since MOPSO-P is an approach that performs a parallel execution of this algorithm, it produces a larger set of these good rules, which is the sum of the rules of the nodes. Fig. 8.5 presents the results of both algorithms for the Nursery dataset. The dark points represent rules generated by MOPSO-P (one execution). The lighter points are the rules generated by MOPSO-N. Note that MOPSO-P generated a larger number of rules with good values of sensitivity and specificity. However, not all of them are non-dominated solutions.

MOPSO-P presented good results in terms of AUC, but the main goal of the proposed algorithm is the reduction of the execution time, specially for large datasets. Table 8.7 presents the average time for the 10 executions of both algorithm in all the datasets, and Table 8.8 presents the average time of the execution on only one partition. Analyzing the execution of the partitions, the parallel algorithm executes faster than MOPSO-N. At the Digits dataset, all partitions are executed in a total time of 223.36 seconds. So, the time of the splitting and the

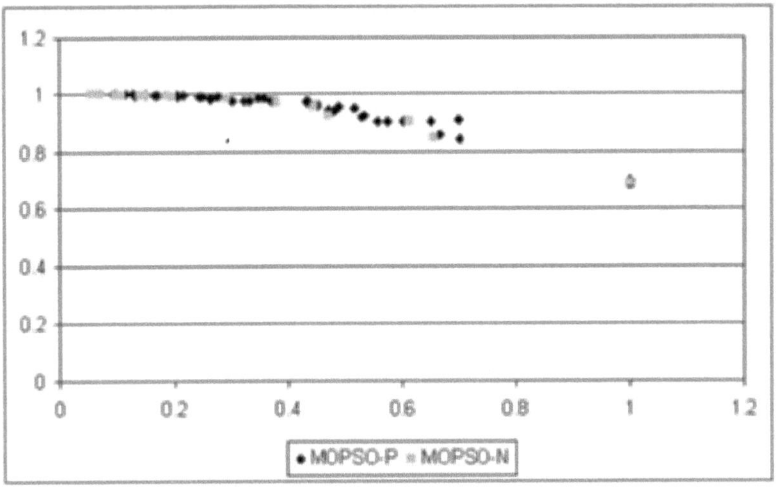

Fig. 8.5. Rules set for MOPSO-P and MOPSO-N, Nursery Dataset

Table 8.7. Execution Mean Time, MOPSO-P and MOPSO-N

Datasets	MOPSO-P	MOPSO-N
adult	606.71 sec	584.23 sec
digits	249.88 sec	166.47 sec
lettera	364.47 sec	251.20 sec
nursery	166.81 sec	166.47 sec

Table 8.8. Execution Mean Time, A Partition of MOPSO-P

Datasets	MOPSO-P
adult	101.00 sec
digits	55.84 sec
lettera	71.89 sec
nursery	41.65 sec

joining procedures was 26.52 seconds. For this dataset, the execution of one partition plus the rule learning algorithm was 82.36. So, if the algorithm is executed in a parallel fashion, the MOPSO-P spends less time than MOPSO-N.

8.7 Conclusions

This work explores Multi-objective Particle Swarm Optimization in a rule learning context. The algorithm MOPSO-N, which is based on this philosophy, is discussed. The algorithm has the following properties: First, MOPSO-N uses a Multi-Objective approach that allows us to create classifiers in only one step.

This method works finding the best non-dominated rules of a problem, by selecting them in the rules generation process. It is a simplified way to obtain more intelligible rules from a database and gives certain freedom to choose new objectives like interest or novelty. Second, the algorithm deals with both numerical and discrete data, and third, the induced classifiers present good results in terms of the AUC measure.

Besides the good properties presented by MOPSO-N, this work focuses on some enhancements to the algorithm. First, it was proposed a new procedure that reduces the number of rules without losing quality in the classification. This procedure removes from the approximation of the Pareto Front all the specific rules that cover the same set of instances than a more generic rule. Second, The parameters of the MOPSO approach were adjusted and a procedure that allows the algorithm to run with incomplete datasets was implemented. Furthermore, a wide set of experiments was performed. MOPSO-N was empirical evaluated aiming to confirm the initial results presented in [8] and to validate the proposed enhancements. MOPSO-N was compared with four well-known algorithms from the literature: C4.5, C4.5 No Pruning, RIPPER and NNge, in terms of AUC. The results confirm that MOPSO-N is very competitive with respect to the other algorithms. Furthermore, the procedure of removing the specific rules does not reduce the classification quality of MOPSO-N.

In addition to the enhancements of MOPSO-N, the main goal of this chapter was to introduce MOPSO-P, a parallel version of the MOPSO-N algorithm to deal with large datasets. MOPSO-P reduces the execution time of MOPSO-N without losing quality on the classification results. In order to do this, two procedures were implemented: a procedure that splits the training dataset and calls multiple versions of the MOPSO-N algorithm, and a procedure that joins all the rules that are generated. MOPSO-P has been evaluated through an experiment where the results of MOPSO-N and MOPSO-P were compared using the AUC measure. Furthermore, the execution time of both algorithm was also confronted. This study showed that MOPSO-P has good results in terms of AUC, in most cases higher than MOPSO-N. These better results occur because MOPSO-P generates a larger set of rules than MOPSO-N. The execution time comparison showed that the parallel approach is very efficient.

Future works include: the execution of a greater number of experiments to validate the initial results of MOPSO-P, now working on a cluster. We also plan to execute MOPSO-P in a parallel domain and implement some schemas for removing rule redundancy. We are also interested in the comparison of our approach with other parallel algorithms found in the literature. Finally, we believe that the algorithm proposed has a lot of possible applications for diverse area such as Software Engineering, Medicine, etc., and they should be explored.

Acknowledgment

This work was supported by the CNPQ (project: 471119/2007-5).

References

[1] Asuncion, A., Newman, D.: UCI Machine Learning Repository. University of California, School of Information and Computer Science, Irvine (2007), `http://www.ics.uci.edu/~mlearn/MLRepository.html`

[2] Bacardit, J., Garrell, J.M.: Bloat control and generalization pressure using the minimum description length principle for a pittsburgh approach learning classifier system. In: Kovacs, T., Llorà, X., Takadama, K., Lanzi, P.L., Stolzmann, W., Wilson, S.W. (eds.) IWLCS 2003. LNCS (LNAI), vol. 4399, pp. 59–79. Springer, Heidelberg (2007)

[3] Baronti, F., Starita, A.: Hypothesis Testing with Classifier Systems for Rule-Based Risk Prediction, pp. 24–34 (2007), `http://dx.doi.org/10.1007/978-3-540-71783-6_3`

[4] Bradley, A.P.: The use of the area under the ROC curve in the evaluation of machine learning algorithms. Pattern Recognition 30(7), 1145–1159 (1997)

[5] Bratton, D., Kennedy, J.: Defining a standard for particle swarm optimization. In: Swarm Intelligence Symposium, pp. 120–127 (2007)

[6] Clark, P., Niblett, T.: Rule induction with CN2: Some recent improvements. In: ECML: European Conference on Machine Learning. Springer, Berlin (1991)

[7] Cohen, W.W.: Fast effective rule induction. In: Proceedings of the Twelfth International Conference on Machine Learning, pp. 115–123 (1995)

[8] de Carvalho, A.B., Pozo, A.: Non-ordered data mining rules through multi-objective particle swarm optimization: Dealing with numeric and discrete attributes. In: Poceedings of Hybrid Intelligent Systems, 2008. HIS 2008. Eighth International Conference, pp. 495–500 (2008)

[9] de Carvalho, A.B., Pozo, A., Vergilio, S., Lenz, A.: Predicting fault proneness of classes trough a multiobjective particle swarm optimization algorithm. In: Poceedings of 20th IEEE International Conference on Tools with Artificial Intelligence (2008)

[10] Egan, J.: Signal detection theory and ROC analysis. Academic Press, New York (1975)

[11] Fawcett, T.: Using rule sets to maximize ROC performance. In: IEEE International Conference on Data Mining, pp. 131–138. IEEE Computer Society Press, Los Alamitos (2001)

[12] Fawcett, T.: Roc graphs: Notes and practical considerations for researchers (2004)

[13] Ferri, C., Flach, P., Hernandez-Orallo, J.: Learning decision trees using the area under the ROC curve. In: Sammut, C., Hoffmann, A. (eds.) Proceedings of the 19th International Conference on Machine Learning, pp. 139–146. Morgan Kaufmann, San Francisco (2002)

[14] Hanley, McNeil: The meaning and use of the area under a receiver operating characteristic (ROC) curve. Radiology 143(1), 29–36 (1982)

[15] Kennedy, J., Eberhart, R.C.: Swarm intelligence. Morgan Kaufmann Publishers Inc., San Francisco (2001)

[16] Knowles, J., Thiele, L., Zitzler, E.: A Tutorial on the Performance Assessment of Stochastic Multiobjective Optimizers. 214, Computer Engineering and Networks Laboratory (TIK), ETH Zurich, Switzerland (February 2006) (revised version)

[17] Martin, B.: Instance-Based learning: Nearest Neighbor With Generalization. PhD thesis, Department of Computer Science, University of Waikato, New Zealand (1995)

[18] Azé, J., Sebag, M., Lucas, N.: ROC-based evolutionary learning: Application to medical data mining. In: Liardet, P., Collet, P., Fonlupt, C., Lutton, E., Schoenauer, M. (eds.) EA 2003. LNCS, vol. 2936, pp. 384–396. Springer, Heidelberg (2004)
[19] Mostaghim, S., Teich, J.: Strategies for finding good local guides in multi-objective particle swarm optimization. In: Proceedings of the 2003 IEEE Swarm Intelligence Symposium, SIS 2003 Swarm Intelligence Symposium, pp. 26–33. IEEE Computer Society, Los Alamitos (2003)
[20] Provost, F., Fawcett, T.: Robust classification for imprecise environments. Machine Learning 42(3), 203 (2001)
[21] Provost, F., Fawcett, T., Kohavi, R.: The case against accuracy estimation for comparing induction algorithms. In: Proceedings 15th International Conference on Machine Learning, pp. 445–453. Morgan Kaufmann, San Francisco (1998)
[22] Provost, F.J., Fawcett, T.: Analysis and visualization of classifier performance: Comparison under imprecise class and cost distributions. In: KDD, pp. 43–48 (1997)
[23] Quinlan, R.: C4.5: Programs for Machine Learning. Morgan Kaufmann Publishers, San Mateo (1993)
[24] Rakotomamonjy, A.: Optimizing area under roc curve with SVMs. In: Hernández-Orallo, J., Ferri, C., Lachiche, N., Flach, P.A. (eds.) ROCAI, pp. 71–80 (2004)
[25] Reyes-Sierra, M., Coello, C.A.C.: Multi-objective particle swarm optimizers: A survey of'the state-of-the-art. International Journal of Computational Intelligence Research 2(3), 287–308 (2006)
[26] Wilson, S.W.: Classifier fitness based on accuracy. Evolutionary Computation 3(2), 149–175 (1995)
[27] Witten, I.H., Frank, E.: Data Mining: Practical machine learning tools and techniques, 2nd edn. Morgan Kaufmann, San Francisco (2005)
[28] Yanbo, Q.X., Wang, J., Coenen, F.: A novel rule ordering approach in classification association rule mining. International Journal of Computational Intelligence Research 2(3), 287–308 (2006)

9

The Basic Principles of Metric Indexing

Magnus Lie Hetland

Norwegian University of Science and Technology
Department of Computer and Information Science
Sem Sælands vei 7-9
NO-7491 Trondheim, Norway
mlh@idi.ntnu.no

Summary. This chapter describes several methods of similarity search, based on metric indexing, in terms of their common, underlying principles. Several approaches to creating lower bounds using the metric axioms are discussed, such as pivoting and compact partitioning with metric ball regions and generalized hyperplanes. Finally, pointers are given for further exploration of the subject, including non-metric, approximate, and parallel methods.

Introduction

This chapter is a tutorial – a brief introduction to the field of metric indexing, which is a way of optimizing certain kinds of similarity retrieval. While there are several excellent publications that cover this area, this chapter takes a slightly different approach in several respects.[1] Primarily, it focuses on giving a concise explanation of underlying principles rather than a comprehensive survey of specific methods (which means that some methods are explained in ways that differ significantly from the original publications). Also, the main effort has been put into explaining these principles clearly, rather than going in depth theoretically or covering the full breadth of the field.

The first two sections of this chapter provide an overview of the main goals and principles of similarity retrieval in general. Section 9.3 discusses the specifics of metric spaces, and the following sections deal with three approaches to metric indexing: pivoting, ball partitioning, and generalized hyperplane partitioning. Section 9.7 summarizes some methods and issues that aren not dealt with in detail elsewhere, and finally the appendices give some mathematical details, as well as a listing of all the specific indexing methods discussed in the chapter. To enhance the flow of the text, many technical details have been relegated to end notes, which can be found on pp. 220–225.

9.1 The Goals of Distance Indexing

Similarity search is a mode of information retrieval where the query is a sample object, and the desired result is a set of objects deemed to be similar – in

C.A. Coello Coello et al. (Eds.): Swarm Intel. for Multi-objective Prob., SCI 242, pp. 199–232.
springerlink.com

some sense – to the query. The similarity is usually formalized (inversely) as a so-called *distance function*,* and indexing is any form of preprocessing of the data set designed to facilitate efficient retrieval.[2] The applications range from entertainment and multimedia (such as image or music search) to science and medicine (such as data mining or matching biological sequences), or anything else that requires efficient query-by-example, but where traditional (coordinate-based) spatial access methods cannot be used. Beyond direct search, similarity retrieval can be used as an internal component in a wide spectrum of systems, such as nearest neighbor classification (with large sample sets), compressed video streaming (to reuse image patches) and multiobjective optimization (to avoid near-duplication of solutions).[3]

Under the distance function formalism, several query types may be formulated. In the following, the author focuses on one of the basic kinds, so-called *range queries*, where all objects that fall within a given distance of the sample are returned. In other words, for a distance function d, a query object q, and a *search radius* r, objects x for which $d(q,x) \leq r$ are returned (see Fig. 9.1). While alternatives are discussed in the literature (most notably returning the nearest object, or the k nearest objects) it can be argued that range queries are fundamental, and virtually all published metric indexing methods support them. (For more information about other query types, see Sect. 9.7.)

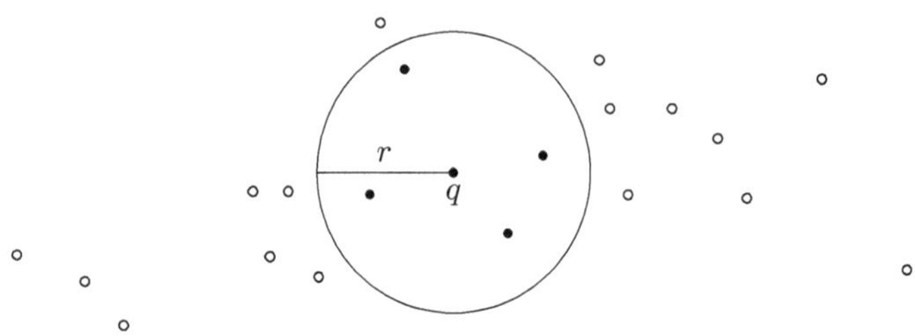

Fig. 9.1. Visualization of a range query in two-dimensional Euclidean space. The small circles are objects in the data set, while the filled dots are returned as a result of the query with sample object q and search radius r

Index structures (such as the inverted files traditionally used in text search) are structures built over the given data set in order to speed up queries. The time cost involved in building the index is amortized over the series of queries,

*A distance function (or simply a *distance*) d is a non-negative, real-valued, binary function that is reflexive ($d(x,x) = 0$) and symmetric ($d(x,y) = d(y,x)$). The function is defined over some universe of possible objects, $\mathbb{U}$(that is, it has the signature $d : \mathbb{U}^2 \rightarrow \mathbb{R}_0^+$). Both here and later on in the chapter, constraints are implicitly quantified over $\mathbb{U}$; for example, symmetry implies that $d(x,y) = d(y,x)$ *for all objects* x, y in $\mathbb{U}$. When discussing queries, it is assumed that query objects may be arbitrarily chosen from $\mathbb{U}$, while the returned objects are taken from some finite subset $\mathbb{D} \subseteq \mathbb{U}$ (the *data set*).

and is usually ignored when considering search cost. The main goal, then, of an index method is to enable efficient search, either asymptotically or simply in real wall-clock time. However, in any specialized form of search there may be a few wrinkles to the story; for example, in examining an index structure theoretically, some basic operations of the search algorithm may completely dominate others, but it may not be entirely clear which ones are the more costly, and there may be different constraints (such as memory use) or required pieces of functionality (such as being able to accomodate new objects) that influence what the criteria of optimization are. The three most important measures of quality used in the literature are:

- The number of distance computation needed during a query;
- The number of I/O operations (disk accesses) needed; and
- The CPU time used beyond distance computations or I/O operations.

Of these three, the first is of primary importance, mainly because it is generally assumed that the distance computations (which may involve comparing highly complex objects, such as video clips) are expensive enough to completely dominate the running time. Beyond this, some (mostly more recent) methods provide mechanisms for going beyond main memory without incurring an inordinate number of disk accesses, and many methods also involve "tricks" for cutting down on the CPU time. It is quite clear, though, that an underlying assumption for most of the field is that minimizing the number of distance computation is the main goal.[4]

9.2 Domination, Signatures, and Aggregation

Without even considering the specifics of metric spaces, it is possible to outline some mechanisms that can be of use when implementing distance indices. This section discusses some fundamental principles, which may then be implemented, so to speak, using metric properties (dealt with in the following section), as shown in Sects. 9.4 through 9.6.

The most basic algorithm for similarity retrieval (or any form of search) is the *linear scan*: Traverse the entire data set, examining each object for eligibility.[5] In order to improve upon the linear scan, we must somehow infer that an object x can be included in, or excluded from, the search result *without* calculating $d(q, x)$. The way to go is to find a cheap way of *approximating* the distance. In order to avoid wrongfully including or excluding objects, we need an approximation with certain properties.[6] In particular, the approximated distance must either overestimate or underestimate the actual distance, and must do so consistently. If the distance function $\hat{d}$ consistently yields higher values than another distance function $\check{d}$ (over the same set of objects), we say that $\hat{d}$ *dominates* $\check{d}$.

Let $\hat{d}$ be a distance that dominates the actual distance d (i.e., it forms an upper bound to d), and let d dominate another distance $\check{d}$ (a lower bound). These estimates may then be used to partition the data set into three categories: no, yes, and maybe. If $\check{d}(q, o) > r$, the object o may be safely excluded (because no objects

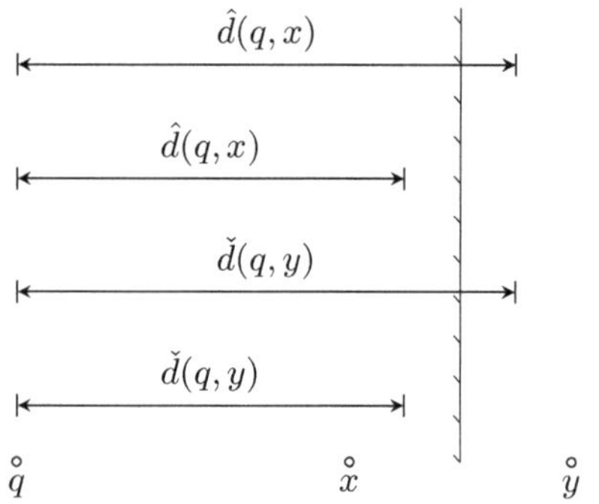

Fig. 9.2. Filtering through distance estimates. The vertical hatched line represents the range of the query; if an overestimated distance ($\hat{d}$) falls within the range, the object is included, and if an underestimated distance ($\check{d}$) falls beyond it, the object is discarded. Otherwise, the true distance must be computed

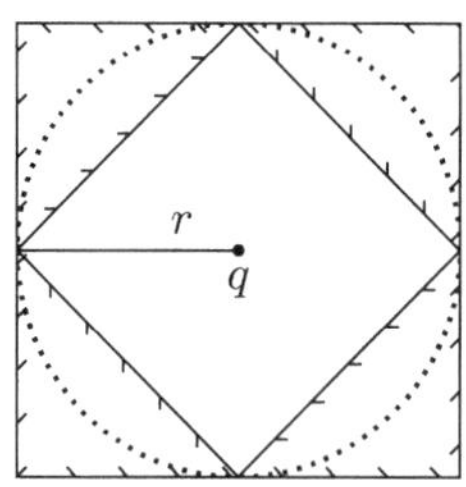

Fig. 9.3. An example of filtering through domination, where $\hat{d}$, d, and $\check{d}$ are all Minkowski metrics, $L_p(x,y) = \sqrt[p]{\sum_i |x_i - y_i|^p}$, with $p = 1$, 2, and ∞, respectively. The circle is the set of points x for which $d(q,x) = r$. The outer square and the inner diamond represent similar regions for $\check{d}$ and $\hat{d}$. Objects inside the diamond are safe to include, objects outside the square may be safely excluded, while the region between them (indicated by the hatching) contains the "maybe" objects

may, under $\check{d}$, be "closer than they appear"). Conversely, if $\hat{d}(q,o) \le r$, the object o may be safely included. The actual distance $d(q,o)$ must then be calculated for any objects that do not satisfy either criterion. See Fig. 9.2 above and Fig. 9.3 on this page for illustrations of this principle. The more closely $\hat{d}$ and $\check{d}$ approximate d, the smaller the maybe set will be; however, there will normally be a tradeoff between approximation quality and cost of computation. One example of this kind of tradeoff is when there are several sources of knowledge available, and thus several upper and lower bounds; these may easily be combined, by letting $\hat{d}$ be the minimum of the upper bounds and $\check{d}$, the maximum of the lower bounds.

Note that under the (quite reasonable) assumption that most of the data set will *not* be returned in most queries, the notion of lower bounds and automatic exclusion is more important than that of upper bounds and automatic inclusion.

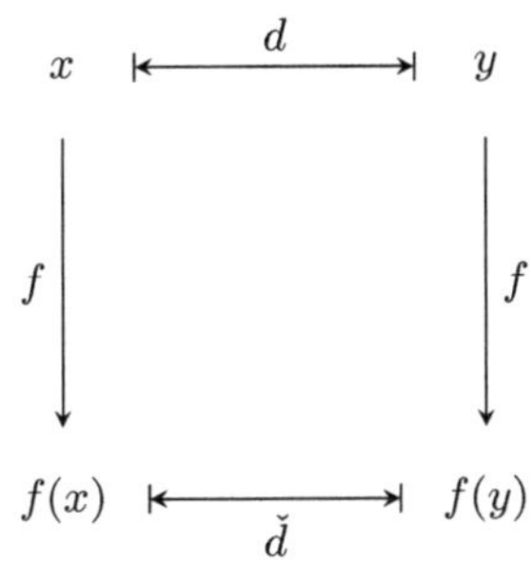

Fig. 9.4. The non-expansive mapping f maps objects from a a universe of objects, $\mathbb{U}$, to a signature space, $\mathbb{P}$. The distance d is defined on $\mathbb{U}$, while $\check{d}$ is defined on $\mathbb{P}$. For all x, y in $\mathbb{U}$, we have that $d(x,y) \geq \check{d}(f(x), f(y))$

Thus, a major theme in the quest for better distance indexing structures is the search for ever more accurate, yet still cheap, lower-bounding estimates.

One way of viewing such lower bounds is in terms of so-called *non-expansive mappings*: All objects are mapped to new objects, or *signatures*, and the *signature distance* becomes an underestimate for the original distance (see Fig. 9.4).[7] If one has knowledge about the internal structures of the objects, such signature spaces may be defined quite specifically using this domain knowledge.[8] In the general case of distance indexing, however, this is not possible, and the geometry of the space itself must be exploited. For metric spaces, certain applications of pivoting (see Sect. 9.4) make the so-called pivot space quite explicit.

Many indexing methods add another mechanism to the (possibly signature-based) bounds, namely *aggregation*: The search space is partitioned into regions, and all objects within a region are handled collectively. At the level of object filtering, this does not give us any obvious advantages. In order to find a common lower bound that lets us discard an entire region, the bound must be defined as the minimum over all possible objects in that region – clearly not an improvement in accuracy. (Another take on this lower bound is to see it as a non-expansive mapping to $\mathbb{R}$, where no objects in the region have positive signatures; see Fig. 9.5 on the following page.)[9]

There are, however, several reasons for using aggregation in this way, related to the goals discussed previously (see Sect. 9.1).* Reducing extra CPU time has been an implicit goal since the earliest metric indexing structures. Hierarchical decomposition of the search space makes it possible to filter out regions of data at increasing levels of precision, somewhat like what is done in search trees over ordered data.[10] When it comes to reducing the number of disk accesses, which is an important consideration for several recent methods, it is also clear that being able to exclude entire disk blocks without examining them (i.e., without reading them into main memory) is vital, and in order to manage this, some form of

*These reasons, or the fact that aggregation in itself can only reduce filtering power, do not seem to be addressed by most authors.

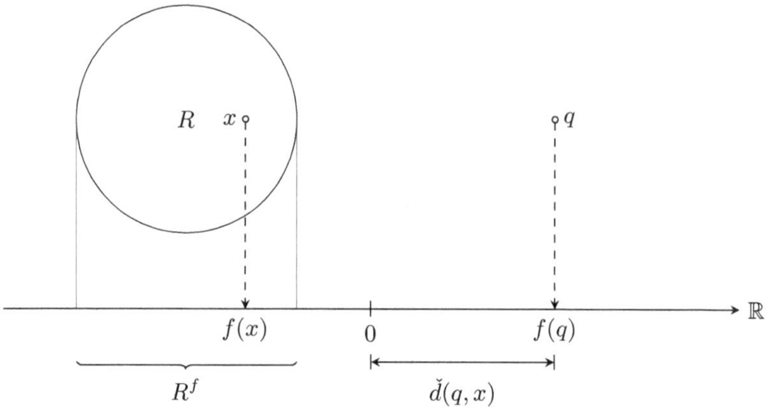

Fig. 9.5. Illustration of a region R with a non-expansive *certification function* (that is, a mapping to the real numbers, $\mathbb{R}$), where $f(x) \leq 0$ for all x in R. R^f is the image of R, mapped to $\mathbb{R}$ through f. As can be seen in this example, because R^f is required to lie below zero, $\check{d}(q,x) = f(q) \leq d(q,x)$, for all x in R. Note that if $f(q) \leq 0$, the relationship is trivially satisfied, as d is non-negative, by definition

aggregation is needed.[11] However, even with a single-minded focus on reducing the number of distance computations, aggregation may be an important factor. Although using regions rather than individual objects reduces the precision of the distance estimates (and, hence, the filtering power), it may also reduce the amount of memory needed for the index structure; if this is the case, methods using aggregation may, given the same amount of memory, actually be able to *improve* upon the filtering power of those without it.[12]

9.3 The Geometry of Metric Spaces

In order to qualify as a distance function, a function must normally be symmetric and reflexive. If d is a distance defined on the universe $\mathbb{U}$, the pair $(\mathbb{U}, d)$ is called a *distance space*. A *metric space* is a distance space where non-identical objects are separated by positive distances and where there are no "short-cuts": The distance from x to z cannot be improved by going via another point (object) y.*
The distance function of a metric space is called (naturally enough) a *metric*. One important property of metric spaces is that subsets (or subspaces) will also be metric, so if a metric is defined over a given data type, a finite data set (and relevant queries) of that data type will form a (finite) metric space, subject to the same constraints.

Metric spaces are a generalization of Euclidean space, keeping some of its well-known geometric properties. These properties allow us to derive certain

*More precisely, a distance function satisfies $d(x,y) = d(y,x)$ and $d(x,x) = 0$, while a metric also satisfies $d(x,y) > 0$, unless $x = y$, as well as the *triangle inequality*, $d(x,z) \leq d(x,y) + d(y,z)$.

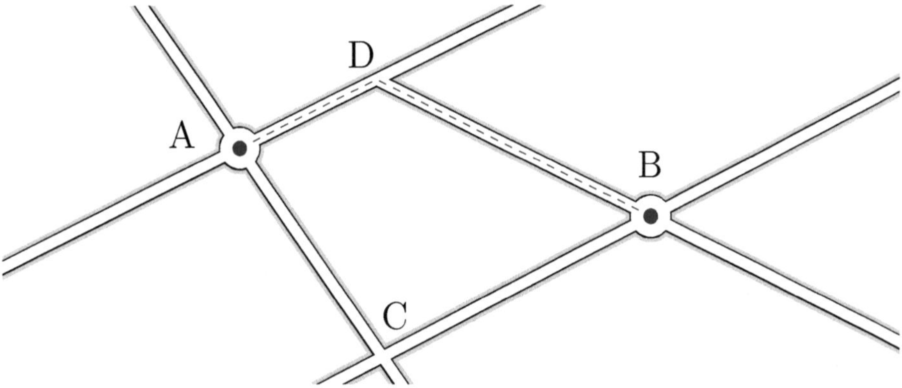

Fig. 9.6. Triangularity in a road network. The distance from A to B is determined by the shortest route, which goes via D, and it cannot be improved by going via C. More importantly, it cannot be improved by going via D either (because it is already part of the shortest path) – or any other point, for that matter; that is, $d(\mathrm{A},\mathrm{B}) \leq d(\mathrm{A},\mathrm{X}) + d(\mathrm{X},\mathrm{B})$, for any point X in the network

facts (and from them, upper and lower bounds) without knowing the exact form of the distance in question.[13]

While the metric properties may not be a perfect fit for modelling our intuition of similarity,[14] nor applicable to all actual similarity retrieval applications,[15] they do capture some notions that seem essential for the intuitive idea of geometric distance. One way to see this is through the metaphor of road networks. The geometry of roads relaxes some of the requirements of Euclidean space, in that the shortest path between two points no longer needs to follow a straight line, but distances still behave "correctly": Unless we have one-way streets, or similar artificial constraints, distance in a road network follows the metric properties exactly. The most interesting of these in this context is, perhaps, triangularity: In finding the distance from A to B we would never follow a more winding road than necessary – we invariably choose the shortest route. Thus, first going from A to C, and subsequently to B, could never give us a smaller sum (that is, a shorter route) than the distance we have defined from A to C. (see Fig. 9.6). While the distance may not be measured along a straight line, we still measure it along the shortest path possible.[16]

One way of interpreting triangularity is that the concept of overlap becomes meaningful for naturally defined regions such as *metric balls* (everything within a given distance of some center object). If we're allowed to violate triangularity, two seemingly non-overlapping balls could still share objects (see Fig. 9.7 on the next page).

If we assume that our dissimilarity space is, in fact, a metric space, the essential next step is to construct bounds (as discussed previously, in Sect. 9.2). Two lemmas constructing quite general bounds using only the metric properties can

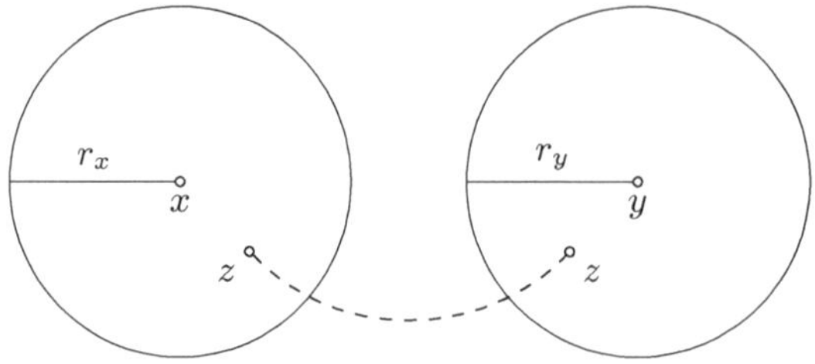

Fig. 9.7. Illustration of a non-triangular space. Two regions defined as balls (everything within a distance r_x and r_y of x and y, respectively) seemingly don't overlap, as $d(x, y) > r_x + r_y$. Even so, there may be objects (such as z) that are found in both regions

be found in Appendix 9.7. The techniques presented in Sects. 9.4 through 9.6 may be seen as special cases of these lemmas, but it should be possible to follow the main text even if the appendix is skipped.

9.4 Pivoting and Pivot Space

The indexing approaches, and their distance bounds, are generally based on selecting sample objects, also known as *pivots* (or, for the regions discussed in Sects. 9.5 and 9.6, sometimes called *centers*). Imagine having selected some pivot object p from your data set, and having pre-computed the distance $d(p, o)$ from the pivot to every other indexed object. This collection of pre-computed distances then becomes your index. When performing a query, you also compute the distance from query to pivot, $d(q, p)$. It is quite easy to see (through some shuffling of the triangle inequality)* that the distance $d(q, o)$ can be lower-bounded as follows (see Fig. 9.8 on the facing page):

$$\check{d}_p(q, o) = |d(q, p) - d(p, o)| \tag{9.1}$$

A search may then be performed by scanning through the data set, filtering out objects based on this lower bound. While this does not take into account CPU costs or I/O (see Sect. 9.1), it does (at least potentially) reduce the total number of distance computations.†

Perhaps the most obvious way of improving this bound is to combine several bounds – using more than one pivot. For any query, the maximum over all the bounds can then be chosen (as the real distance must, of course, obey all the bounds, and the maximum gives us the closest estimate and therefore the most pruning power). For a set of pivots, P, the lower bound can be defined as follows:

* $d(q, o) + d(p, o) \geq d(q, p) \Rightarrow d(q, o) \geq d(q, p) - d(p, o)$. Equivalently, $d(q, o) \geq d(p, o) - d(q, p)$, and hence, $d(q, o) \geq |d(q, p) - d(p, o)|$. This is a special case of Lemma 1 in Appendix 9.7.

† An obvious related upper bound, which can be used for inclusion, is $\hat{d}(q, o) = d(q, p) + d(p, o)$, which follows directly from the triangular inequality.

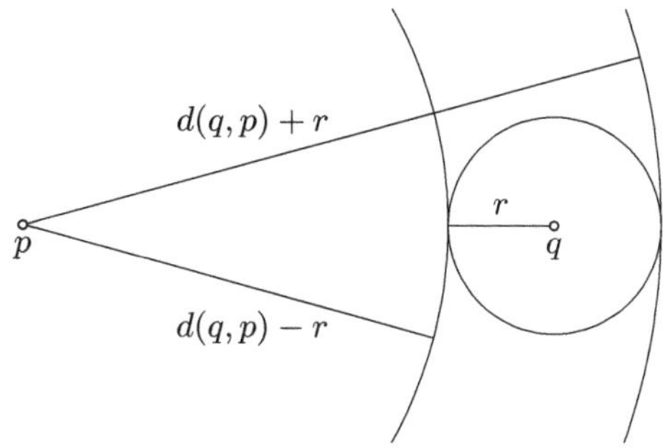

Fig. 9.8. An illustration of basic pivot filtering, based on (9.1). For all objects inside the inner ball around p, we have that $d(q,p) - d(p,o) > r$ and for those outside the outer ball, $d(p,o) - d(q,p) > r$. The "maybe" region is the shell between the two, where $|d(q,p) - d(p,o)| \leq r$. Using several lower bounds (that is, pivots) is equivalent to intersecting several shells, more closely approximating the query ball

$$\check{d}(q,o) = \max_{p \in P} \check{d}_p(q,o) \tag{9.2}$$

One interesting way of viewing this is as a non-expansive mapping to a vector space, often called *pivot space*. We can define a mapping $f(x) = \langle d(p,x) \rangle_{p \in P}$. Our lower bound then, in fact, becomes simply L_∞.

This, in fact, is the gist of one of the basic metric indexing methods, called LAESA.[17]In LAESA, a set of m pivots is chosen from the data set of size n, and an $n \times m$ matrix is filled with the object-to-pivot distances in a preprocessing step. Searching becomes a linear scan through the matrix, filtering out rows based on the lower bound (9.2). As the focus is entirely on reducing the number of comparisons, the linear scan is merely seen as acceptable extra CPU cycles. Some CPU time can be saved by storing columns separately, sorted by distance from its pivot. In each column, the set of viable objects will be found in a contiguous interval, and this interval can be found by bisection. One version of the structure even maintains pointers from cells in one column to the next, between those belonging to the same data object, permitting more efficient intersection of the candidate sets.[18] The basic idea of LAESA was rediscovered by Filho et al. (who call it the Omni method), who also index the pivot space using B-trees and R-trees, to reduce CPU time and I/O.[19]

The choice of pivots can have quite an effect on the performance of this basic pivoting scheme. The simplest approach – simply selecting at random – does work, and several heuristics (such as choosing pivots that are far apart or that have similar distances to each other*) have been proposed to improve the filtering power. One approach, in particular, is to heuristically maximize the lower bound directly. Pivots are added to the pivot set one at a time. For each iteration, the choice stands between a set of randomly sampled candidate pivots. In order to evaluate these, a set of *pairs* of objects is sampled randomly from the data,

*The latter is the approach used by the Omni method.

and the average pivot distance (using the tentative pivot set, including the new candidate in question) is computed. The pivot that gets the highest average is chosen, and the next iteration begins. [20]

The LAESA method is, in fact, based on an older method, called AESA.[21] In AESA, there is no separate pivot set; instead, any data object may be used as a pivot, and the set of pivots used depends on the query object. In order to achieve this flexibility (and the resulting unsurpassed pruning power) one needs to store the distances between all objects in the data set in an $n \times n$ matrix (or, rather, half of that, because of symmetry). In the first iteration, a pivot is chosen arbitrarily, and the data set is filtered. (As we now have the actual distance to the pivot, it can be either included in or excluded from the final result set.) In subsequent iterations, a pivot is chosen among the remaining, unfiltered objects. This object should – in order to maximize pruning power – be as close as possible to the query. This distance is approximated with the pivot distance (9.2), using the pivot set built so far.[22]

In addition to these main algorithms based on pivoting, and their variations, the basic principles are used as components in more complex, hybrid structures.[23]

9.5 Metric Balls and Shells

The pivoting scheme can give quite accurate lower bounds, given enough pivots. However, that number can be high – in many cases leading to unrealistically high space requirements.[24] One possible tradeoff is to reduce the number of pivots below the optimal. As noted in Sect. 9.2, another possibility is to reduce the information available about each object, through aggregation: Instead of storing the exact distance from an object to a pivot, the object is merely placed in a *region*, somehow defined in terms of some of the pivots. This also implies that the relationship between some objects and some pivots can be left completely undefined.

One of the most obvious region choices in a metric space is, perhaps, metric balls. A ball region $[p]_r$ is defined by one pivot p (often referred to as its *center*), along with a so-called *covering radius* r, an upper bound to the distance from the pivot to any object in the region. There are two common ways of partitioning the data using metric ball regions (see Fig. 9.9 on the next page). The first, found in the VP-tree (*vantage point tree*), among others, uses a single ball to create two regions: one inside the ball, and one outside it. The other, found in the BS-tree (*bisector tree*), for example, uses a ball for each region. These two basic partitioning schemes are described in more detail in the following.[25]

The VP-tree is a static balanced binary tree structure built as follows, from a given data set: Choose a pivot (or *vantage point*) p and compute the median distance d_m from p to the rest of the data set. Keep p and d_m in the root node, and recursively construct left and right subtrees from the objects that fall inside and outside d_m, respectively.[26] When searching the tree with a query q and a search radius r, the search ball is examined for overlap with the inside and outside regions (there might, of course, be overlap with both), and the

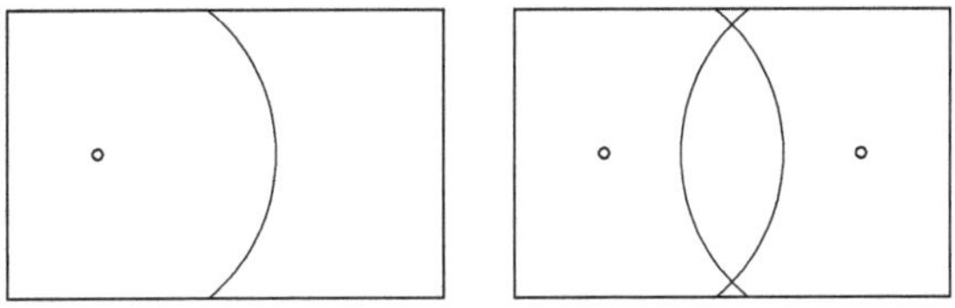

(a) VP-style (b) BS-style

Fig. 9.9. An illustration of the two basic ball partitioning principles. In VP-style partitioning, the two regions correspond to inside and outside one ball, while in BS-style, each region has its own ball. Multiway versions of both are possible

overlapping regions are examined recursively. The respective lower bounds for objects in the inside and outside regions are, of course, $d(q,p) - d_m$ and $d_m - d(q,p)$ (see Fig. 9.10 on the following page).*

While the property of balance might intuitively seem to be an obvious benefit, this may not always be the case. Clearly, the balance does not give us any guaranteed upper bounds on the search time, as we may need to visit multiple (or, indeed, all) subtrees at any point. There may, in fact, be reason to think that unbalanced structures are superior for metric indexing in certain cases. One structure based on this idea is LC (list of clusters).[27] Simply put, LC is a highly unbalanced VP-tree, with each left (inside) branch having a fixed size (either in terms of the radius or the number of items). The right (outside) branch contains the rest of the data, and functions almost as a next-pointer in a linked list consisting of the inside region clusters (hence the name). Search in LC is performed in the same manner as in the VP-tree, but it has one property that can give it an edge in some circumstances: If, at any point, the query is completely contained within one of the cluster balls, the entire tail of the list (the outside) can be discarded.[28]

The VP-style partitioning can be generalized to multiway trees in at least two ways: Either, one can use multiple radii and partition the data set into shells, or bands, between the radii† (either with a fixed distance between them, or with a fixed number of object in each region), or one can use several pivots in each node. The *multi-vantage point* (MVP) tree uses both of these techniques: Each node in the tree has two pivots (more or less corresponding to two levels of a VP-tree collapsed into a single node) as well as (potentially) several radii for each pivot.[29]

The original BS-tree is also a binary tree, but it is dynamic and potentially unbalanced, unlike the VP-tree. The objects in a subtree are contained in a metric ball specific to that tree. The pivot and radius of this ball are kept in the parent node, so each node contains up to two pivots (and corresponding radii). When an object is inserted into a node with zero or one pivots, it is added as a

*See also Fig. 9.5 on page 204, with $f(q) = d(q,p) - d_m$ for the inside region, and $f(q) = d_m - d(q,p)$ for the outside region.

†For more on using shells in indexing, see Fig. 9.11 on page 211 and the discussion of GNAT in Sect. 9.6.

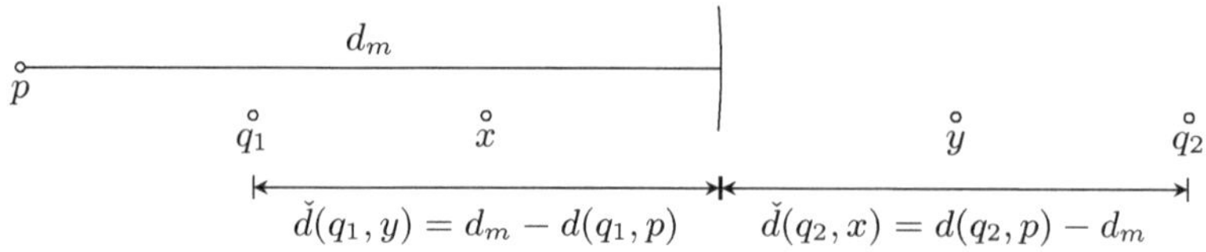

Fig. 9.10. The lower bounds based on VP-style partitioning: The inside region must be checked if $d(q,p) - d_m \leq r$, and the outside region, if $d_m - d(q,p) \leq r$. If q is in the same region as an object, the lower bound becomes negative

pivot to that node (with an empty subtree). If a node is already full, the pivot is inserted into the subtree of the nearest pivot.[30]

The BS-tree has some closely related descendants[31] but one of the most well-known structures using BS-style ball partitioning is rarely presented as one of them: The M-tree.[32]

The structures in the M-tree family are, at their core, multiway BS-trees, with each subtree contained in a metric ball, and new objects inserted into the closest ball (if inside, the one with the closest pivot; if outside, the one whose radius will increase the least). While more recents structures based on the M-tree may have extra features (such as special heuristics for insertion or balancing, or AESA-like pivot filtering in each node) this basic structure is common to them all. The main contribution of the M-tree, however, is not simply the use of a multiway tree structure; it is the way it is implemented – as a balanced, disk-based tree. Each node is a disk block, and balancing is achieved through algorithms quite similar to those of the B-tree family (or the spatial index structures of the R-tree family).[33] This means that it is dynamic (with support for insertions and deletions) and that it is usable for large, disk-based data sets.

The iDistance method is a somewhat related approach. Like in the M-tree, the data set is partitioned into ball regions, and one-dimensional pivot filtering is used within each region. However, instead of using a custom data structure, all information about regions and pivot distances is stored in a B^+-tree. Each object receives a key that, in effect, consists of two digits: The first is the number of its region, and the second is its distance to the region pivot. A metric search can then be reduced to a set of one-dimensional searches in the B^+-tree – one for each region that overlaps the query.[34]

A region type closely related to the ball is the *shell*. It is the set difference $[p]_r \setminus [p]_s$ of two metric balls (where $r \geq s$), and a shell region R can be represented with single center, p, and two radii: the *inner* radius, $d^-(p, R)$, and the *outer* radius, $d^+(p, R)$. These will then be lower and upper bounds on the distance $d(p, o)$ for any object o in the region R. Of course, we have similar lower and upper bounds for any object in the query region as well: $d(q,p) - r$ and $d(q,p) + r$. Only if these two distance intervals overlap can there be any hits in R. To put things differently, for each pivot p and corresponding shell region R, we have two lower bounds:

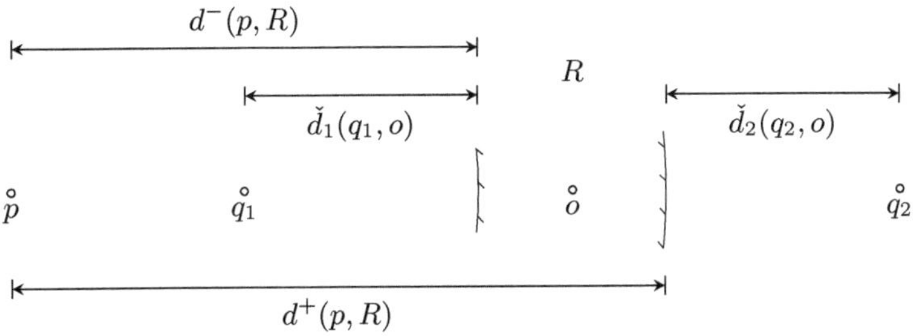

Fig. 9.11. Lower bounds for a shell region R. As discussed in the text, the shell region gives rise to two bounds: $\check{d}_1(q,o) = d^-(p,R) - d(q,p)$ and $\check{d}_2(q,o) = d(q,p) - d^+(p,R)$. The bound $\check{d}_1$ is only useful if the query falls inside the shell, and $\check{d}_2$ only when it falls outside

$$\check{d}_1(q,o) = d^-(p,R) - d(q,p)$$
$$\check{d}_2(q,o) = d(q,p) - d^+(p,R)$$

Here, $\check{d}_1$ is non-negative if the query falls inside the shell and $\check{d}_2$ is non-negative if it falls outside* (see Fig. 9.11).[35]

For an example of a structure using shell regions, see the discussion of GNAT in the following section.

9.6 Metric Planes and Dirichlet Domains

A less obvious partitioning choice than metric balls is the generalized hyperplane, also known as the *midset*, between two pivots in a metric space – the set of objects for which the pivots are equidistant. The two regions generated by this midset consist of the objects closer to one pivot or the other. In the Euclidean plane, this can be visualized by the half-plane on either side of the bisector between two points (see Fig. 12(a) on the following page).[36] This space bisection is used by the GH-tree, and the multiway equivalent is used by the structure called the Geometric Near-neighbor Access Tree, or simply GNAT (see Fig. 12(b)).[37]

The GH-tree is a close relative of the other basic metric trees – especially the BS-tree. In fact, you might say it's simply a BS-tree without the covering radii. It's a dynamic binary structure where every node has (up to) two pivots and every new object is recursively inserted into the subtree of the closest pivot, just as in the BS-tree. The difference is that no covering radius is maintained or used for searching. Instead, the hyperplane criterion is used directly: If the object o is known to be closer to pivot v than of u, then $(d(q,v) - d(q,u))/2$ is a lower bound on $d(q,o)$, and can be used to filter out the v branch.[38]

At core, GNAT is simply a generalization of the GH-style partitioning to multiway trees, partitioning the space into so-called Dirichlet domains (a generalization of Voronoi partitioning). While GNAT is often classified into the

*If the query falls inside R, both bounds are negative.

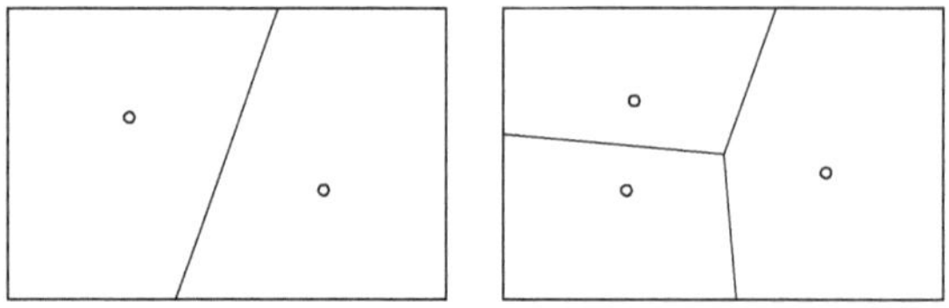

(a) GH-style (b) GNAT-style

Fig. 9.12. Illustration of two-way and multiway generalized hyperplane partitioning. In two-way (GH-style) partitioning, two half-spaces are formed, each consisting of objects closer to one of the two pivots. In multiway (GNAT-style) partitioning, this is simply extended to multiple pivots. The resulting regions are called Dirichlet domains

"hyperplane family" of metric index methods, this is a bit misleading, as it is almost a closer relative of the BS-tree (and, indeed, the M-tree family) than to the GH-tree, for example. It is built by selecting a set of pivots/centers (heuristically, so they are reasonably distant from each other), allocating objects to their closest pivots, and then processing each region recursively (with a number of pivots proportional to the number of objects in that region, to balance the tree). However, the Dirichlet domains are not used for searching; instead, for each region, a set of covering shells (distance ranges) are constructed – one for each other pivot in that node. If the query ball does not intersect with each of these shells, the given region is discarded. (For a discussion on shell regions, see Sect. 9.5.)

Recently, Uribe et al. have developed a dynamic version of GNAT called the Evolutionary Geometric Near-neighbor Access Tree, or EGNAT, which is also well-suited for secondary memory and parallelization.[39] EGNAT is, at heart, simply a GNAT, but it has one addition: All nodes are initially simply buckets (with a given capacity), where the only information stored is the distance to the parent pivot. When a bucket becomes full, it is converted into a GNAT node (with buckets as its children). Searching a GNAT node works just like with GNAT, while searching a bucket uses plain pivoting (using the single pivot available).

Another related structure is the Spatial Approximation (SA) tree.[40] The SA-tree approximates the metric space similarly to a GNAT node – that is, by a partition into Dirichlet domains – but the tree structure is quite different. Rather than representing a hierarchical decomposition, the tree edges simply connect adjacent regions (with one pivot per node).[41] Search is then performed by traversing this tree, moving from region to region.

The SA-tree is built as follows. First a random object is selected as the root, and a set of suitable neighbors are chosen: Every neighbor is required to be closer to the root than to all other neighbors. Also, all *other* objects are closer to at least one of the neighbors than to the root (otherwise they would simply be included as neighbors).[42] Each remaining (non-neighbor) object is assigned to a subset associated with the closest of the neighbors (and subtrees for these subsets are constructed recursively). As a consequence, if we start our traversal

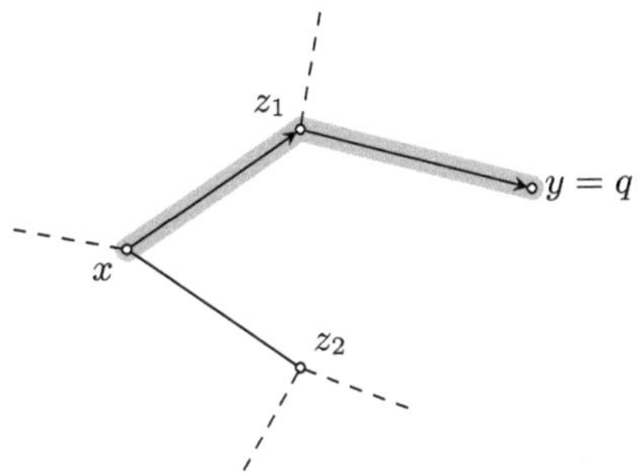

$$d(z_1, z_2) \geq d(z_1, x), d(z_2, x)$$
$$d(y, x) \geq d(y, z_1)$$
$$d(z_1, y) \leq d(z_2, y)$$

Fig. 9.13. For any objects x and y in an SA-tree, y is either a neighbor of x or it is closer to one of x's neighbors than to x. Thus, when the query q is in the data set, we can move gradually closer along the edges. For more realistic searches (involving multiple objects or potential dead ends), backtracking is needed

at the root, we can always get closer to any object by moving to the neighbor that is closest to our destination (see Fig. 9.13).

The basic traversal strategy only works for single objects, and only when searching for objects taken from the data set. If we search for multiple objects (as in a range query) we may need to retrieve objects from multiple branches, and if the query object is unknown, we may hit dead ends that don't end in the desired objects. In either case, we need to add backtracking to our search; or, to put it in the terms used for the other tree-based algorithms: We must traverse the tree, discarding subtrees when possible, using a lower bound on the distance.

The lower bound is twofold: First, a covering radius is kept for each node, and BS-style filtering is used as an initial step (see Sect. 9.5 for details). Second, the the generalized hyperplane bound (the same as in the GH-tree) is used: $\check{d}(q, o) = (d(q, v) - d(q, u))/2$, for objects closer to (that is, in the subtree of) v.[43] Note that when considering the branch of node v, we could construct a separate bound for each other neighbor, u. Instead, we can get the same filtering power by simply using the u for which $d(q, u)$ is lowest. In fact, at any state of the search, we can choose this u among all the ancestors of the current root, as well as their neighbors, because we know the object o will be closer to v than any of them – and we will already have computed $d(q, u)$ for all of these. The minimum of these distances can be passed along as an extra parameter to the recursive search (updated at each node).[44]

9.7 Further Approaches and Issues

Although an attempt has been made in this chapter to include the basic principles of all the central approaches to metric indexing, there are, of course, potentially relevant methods that have not been covered, and interesting topics that have not been discussed. As mentioned in the introduction, there are some surveys and textbooks that cover the field much more thoroughly; the remainder of this section elaborates on the scope of the chapter, and points to some sources of further information.

Other query types

The range search is, in many ways, the least complex of the distance based query modes. Some other types include the following [see 3].

- Nearest or farthest neighbors: Find the k objects that are nearest to or furthest from the query object.
- Closest or most distant pair: Find the two object that are closest to or most distant from each other. Sometimes known as similarity self-join.
- Ranking and inverse ranking: Traverse the objects in (inverse) order of distance from the query.
- Combinations, such as finding the (at most) k nearest objects within a search radius of r.

Of these, the k-nearest-neighbor (kNN) search (possibly combined with range search) is by far the most common, and perhaps the most intuitive from the user and application point of view. Some principles discussed in this chapter – such as domination and non-expansive mappings – do not transfer directly to the case of kNN search. In order for a transform between distance spaces to preserve the k nearest neighbors, it must preserve the relative distances from the query to these objects; simply underestimating them is not enough to guarantee a correct result. Even so, the techniques behind range search can be used quite effectively to perform kNN searches as well.

Although there are special-purpose kNN mechanisms for some index structures, the general approach is to perform what amounts to a branch-and-bound search: If a given object (or set of objects) cannot improve upon the solution we have found so far, it is discarded. More specifically, we maintain a candidate set as the index is examined, consisting of the (at most) k nearest neighbors found so far, along with a dynamic query radius that tightly covers the current candidates.*

As the search progresses, we want to know if we can find any objects that will improve our candidate set; any such objects would have to fall within the current query radius (as they would have to be closer to the query object than the furthest current candidate). Thus, we can use existing radius-based search techniques, shrinking the radius as new candidates are found. The magnitude of the covering radius will then be crucial for search efficiency. It can be kept low for most of the search by heuristically seeking out good candidates early on, for example by visiting regions that are close to the query before those that are further away [3].

Take one of the simplest index structures, the BS-tree (see p. 208). For a range query, subtrees are discarded if their covering balls don't intersect the query. For

*Until we have k candidates, the covering radius is infinite.

a kNN search, we use the candidate ball instead: Only if it is intersected is there any hope of finding a closer neighbor.

Another perspective can be found in the original 1-NN versions of pivoting methods such as AESA and its relatives (see p. 208). They alternate between two steps:

1. Heuristically choose a promising neighbor and compute the actual distance to it from the query.
2. Use the pivoting lower bound to eliminate all objects that are further away than the current candidate.

Both the heuristic and the lower bound are based on the set of candidates chosen so far (that is, they are used as pivots). While it might seem like a different approach, this too is really just a matter of maintaining a shrinking query ball with the current candidates, and the method readily generalizes to kNN for $k > 1$.

Other indexing approaches

While the principles of distance described in Sect. 9.2 hold in general, in the discussions of particulars in the previous sections, several approaches have deliberately been left out. Some important ones are:

- Coordinate-based, or spatial, access methods (see the textbook on the subject by Samet [4] for extensive information on multidimensional indexing). This includes methods in the borderland between purely distance-based methods and spatial ones, such as the M^+- and BM^+-trees of Zhou et al. [5, 6].
- Coordinate-based vector space embeddings, such as FastMap [7]. While pivot space mapping (as used in LAESA [8] and Omni [1]) is an embedding into a vector space ($\mathbb{R}^k$ under L_∞) it is not based on the initial objects having coordinates.
- Approximate methods, such as approximate search with the M-tree [9], MetricMap [10], kNN graphs [see, e.g., 4, pp. 637–641], SASH [11], the proximity preserving order of Chávez et al. [12] or several others [13–15].
- Methods based on stronger assumptions than the metric axioms, such as the growth-restricted metrics of Karger and Ruhl [16].
- Methods based on weaker assumptions than the metric axioms, such as the TriGen method of Skopal [17, 18].*
- Methods exploiting the properties of discrete distances (or discrete metrics, in particular), such as the Burkhart-Keller tree, and The Fixed-Query Tree and its relatives.[45]
- Methods dealing with query types other than range search,† such as the similarity self-join (nearest pair queries) of the eD-index [19], the multi-metric

*Some approximate methods, such as SASH, kNN graphs, and the proximity preserving order of Chávez et al. also waive the metric axioms in favor of more heuristic ideas of how a distance behaves.

†While kNN is only briefly discussed in this chapter, most of the methods discussed support it.

or complex searches of M^3-tree [20] and others [21–24], searching with user-defined metrics, as in the QIC-M-tree [25], or the incremental search of Hjaltason and Samet [26].
- Parallel and distributed methods, using several processors (beyond simple data partitioning), such as GHT* [27], MCAN [28], parallel EGNAT [29] and several other approaches [25, 28, 30–32].

Metric space dimension

It is well known that spatial access methods deal more effectively with low-dimensional vector spaces than high-dimensional ones (one aspect of the so-called *curse of dimensionality*).[46] Some attempts have been made to generalize the concept of dimension to metric spaces as well, in order to measure how difficult a given space would be to index.

One thing that happens when the dimension increases in Euclidean space, for example, is that all distances become increasingly similar, offset somewhat by any clustering in the data. One hypothesis is that it is this convergence of distances that makes indexing difficult, and this idea is formalized in the *intrinsic dimensionality* of Chávez et al. [33]. Rather than dealing with coordinates, this measure is defined using the distance distribution of a metric (or, in general, distance) space. The intrinsic dimensionality is defined as $\mu^2/2\sigma^2$, where μ and σ^2 are the mean and variance of distribution, respectively. For a set of uniformly random vectors, this is actually proportional to the dimension of the vector space. Note that a large spread of distances results in a lower dimensionality, and vice versa.

An interesting property of distance distributions with a low spread can be seen when considering the cumulative distribution function $N(r)$, the average number of objects returned for a range query with radius r. For r-values around the mean distance, the value of N will increase quite rapidly – the narrower the distribution, the more quickly it will increase. Traina Jr. et al. [34], building on the work on fractal dimensions by Belussi and Faloutsos [35], show that for many synthetic and real-world metric data sets, for suitable ranges of r, the cumulative distribution follows a power-law; that is, $N(r) \in \Theta(r^{\mathfrak{D}})$, for some constant $\mathfrak{D}$, which they call the *distance exponent*.[47] They show how to estimate the distance exponent, and demonstrate that it is closely related to, among other things, the number of disk accesses needed when performing range queries using the M-tree.

Another, rather different, hypothesis is that the problem with high dimensionality in vector spaces is the abundance of emptiness: As the dimension grows, the ratio of data to the sheer volume of space falls exponentially. It becomes hard to create tight regions around subsets of the data, and region overlaps abound. The *ball-overlap factor* (BOF) of Skopal [18] tackles this problem head-on, and measures the relative frequency of overlap between suitably chosen ball regions. BOF_k is defined as the relative frequency of overlap between balls that each cover an object and its k nearest neighbors. In other words, it predicts the likelihood that two rather arbitrary ball-shaped regions will overlap. This factor can describe both the degree of overlap by distinct regions in an index structure, and the probability that a query will have to investigate irrelevant ball regions (only non-overlapping regions can be discarded).

Method quality

In the introduction, the main goals and quality measures used in the development of metric indexing methods were briefly discussed, but the full picture is rather complex. The theoretical analysis of these structures can be quite difficult, so experimental validation becomes essential. There are some theoretically defined measures such as the fat and bloat factors and prunability of Traina Jr. et al. [36, 37], but even such properties are generally established empirically for a given structure.[48] Moret [38] gives some reasons why asymptotic analysis alone may not be enough for algorithm studies in general (the worst-case behavior may be restricted to a very small subset of instances and thus not be at all characteristic of instances encountered in practice, and even in the absence of these problems, deriving tight asymptotic bounds may be very difficult). Given a sufficiently problematic metric space, a full linear scan can never be avoided, so the non-asymptotic (primarily empirical) analysis may be particularly relevant for metric indexing.

Even beyond the specifics of a given data set (including measures such as intrinsic dimensionality, fractal dimension, and ball-overlap factor) there are, of course, the real-world issues that plague all theoretical studies of algorithms, such as caching and the memory hierarchy – issues that are not easily addressed in terms of basic principles (at least not given the current state of the theory of metric index structures) and therefore have been omitted from this chapter.

Appendix A: Bounding Lemmas, with Proofs

The following two lemmas give us some useful bounds for use in metric indexing. For a more detailed discussion, see, for example, the survey by Hjaltason and Samet [3] or the textbook by Zezula et al. [39]. Lemmas 1 and 2 are illustrated in figures 9.14 and 9.15, respectively.

Lemma 1 (Ball lemma). *Let o, p and q be objects in $\mathbb{U}$, and let d be a metric over $\mathbb{U}$. For any objects u, v in $\mathbb{U}$, assume that the the value of $d(u,v)$ is known to be in the range $[d^-_{uv}, d^+_{uv}]$. The value of $d(q,o)$ may then be bounded as follows:*

$$\max\{0, d^-_{po} - d^+_{pq}, d^-_{qp} - d^+_{op}\} \leq d(q,o) \leq d^+_{qp} + d^+_{po}\,.$$

Proof. From the triangle inequality we have $d(p,o) \leq d(p,q) + d(q,o)$, which gives us $d(q,o) \geq d(p,o) - d(p,q) \geq d^-_{po} - d^+_{pq}$. Similarly, we have $d(q,p) \leq d(q,o) + d(o,p)$, which gives us $d(q,o) \geq d(q,p) - d(q,p) \geq d^-_{qp} - d^+_{op}$. Finally, the upper bound follows directly from the triangle inequality: $d(q,o) \leq d(q,p) + d(p,o) \leq d^+_{qp} + d^+_{po}$.

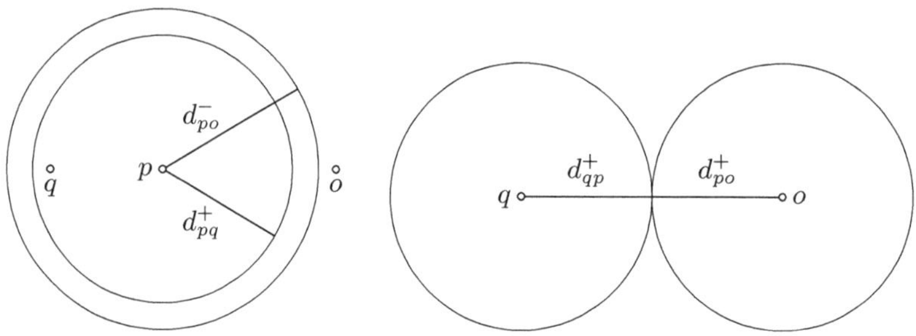

(a) $\check{d}(q,o) = d^-_{po} - d^+_{pq}$ (b) $\hat{d}(q,o) = d^+_{qp} + d^+_{po}$

Fig. 9.14. Illustration of Lemma 1. (a) The bounds on $d(o,p)$ and $d(q,p)$ define two shell regions around p: one containing o and one containing q, and the underestimates are given by the distances from the outer one to the inner one. (b) The upper bound is also given by two shell regions: one around o and one around q, both containing p; because these must overlap in at least one point (that is, p), the overestimate is simply the sum of their (outer) radii

Lemma 2 (Plane lemma). *Let o, q, u and v be objects in $\mathbb{U}$and let $d(o,v) \leq d(u,o)$. We can then bound $d(q,o)$ as follows:*

$$\max\{(d(q,v) - d(q,u))/2, 0\} \leq d(q,o)\,.$$

Proof. From the triangle inequality, we have $d(q,v) \leq d(q,o) + d(o,v)$, which yields $d(q,v) - d(q,o) \leq d(o,v)$. When combined with $d(u,o) \leq d(q,u) + d(q,o)$ (from the triangle inequality) and $d(o,v) \leq d(u,o)$, we obtain $d(q,v) - d(q,o) \leq d(q,u) + d(q,o)$. Rearranging yields $d(q,v) - d(q,u) \leq 2d(q,o)$, which yields the first combined of the lower bound, the second component being furnished by non-negativity.

Appendix B: An Overview of the Indexing Methods

Many methods are discussed in this chapter, either in the main text or in the end notes. Some important ones are summarized in Table 9.1 on the next page, with a brief indication of their structure and functionality. The structural properties indicate the use of pivoting (P), BS-style ball partitioning (BS), VP-style ball partitioning (VP), generalized hyperplane partitioning (GH), and multiway partitioning (MW). The functional features indicate whether the method is dynamic (D), whether it is designed to reduce extra CPU cost (CPU) or memory use and/or disk accesses (I/O). A bullet (•) indicates the presence of a property, while a circle (○) indicates partial (or implicit) presence. This table, of course, only gives the coarsest of overviews of the differences between the methods, certainly does not cover all their unique features, and is not meant as a comparison of their relative merits.

Table 9.1. Some structural properties and functional features of some of the metric indexing methods discussed in this chapter. The numbers refer to the bibliography

		Structure				Functionality			
Method		P	BS	VP	GH	MW	D	CPU	I/O
AESA	62, 67, 68	●							
AHC	83	●		●				●	
Antipole Tree	93	●	●		○	○		●	
BS-Tree	44		●		○		●	●	
BU-Tree	90		●		○			●	
CM-Tree	45	●	●		○	●	●	●	●
DBM-Tree	84, 85	○	●		○	●	●	●	●
DF-Tree	37	●	●		○	●	●	●	●
D-Index	53, 55–57	●		●		●	○	●	●
DSA-Tree	102–104				●	●	●	●	○
DVP-Tree	79			●		●	●	●	●
EGNAT	2	○		●	○	●	●	●	●
EHC	83			●		○		●	
EM-VP-Forest	78			●				●	
GH-Tree	75, 98				●		●	●	
GNAT	99			●	○	●		●	
HC	82, 83			●				●	
HSA-Tree	105–107	●			●	●	●	●	○
iAESA	69	●							
iDistance	96	○	●		○	●	●	●	●
LAESA	8, 62, 64, 65	●					●		○
LC	80, 81			●				●	○
Linear Scan	47						●		
MB$^+$-Tree	97			●		●	●	●	●
MB*-Tree	88, 89		●		○			●	
M*-Tree	94	●	●		○	●	●	●	●
M-Tree	49–51, 91	○	●		○	●	●	●	●
MVP-Tree	52, 73	●		●		●	●	●	●
Omni	1	●					●	●	●
Opt-VP-Tree	77			●		●	●	●	●
PM*-Tree	94	●	●		○	●	●	●	●
PM-Tree	74	●	●		○	●	●	●	●
ROAESA	70	●						●	
SA-Tree	100, 101				●	●		●	
Slim-Tree	36, 92	○	●		○	●	●	●	●
Spaghettis	63	●					●	●	
SSS-Tree	46		●		●	●		●	
SSS-LC	66	●		●			○	●	●
TLAESA	71, 72	●					●	●	
Voronoi-Tree	86		●		○	○	●	●	
VP-Tree	75, 76, 98			●				●	

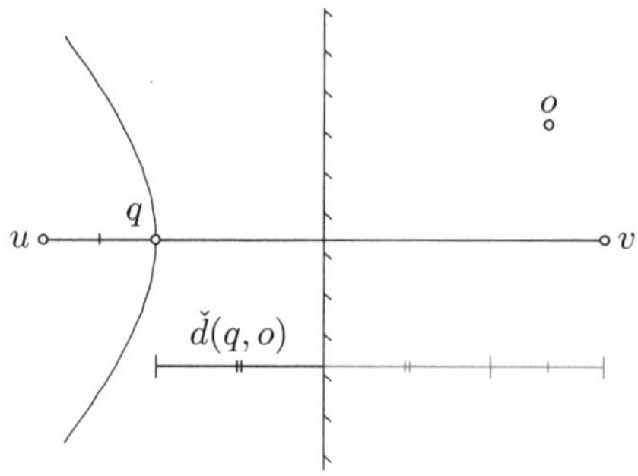

Fig. 9.15. Illustration of Lemma 2 ($\check{d}(q,o) = (d(q,v) - d(q,u)/2)$). The lower bound may intuitively be seen as a bound on the distance from q (on the u-side) to the "plane" of points midway from u to v (note that no such points may actually exist). As the object o is on the v-side, this is a lower bound on $d(q,o)$. The figure shows the case where q is on the metric interval between u and v (that is, $d(u,v) = d(u,q) + d(q,v)$); other placements of q for the same bound (the hyperbola in the figure) would be further away from the v-region

Acknowledgements

The author would like to thank Simon Jonassen for useful and constructive comments.

Notes

1. Chávez et al. [33] have written a very thorough and rather theoretical survey paper on the subject. In addition to describing the methods available at the time, they construct a taxonomy of the methods and derive theoretical bounds for their complexity. They also introduce the concept of intrinsic dimensionality (see Sect. 9.7). The later survey by Hjaltason and Samet [3] takes a somewhat different approach, and uses a different fundamental formalism, but is also quite comprehensive and solid, and complements the paper by Chávez et al. nicely. In recent years, a textbook on the subject by Zezula et al. has appeared [39], and Sect. 4.5 of Samet's book on multidimensional and metric data structures [4] is also devoted to distance based methods. The paper by Pestov and Stojmirović [40] is specifically about a model for similarity search, but in many ways provides a brief survey as well. The encyclopedia entry by Chávez and Navarro [41] is another example of a brief introduction. In addition to publications that specifically set out to describe the field, there are also publications, such as the PhD thesis of Skopal [42], with more specific topics, that still have substantial sections devoted to the general field of metric indexing.

2. Note that the terminology is not entirely consistent across the literature. The use of 'distance' here conforms with the usage of Deza and Deza [43].

3. Many of the surveys mentioned previously [3, 4, 33, 39–42] discuss various applications, as do most publications about specific metric indexing methods.

Spatial access methods [see, e.g., 4] can also be used for many query-by-example applications, but they rely on the specific structure of the problem – the fact that all objects are represented as vectors of a space of fixed, low dimensionality. Metric indexing is designed for cases where less is known about the space (i.e., where we only know the distances between objects, and those distances satisfy the metric axioms), and thus have a different, possibly wider, field of applications. Because the metric indexing methods disregard the number of dimensions of a vector space and are only hampered by the innate complexity of the given distance (or the distribution of objects), they may also be better suited to indexing high-dimensional traditional vector spaces (see Sect. 9.7).

4. The field in question is defined by in excess of fifty published methods for metric indexing, stretching back to the 1983 paper of Kalantari and McDonald [44] (with more recent publications including ones by Aronovich and Spiegler [45] and Brisaboa et al. [46], for example), not counting several methods aimed specifically at discrete distance measures (surveyed along with more general methods by Chávez et al. [33]; see also Sect. 9.7).

5. While a linear scan is very straightforward, and seemingly quite inefficient, it may serve as an important "reality check," particularly for complex index structures (especially involving disk access) or high-dimensional data. The extra work in maintaining and traversing the structures may become so high, in some cases, that it swamps the reduction in distance computations [see, e.g., 47].

6. As discussed in Sect. 9.7, there are approximate retrieval methods as well, where some such errors are permitted. In these cases, looser notions of distance approximation may be acceptable.

7. Such mappings are normally defined between metric spaces, and are known by several names, including *metric mappings* and 1-*Lipschitz mappings*.

8. One example of this technique is similarity indexing for time series, where many signature types have been suggested, normally in the form of fixed-dimensional vectors that may be indexed using spatial access methods. Hetland [48] gives a survey of this application.

9. Hjaltason and Samet [3] use hierarchical decompositions of the space into regions (represented as tree structures) as their main formalism when discussing indexing methods. In order to discard such a region R from a search, a lower bound on the point-set distance $d(q, R) = \inf_{x \in R} d(q, x)$ must be defined; this bound is a characteristic of the region type used. Pestov and Stojmirović [40] use basically the same formalism, although presented rather differently. Instead of equipping the tree nodes corresponding to regions with lower-bounding distance estimates, they give them *certification functions*, which are non-expansive (1-Lipschitz) mappings $f : R \to \mathbb{R}$, where $f(x) \leq 0, \forall x \in R$. While not discussed by the authors, it should be clear that these certification functions are equivalent to lower-bounding estimates of the point-set distance (see Fig. 9.4 on page 203). For a rather different approach, see the survey by Chávez et al. [33]. They base their indexing formalism on the hierarchical decomposition of the search space into

equivalence classes, and discuss overlapping regions without directly involving lower bounds in the fundamental model.

10. Indeed, it may seem like some structures have rather uncritically transplanted the ideas behind search trees to the field of similarity search, even though the advantages of this adaptation are not as obvious as they might seem. In addition to the varying balance between distance computations, CPU time, and I/O, there are such odd phenomena as unbalanced structures being superior to balanced ones in certain circumstances (see the discussion of LC in Sect. 9.7).

11. Examples of methods that rely on this are the M-tree [49–51] (and its descendants), the MVP-tree [52] and D-index [53–57].

12. Chávez et al. [33] discuss this in depth for the case of metric indexing, and the use of compact regions versus pivoting (see Sects. 9.4 through 9.6), and show that, in general, if one has an unlimited amount of memory, pivoting will be superior, but that the region-based methods will normally utilize limited memory more efficiently.

13. The theory of metric spaces is extensive (see the tutorial by Semmes [58] for a brief introduction, or the book by Jain and Ahmad [59] for a more thorough treatment), but the discussion in this chapter focuses on the basic properties of metric spaces, and how they permit the construction of bounds usable for indexing.

14. See, for example, the discussion of Skopal [18] for some background on this issue. One example he gives, illustrating broken triangularity, is that of comparing humans, horses and centaurs. While a human might be quite similar to a centaur, and a centaur looks quite a bit like a horse, most would think a human to be completely different from a horse.

15. One example of a decidedly non-metric distance is the one used in the query-by-whistling system of Arentz et al. [60]. While this distance is non-negative and reflexive, it is neither symmetric nor triangular, and it does not separate non-identical objects.

16. This is, of course, simply a characterization of metric spaces in terms of (undirected) graphs. It is clear that shortest paths (geodesics) in finite graphs with positive edge weights invariably form metric spaces, but it is interesting to note that the converse is also true: Any finite metric space may be realized as such a graph geodesic [see, e.g., Lemma 3.2.1 of 61, p. 62]. It is also interesting to note that the triangularity of geodesics is preserved in the presence of negative edge weights (that is, shortest paths still cannot be shortened). Similarly, directed triangularity holds for directed graphs.

17. Linear Approximating and Eliminating Search Algorithm [8, 62].

18. This is the Spaghettis structure, described by Chávez et al. [63].

19. See the paper by Filho et al. [1] for more information.

20. For more details on this approach, and a comparison between it and a few similar heuristics, see the paper by Bustos et al. [64]. A recent variation on choosing pivots that are far apart is called Sparse Spatial Selection [65], and it is used in the so-called SSS-Tree [46] and in SSS-LC [66].

21. Approximating and Eliminating Search Algorithm [67, 68].

22. Actually, a recent algorithm called iAESA [69], managed, as the first method in twenty years, to improve upon AESAs search performance. iAESA uses the same basic strategy as AESA, but uses a different heuristic for selecting pivots, involving the correlation between distance-based permutations of *another* set of pivots. In theory, any cheap and accurate approximation could be used as such a heuristic, potentially improving the total performance.

23. A couple of variations of interest are ROAESA, or Reduced Overhead-AESA [70], and TLAESA, or Tree-LAESA [71, 72], which reduce the CPU time of AESA and LAESA, respectively, to sublinear. Applications of pivoting in hybrid structures include MVP-tree [52, 73], D-index [53, 53–57] and the related, DF-tree [37], PM-tree [74], and CM-tree [45].

24. Chávez et al. [33] show that the optimal number is logarithmic in the size of the data set ($\Theta(\lg n)$, for n objects). Filho et al. [1], on the other hand, claim that the required number of pivots is proportional to the fractal (which they also call intrinsic) dimensionality of the data set, and that using pivots beyond this is of little use. In other words, this means that the optimal number of pivots is not necessarily related to the size of the data set (that is, it is $\Theta(1)$). See Sect. 9.7 for a brief discussion of fractal and intrinsic dimensionality.

25. For more information about the VP-tree, see the papers by Uhlmann [75] (who calls them simply *metric trees*) and Yianilos [76] (who rediscovered them, and gave them their name). The VP partitioning scheme has been extended in, for example, the Optimistic VP-tree [77] and the MVP-tree [52, 73]. It is extended with a so-called *exclusion zone* in the Excluded Middle Vantage Point Forest [78] (which is in many ways very similar to the more recent D-index, discussed later). The BS-tree was first described by Kalantari and McDonald [44], and the BS partitioning scheme has been adopted by several subsequent structures, including the M-tree [49–51] and its descendants. A recent addition to the BS-tree family is the SSS-Tree [46].

26. A dynamic version of the VP-tree has been proposed by chee Fu et al. [79].

27. Chávez and Navarro [80, 81] give a theoretical analysis of high-dimensional metric spaces (in terms of intrinsic dimensionality, as explained in Sect. 9.7) as support for their assumption that unbalanced structures may be beneficial. Fredriksson [83] describes several extended versions of the LC structure (HC, AHC and EHC). Marin et al. [66] describe a hybrid of LC and LAESA, using the SSS pivot selection strategy. Another structure that relaxes the balance requirement and achieves improved performance is the DBM-tree [84, 85]

28. An interesting point here is that one can use other metric index structures to speed up the building of LC, because one needs to find which objects (of those remaining) are inside the cluster radius, or which k objects are the nearest neighbors of the cluster center – and both of these operations are standard metric queries.

29. The MVP-tree is discussed in detail by Bozkaya and Özsoyoglu [52, 73]. In addition to the basic partitioning strategy, each leaf also contains a LAESA

structure, filtering with the pivots found on the path from the root to that leaf (thereby reusing distances that have already been computed).

30. Note that this decision is actually based on a generalized hyperplane, or bisector – hence the name "bisector tree." Hyperplanes are discussed in more detail in Sect. 9.6.

31. Perhaps the most closely related structure is the Voronoi-tree [86, 87], which is essentially a ternary BS-tree with an additional property: When a new leaf is created, the parent pivot (the one closest to the new object) is also added to the leaf. This guarantees that no node can have a greater covering radius than its parent. Another relative is the Monotonous Bisector* Tree (MBS*-tree) [88, 89]. A more recent relative is the BU-tree [90], which is simply a static BS-tree, built bottom-up (hence the name), using clustering techniques.

32. For more details on the M-tree, see the papers by Zezula et al. [49] and Ciaccia et al. [50, 51, 91], for example. Some recent structures based on the M-tree include the QIC-M-tree [25], the Slim-tree [92], the DF-tree [37], the PM-tree [74], the Antipole Tree [93], the M*-tree [94], and the CM-tree [45].

33. The B-tree and its descendants are described in several basic textbooks on algorithms; a general structure called GiST (Generalized Search Tree, available from `http://gist.cs.berkeley.edu`) implements disk-based B-tree-style balancing for use in index structures in general, and has been used in several published implementations (including the original M-tree). For more information on recent developments in the R-tree family, see the book by Manolopoulos et al. [95].

34. In practice, the object key is a single value that is constructed by multiplying the region number with a sufficiently large constant, and adding the two. Note that the original description of iDistance focuses on kNN search, but the method works equally well for range queries. For more information on iDistance, including partitioning heuristics, see the paper by Yu et al. [96]. For another way of mapping a metric space onto a B^+-tree keys, see the MB^+-tree of Ishikawa et al. [97].

35. For more on why this is correct, see Lemma 1 on page 217.

36. Note that in general, the midset itself may be empty.

37. For more information on the GH-tree, see the papers by Uhlmann [75, 98]; GNAT is described in detail by Brin [99].

38. See Lemma 2 on page 218 for an explanation of this bound. Also note that there is no reason why this criterion couldn't be used in *addition* to a covering radius.

39. For more information on EGNAT, see the paper by Uribe et al. [2].

40. For more information on the SA-tree, see the papers by Navarro [100, 101]. Though the originally published structure is static, it has since been extended to admit insertions and deletions [102–105] and combined with pivoting [106, 107].

41. If you create a graph from the pivots by adding edges between neighboring Dirichlet domains, you get what is often known as the Delaunay graph (or, for

the Euclidean plane, the Delaunay triangulation) of the pivot set. The SA-tree will be a spanning tree of this graph.

42. Note that the neighbor set is defined recursively, and it is not necessarily entirely obvious how to construct it. In fact, there can be several sets satisfying the definition. As finding a minimum set of neighbors it not a trivial task, Navarro [101] uses an incremental heuristic approach: Consider nodes in order of increasing distance from the root, and add them if they are closer to the root than all neighbors added previously. While the method does not guarantee neighbor sets of minimum size, the sets are correct (and usually rather small): Clearly all object that are closer to the root than the other neighbors have been added. Conversely, let's say consider a root r and two neighbors u and v, added in that order. By assumption, $d(u,r) \leq d(v,r)$. We could only have added v if $d(v,r) < d(u,v)$, so both must be closer to r than to each other.

43. See Lemma 2 on page 218 for proof.

44. Although the description of the search algorithm here is formally equivalent to that of Navarro [101], the presentation differs quite a bit. He gives two explanations for the correctness of the search algorithm: one involving a lower bound (similar to the hyperplane argument given here), and one based on the notion of traversing the tree, moving toward a hypothetical object within a distance r of the query. The final result is the same, of course.

45. These relatives include FHQT, FQA/FMVPA, FHQA, and FQMVPT; see the survey by Chávez et al. [33] for details on these discrete-metric methods.

46. For a discussion of how to deal with high-dimensional vector spaces, see the survey by Hetland [48]. It is interesting to note that the intrinsic dimensionality of a vector data set may be lower than its representational dimensionality; that is, the vectors may be mapped faithfully to a lower-dimensional vector space without distorting the distances much. In this case, using a distance-based (metric) index may be quite a bit more efficient than indexing the original dimensions directly with a spatial access method. (See the discussion of intrinsic and fractal dimension in Sect. 9.7.)

47. This $\mathfrak{D}$ plays a role similar to that of the 'correlation' fractal dimension of a vector space, also known as D_2 [35]. The fractal dimension of a metric space may be found in quadratic time [see, e.g., 108], but the method of Traina Jr. et al. [34] is based on a linear time approximation. Filho et al. [1] use the concept of fractal space in their analysis of the required number of pivots (or foci) for optimum performance of the Omni method (see note 24).

48. Note that Chávez et al. [33] give some general bounds for pivoting and region-based methods based on intrinsic dimensionality, but these are not really fine-grained enough to give us definite comparisons of methods.

References

[1] Santos Filho, R.F., Traina, A., Traina Jr., C., Faloutsos, C.: Similarity search without tears: the OMNI-family of all-purpose access methods. In: Proceedings of the 17th International Conference on Data Engineering, ICDE, pp. 623–630 (2001)

[2] Uribe, R., Navarro, G., Barrientos, R.J., Marín, M.: An Index Data Structure for Searching in Metric Space Databases. In: Alexandrov, V.N., van Albada, G.D., Sloot, P.M.A., Dongarra, J. (eds.) ICCS 2006. LNCS, vol. 3991, pp. 611–617. Springer, Heidelberg (2006)

[3] Hjaltason, G.R., Samet, H.: Index-driven similarity search in metric spaces. ACM Transactions on Database Systems, TODS 28(4), 517–580 (2003)

[4] Samet, H.: Foundations of Multidimensional and Metric Data Structures. Morgan Kaufmann, San Francisco (2006)

[5] Zhou, X., Wang, G., Yu, J.X., Yu, G.: M^+-tree: A new dynamical multidimensional index for metric spaces. In: Zhou, X., Schewe, K.-D. (eds.) Proceedings of the 14th Australasian Database Conference, ADC. Conferences in Research and Practice in Information Technology, vol. 17 (2003)

[6] Zhou, X., Wang, G., Zhou, X., Yu, G.: BM^+-tree: A hyperplane-based index method for high-dimensional metric spaces. In: Zhou, L.-z., Ooi, B.-C., Meng, X. (eds.) DASFAA 2005. LNCS, vol. 3453, pp. 398–409. Springer, Heidelberg (2005)

[7] Faloutsos, C., Lin, K.-I.: FastMap: A fast algorithm for indexing, data-mining and visualization of traditional and multimedia. In: Carey, M.J., Schneider, D.A. (eds.) Proceedings of the 1995 ACM SIGMOD International Conference on Management of Data, San Jose, California, pp. 163–174 (1995)

[8] Micó, M.L., Oncina, J.: A new version of the nearest-neighbour approximating and eliminating search algorithm (AESA) with linear preprocessing time and memory requirements. Pattern Recognition Letters 15(1), 9–17 (1994)

[9] Zezula, P., Savino, P., Amato, G., Rabitti, F.: Approximate similarity retrieval with M-Trees. The VLDB Journal 7(4), 275–293 (1998)

[10] Wang, J.T.L., Wang, X., Shasha, D., Zhang, K.: Metricmap: an embedding technique for processing distance-based queries in metric spaces. IEEE Transactions on Systems, Man and Cybernetics, Part B 35(5), 973–987 (2005)

[11] Houle, M.E., Sakuma, J.: Fast approximate similarity search in extremely high-dimensional data sets. In: Proceedings of the 21st IEEE International Conference on Data Engineering, ICDE (2005)

[12] Chávez, E., Figueroa, K., Navarro, G.: Proximity Searching in High Dimensional Spaces with a Proximity Preserving Order. In: Gelbukh, A., de Albornoz, Á., Terashima-Marín, H. (eds.) MICAI 2005. LNCS (LNAI), vol. 3789, pp. 405–414. Springer, Heidelberg (2005)

[13] Amato, G.: Approximate similarity search in metric spaces. PhD thesis, Computer Science Department – University of Dortmund, August-Schmidt-Str. 12, 44221, Dortmund, Germany (2002)

[14] Ciaccia, P., Patella, M.: PAC nearest neighbor queries: Approximate and controlled search in high-dimensional and metric spaces. In: IEEE Computer Society (eds.) Proceedings of the 16th International Conference on Data Engineering, ICDE, pp. 244–255 (2000)

[15] Amato, G., Rabitti, F., Savino, P., Zezula, P.: Region proximity in metric spaces and its use for approximate similarity search. ACM Transactions on Information Systems, TOIS 21(2), 192–227 (2003)

[16] Karger, D.R., Ruhl, M.: Finding nearest neighbors in growth-restricted metrics. In: Proceedings of the 34th annual ACM symposium on Theory of computing, pp. 741–750. ACM Press, New York (2002)

[17] Skopal, T.: On fast non-metric similarity search by metric access methods. In: Ioannidis, Y., Scholl, M.H., Schmidt, J.W., Matthes, F., Hatzopoulos, M., Böhm, K., Kemper, A., Grust, T., Böhm, C. (eds.) EDBT 2006. LNCS, vol. 3896, pp. 718–736. Springer, Heidelberg (2006)

[18] Skopal, T.: Unified framework for exact and approximate search in dissimilarity spaces. ACM Transactions on Database Systems, TODS 32(4) (2007)

[19] Dohnal, V., Gennaro, C., Zezula, P.: Similarity join in metric spaces using eD-index. In: Mařík, V., Štěpánková, O., Retschitzegger, W. (eds.) DEXA 2003. LNCS, vol. 2736, pp. 484–493. Springer, Heidelberg (2003)

[20] Bustos, B., Skopal, T.: Dynamic similarity search in multi-metric spaces. In: Proceedings of the 8th ACM international workshop on Multimedia information retrieval. ACM Press, New York (2006)

[21] Fagin, R.: Combining fuzzy nformation from multiple systems. In: Proceedings of the 15th ACM Symposium on Principles of Database Systems, PODS, pp. 216–226. ACM Press, New York (1996)

[22] Fagin, R.: Fuzzy queries in multimedia database systems. In: Proceedings of the 17th ACM Symposium on Principles of Database Systems, PODS, pp. 1–10. ACM Press, New York (1998)

[23] Ciaccia, P., Montesi, D., Penzo, W., Trombetta, A.: Imprecision and user preferences in multimedia queries: A generic algebraic approach. In: Schewe, K.-D., Thalheim, B. (eds.) FoIKS 2000. LNCS, vol. 1762, pp. 50–71. Springer, Heidelberg (2000)

[24] Ciaccia, P., Patella, M., Zezula, P.: Processing complex similarity queries with distance-based access methods. In: Schek, H.-J., Saltor, F., Ramos, I., Alonso, G. (eds.) EDBT 1998. LNCS, vol. 1377, pp. 9–23. Springer, Heidelberg (1998)

[25] Ciaccia, P., Patella, M.: Searching in metric spaces with user-defined and approximate distances. ACM Transactions on Database Systems (TODS) 27(4), 398–437 (2002)

[26] Hjaltason, G.R., Samet, H.: Incremental similarity search in multimedia databases. Technical Report CS-TR-4199, Computer Science Department, University of Maryland, College Park (2000)

[27] Batko, M., Gennaro, C., Savino, P., Zezula, P.: Scalable similarity search in metric spaces. In: Proceedings of the DELOS Workshop on Digital Library Architectures, pp. 213–224 (2004)

[28] Falchi, F., Gennaro, C., Zezula, P.: Nearest neighbor search in metric spaces through content-addressable networks. Information Processing & Management 43(3), 665–683 (2007)

[29] Marín, M., Uribe, R., Barrientos, R.J.: Searching and updating metric space databases using the parallel EGNAT. In: Shi, Y., van Albada, G.D., Dongarra, J., Sloot, P.M.A. (eds.) ICCS 2007. LNCS, vol. 4487, pp. 229–236. Springer, Heidelberg (2007)

[30] Alpkocak, A., Danisman, T., Ulker, T.: A parallel similarity search in high dimensional metric space using M-tree. In: Grigoras, D., Nicolau, A., Toursel, B., Folliot, B. (eds.) IWCC 2001. LNCS, vol. 2326, p. 166. Springer, Heidelberg (2002)

[31] Zezula, P., Savino, P., Rabitti, F., Amato, G., Ciaccia, P.: Processing M-Trees with parallel resources. In: Proceedings of the 8th International Workshop on Research Issues in Data Engineering, RIDE, pp. 147–154. IEEE Computer Society Press, Los Alamitos (1998)
[32] Novak, D., Zezula, P.: M-Chord: a scalable distributed similarity search structure. In: Proceedings of the 1st International Conference on Scalable Information Systems (2006)
[33] Chávez, E., Navarro, G., Baeza-Yates, R., Marroquín, J.L.: Searching in metric spaces. ACM Computing Surveys 33(3), 273–321 (2001)
[34] Traina Jr., C., Traina, J.M., Faloutsos, C.: Distance exponent: a new concept for selectivity estimation in metric trees. In: Proceedings of the International Conference on Data Engineering, ICDE (2000)
[35] Belussi, A., Faloutsos, C.: Estimating the selectivity of spatial queries using the correlation fractal dimension. In: Proceedings of the International Conference on Very Large Databases, VLDB (1995)
[36] Traina Jr., C., Traina, A.J.M., Seeger, B., Faloutsos, C.: Slim-Trees: High Performance Metric Trees Minimizing Overlap between Nodes. In: Zaniolo, C., Grust, T., Scholl, M.H., Lockemann, P.C. (eds.) EDBT 2000. LNCS, vol. 1777, pp. 51–65. Springer, Heidelberg (2000)
[37] Traina Jr., C., Traina, A., Santos Filho, R.F., Faloutsos, C.: How to improve the pruning ability of dynamic metric access methods. In: Proceedings of the eleventh international conference on Information and knowledge management, pp. 219–226 (2002)
[38] Moret, B.M.E.: Towards a discipline of experimental algorithmics. In: Data Structures, Near Neighbor Searches, and Methodology: Fifth and Sixth DIMACS Implementation Challenges, vol. 59, pp. 197–214. Americal Mathematical Society (2002)
[39] Zezula, P., Amato, G., Dohnal, V., Batko, M.: Similarity Search: The Metric Space Approach. Springer, Heidelberg (2006)
[40] Pestov, V., Stojmirović, A.: Indexing schemes for similarity search: an illustrated paradigm. Fundamenta Informaticae 70(4), 367–385 (2006)
[41] Chávez, E., Navarro, G.: Metric databases. In: Rivero, L.C., Doorn, J.H., Ferraggine, V.E. (eds.) Encyclopedia of Database Technologies and Applications, pp. 367–372. Idea Group, USA (2006)
[42] Skopal, T.: Metric Indexing in Information Retrieval. PhD thesis, Technical University of Ostrava (2004)
[43] Deza, E., Deza, M.M.: Dictionary of Distances. Elsevier, Amsterdam (2006)
[44] Kalantari, I., McDonald, G.: A data structure and an algorithm for the nearest point problem. IEEE Transactions on Software Engineering 9(5), 631–634 (1983)
[45] Aronovich, L., Spiegler, I.: Efficient similarity search in metric databases using the CM-tree. Data & Knowledge Engineering (2007)
[46] Brisaboa, N., Pedreira, O., Seco, D., Solar, R., Uribe, R.: Clustering-based similarity search in metric spaces with sparse spatial centers. In: Geffert, V., Karhumäki, J., Bertoni, A., Preneel, B., Návrat, P., Bieliková, M. (eds.) SOFSEM 2008. LNCS, vol. 4910, pp. 186–197. Springer, Heidelberg (2008)
[47] Beyer, K., Goldstein, J., Ramakrishnan, R., Shaft, U.: When is "nearest neighbor" meaningful? In: Beeri, C., Bruneman, P. (eds.) ICDT 1999. LNCS, vol. 1540, pp. 217–235. Springer, Heidelberg (1998)

[48] Hetland, M.L.: A survey of recent methods for efficient retrieval of similar time sequences. In: Last, M., Kandel, A., Bunke, H. (eds.) Data Mining in Time Series Databases, vol. 2, pp. 23–42. World Scientific, Singapore (2004)

[49] Zezula, P., Ciaccia, P., Rabitti, F.: M-tree: A dynamic index for similarity queries in multimedia databases. Technical Report 7, HERMES ESPRIT LTR Project (1996)

[50] Ciaccia, P., Patella, M., Zezula, P.: M-tree: An efficient access method for similarity search in metric spaces. In: Proceedings of the 23rd International Conference on Very Large Data Bases, VLDB, pp. 426–435. Morgan Kaufmann, San Francisco (1997)

[51] Ciaccia, P., Patella, M., Zezula, P.: A cost model for similarity queries in metric spaces. In: Proceedings of the 17th Symposium on Principles of Database Systems, pp. 59–68 (1998)

[52] Bozkaya, T., Özsoyoglu, M.: Distance-based indexing for high-dimensional metric spaces. In: Proceedings of the, ACM SIGMOD international conference on Management of data, pp. 357–368. ACM Press, New York (1997)

[53] Dohnal, V., Gennaro, C., Savino, P., Zezula, P.: Separable splits of metric data sets. In: Proceedings of the Nono Convegno Nazionale Sistemi Evoluti per Basi di Dati, SEBD (June 2001)

[54] Gennaro, C., Savino, P., Zezula, P.: Similarity search in metric databases through hashing. In: Proceedings of the 2001 ACM workshops on Multimedia: multimedia information retrieval (2001)

[55] Dohnal, V., Gennaro, C., Savino, P., Zezula, P.: D-index: Distance searching index for metric data sets. Multimedia Tools and Applications 21(1), 9–33 (2003)

[56] Dohnal, V.: An access structure for similarity search in metric spaces. In: Lindner, W., Mesiti, M., Türker, C., Tzitzikas, Y., Vakali, A.I. (eds.) EDBT 2004. LNCS, vol. 3268, pp. 133–143. Springer, Heidelberg (2004)

[57] Dohnal, V.: Indexing Structures for Searching in Metric Spaces. PhD thesis, Masaryk University, Faculty of Informatics (2004)

[58] Semmes, S.: What is a metric space? (September 2007) arXiv:0709.1676v1 [math.MG]

[59] Jain, P.K., Ahmad, K.: Metric Spaces, 2nd edn. Alpha Science International Ltd. Pangbourne (2004)

[60] Arentz, W.A., Hetland, M.L., Olstad, B.: Methods for retrieving musical information based on rhythm and pitch correlations. Journal of New Music Research 34(2) (2005)

[61] Jungnickel, D.: Graphs, Networks and Algorithms, 2nd edn. Springer, Heidelberg (2005)

[62] Rico-Juan, J.R., Micó, L.: Comparison of AESA and LAESA search algorithms using string and tree-edit-distances. Pattern Recognition Letters 24(9-10), 1417–1426 (2002)

[63] Chávez, E., Marroquín, J.L., Baeza-Yates, R.: Spaghettis: An array based algorithm for similarity queries in metric spaces. In: Proceedings of the String Processing and Information Retrieval Symposium & International Workshop on Groupware (SPIRE), pp. 38–46. IEEE Computer Society Press, Los Alamitos (1999)

[64] Bustos, B., Navarro, G., Chávez, E.: Pivot selection techniques for proximity searching in metric spaces. Pattern Recognition Letters 24(14), 2357–2366 (2003)

[65] Pedreira, O., Brisaboa, N.R.: Spatial Selection of Sparse Pivots for Similarity Search in Metric Spaces. In: van Leeuwen, J., Italiano, G.F., van der Hoek, W., Meinel, C., Sack, H., Plášil, F. (eds.) SOFSEM 2007. LNCS, vol. 4362, pp. 434–445. Springer, Heidelberg (2007)
[66] Marin, M., Gil-Costa, V., Uribe, R.: Hybrid index for metric space databases. In: Proceedings of the International Conference on Computational Science (2008)
[67] Ruiz, E.V.: An algorithm for finding nearest neighbours in (approximately) constant average time. Pattern Recognition Letters 4(3), 145–157 (1986)
[68] Vidal, E.: New formulation and improvements of the nearest-neighbour approximating and eliminating search algorithm (AESA). Pattern Recognition Letters 15(1), 1–7 (1994)
[69] Figueroa, K., Chávez, E., Navarro, G., Paredes, R.: On the least cost for proximity searching in metric spaces. In: Àlvarez, C., Serna, M. (eds.) WEA 2006. LNCS, vol. 4007, pp. 279–290. Springer, Heidelberg (2006)
[70] Vilar, J.M.: Reducing the overhead of the AESA metric-space nearest neighbour searching algorithm. Information Processing Letters 56(5), 265–271 (1995)
[71] Micó, M.L., Oncina, J., Carrasco, R.C.: A fast branch & bound nearest neighbour classifier in metric spaces. Pattern Recognition Letters 17(7), 731–739 (1996)
[72] Tokoro, K., Yamaguchi, K., Masuda, S.: Improvements of TLAESA nearest neighbour search algorithm and extension to approximation search. In: Proceedings of the 29th Australasian Computer Science Conference, pp. 77–83. Australian Computer Society, Inc (2006)
[73] Bozkaya, T., Özsoyoglu, M.: Indexing large metric spaces for similarity search queries. ACM Transactions on Database Systems, TODS 24(3) (1999)
[74] Skopal, T.: Pivoting M-tree: A metric access method for efficient similarity search. In: Snášel, V., Pokorn'y, J., Richta, K. (eds.) Proceedings of the Annual International Workshop on DAtabases, TExts, Specifications and Objects (DATESO), Desna, Czech Republic, Technical University of Aachen (RWTH), April 2004. CEUR Workshop Proceedings, vol. 98 (2004)
[75] Uhlmann, J.K.: Metric trees. Applied Mathematics Letters 4(5), 61–62 (1991)
[76] Yianilos, P.N.: Data structures and algorithms for nearest neighbor search in general metric spaces. In: Proceedings of the fourth annual ACM-SIAM Symposium on Discrete algorithms, Philadelphia, PA, USA, pp. 311–321. Society for Industrial and Applied Mathematics
[77] Chiueh, T.-c.: Content-based image indexing. In: Proceedings of the Twentieth International Conference on Very Large Databases, VLDB, Santiago, Chile, pp. 582–593 (1994)
[78] Yianilos, P.N.: Excluded middle vantage point forests for nearest neighbor search. Technical report, NEC Research Institute (1999)
[79] chee Fu, A.W., shuen Chan, P.M., Cheung, Y.-L., Moon, Y.S.: Dynamic vp-tree indexing for n-nearest neighbor search given pair-wise distances. The VLDB Journal 9(2), 154–173 (2000)
[80] Chávez, E., Navarro, G.: An effective clustering algorithm to index high dimensional metric spaces. In: Proceedings of the 6th International Symposium on String Processing and Information Retrieval SPIRE, pp. 75–86. IEEE Computer Society Press, Los Alamitos (2000)
[81] Chávez, E., Navarro, G.: A compact space decomposition for effective metric indexing. Pattern Recognition Letters 26(9), 1363–1376 (2005)
[82] Fredriksson, K.: Exploiting distance coherence to speed up range queries in metric indexes. Information Processing Letters 95(1), 287–292 (2005)

[83] Fredriksson, K.: Engineering efficient metric indexes. Pattern Recognition Letters (2006)
[84] Vieira, M.R., Traina Jr., C., Takada Chino, F.J., Traina, A.J.M.: DBM-tree: a dynamic metric access method sensitive to local density data. In: Proceedings of the 19th Brazilian Symposium on Databases (SBBD). UnB (2004)
[85] Vieira, M.R., Traina Jr., C., Takada Chino, F.J., Machado Traina, A.J.: DBM-tree: trading height-balancing for performance in metric access methods. Journal of the Brazilian Computer Society 11(3), 20–39 (2006)
[86] Frank, K.H.A.: Dehne and Hartmut Noltemeier. Voronoi trees and clustering problems. Information Systems 12(2), 171–175 (1987)
[87] Noltemeier, H.: Voronoi trees and applications. In: Proceedings of the International Workshop on Discrete Algorithms and Complexity, pp. 69–74 (1989)
[88] Noltemeier, H., Verbarg, K., Zirkelbach, C.: Monotonous Bisector* trees: a tool for efficient partitioning of complex scenes of geometric objects. In: Monien, B., Ottmann, T. (eds.) Data Structures and Efficient Algorithms. LNCS, vol. 594, pp. 186–203. Springer, Heidelberg (1992)
[89] Noltemeier, H., Verbarg, K., Zirkelbach, C.: A data structure for representing and efficient querying large scenes of geometric objects: MB* trees. In: Farin, G.E., Hagen, H., Noltemeier, H., Knödel, W. (eds.) Geometric Modelling. Computing Supplement, vol. 8. Springer, Heidelberg (1992)
[90] Liu, B., Wang, Z., Yang, X., Wang, W., Shi, B.-L.: A Bottom-Up Distance-Based Index Tree for Metric Space. In: Wang, G.-Y., Peters, J.F., Skowron, A., Yao, Y. (eds.) RSKT 2006. LNCS (LNAI), vol. 4062, pp. 442–449. Springer, Heidelberg (2006)
[91] Ciaccia, P., Patella, M., Rabitti, F., Zezula, P.: Indexing metric spaces with M-tree. In: Cristiani, M., Tanca, L. (eds.) Atti del Quinto Convegno Nazionale su Sistemi Evoluti per Basi di Dati (SEBD 1997), Verona, Italy, June 1997, pp. 67–86 (1997)
[92] Skopal, T., Pokorný, J., Krátký, M., Snášel, V.: Revisiting M-tree building principles. In: Kalinichenko, L.A., Manthey, R., Thalheim, B., Wloka, U. (eds.) ADBIS 2003. LNCS, vol. 2798, pp. 148–162. Springer, Heidelberg (2003)
[93] Cantone, D., Ferro, A., Pulvirenti, A., Recupero, D.R., Shasha, D.: Antipole tree indexing to support range search and k-nearest neighbor search in metric spaces. IEEE Transactions on Knowledge and Data Engineering 17(4), 535–550 (2005)
[94] Skopal, T., Hoksza, D.: Improving the performance of M-tree family by nearest-neighbor graphs. In: Ioannidis, Y., Novikov, B., Rachev, B. (eds.) ADBIS 2007. LNCS, vol. 4690, pp. 172–188. Springer, Heidelberg (2007)
[95] Manolopoulos, Y., Nanopoulos, A., Papadopoulos, A.N., Theodoridis, Y.: R-Trees: Theory and Applications. In: Advanced Information and Knowledge Processing. Springer, Heidelberg (2006)
[96] Yu, C., Ooi, B.C., Tan, K.-L., Jagadish, H.V.: Indexing the distance: An efficient method to KNN processing. In: Proceedings of the 27th International Conference on Very Large Data Bases, VLDB, San Francisco, CA, USA, pp. 421–430. Morgan Kaufmann Publishers Inc, San Francisco (2001)
[97] Ishikawa, M., Chen, H., Furuse, K., Yu, J.X., Ohbo, N.: MB+Tree: A Dynamically Updatable Metric Index for Similarity Search. In: Lu, H., Zhou, A. (eds.) WAIM 2000. LNCS, vol. 1846, pp. 356–373. Springer, Heidelberg (2000)
[98] Uhlmann, J.K.: Satisfying general proximity/similarity queries with metric trees. Information Processing Letters 40(4), 175–179 (1991)

[99] Brin, S.: Near neighbor search in large metric spaces. In: Dayal, U., Gray, P.M.D., Nishio, S. (eds.) Proceedings of 21th International Conference on Very Large Data Bases, VLDB, pp. 574–584. Morgan Kaufman, San Francisco (1995)
[100] Navarro, G.: Searching in metric spaces by spatial approximation. In: Proceedings of the 6th International Symposium on String Processing and Information Retrieval, SPIRE, pp. 141–148. IEEE Computer Society Press, Los Alamitos (1999)
[101] Navarro, G.: Searching in metric spaces by spatial approximation. The VLDB Journal 11(1), 28–46 (2002)
[102] Navarro, G., Reyes, N.: Dynamic spatial approximation trees. In: Proceedings of the XXI Conference of the Chilean Computer Science Society, SCCC, pp. 213–222 (2001)
[103] Navarro, G., Reyes, N.: Fully dynamic spatial approximation trees. In: Laender, A.H.F., Oliveira, A.L. (eds.) SPIRE 2002. LNCS, vol. 2476, pp. 254–270. Springer, Heidelberg (2002)
[104] Navarro, G., Reyes, N.: Improved deletions in dynamic spatial approximation trees. In: Proceedings of the XXIII International Conference of the Chilean Computer Science Society, SCCC (2003)
[105] Arroyuelo, D., Navarro, G., Reyes, N.: Fully dynamic and memory-adaptative spatial approximation trees. In: Proceedings of the Argentinian Congress in Computer Science, CACIC, pp. 1503–1513 (2003)
[106] Arroyuelo, D., Muñoz, F., Navarro, G., Reyes, N.: Memory-adaptative dynamic spatial approximation trees. In: Proceedings of the 10th International Symposium on String Processing and Information Retrieval, SPIRE (2003)
[107] Chavez, E., Herrera, N., Reyes, N.: Spatial approximation + sequential scan = efficient metric indexing. In: Proceedings of the XXIV International Conference of the Chilean Computer Science Society, SCCC (2004)
[108] Schroeder, M.: Fractals, Chaos, Power Laws, 6th edn. W. H. Freeman and Company, New York (1991)

10

Particle Evolutionary Swarm Multi-Objective Optimization for Vehicle Routing Problem with Time Windows

Angel Muñoz-Zavala, Arturo Hernández-Aguirre, and Enrique Villa-Diharce

Centro de Investigación en Matemáticas (CIMAT)
Jalisco s/n, Mineral de Valenciana, C.P. 36240, Guanajuato, Guanajuato, México
aemz@cimat.mx,artha@cimat.mx,villadi@cimat.mx

Summary. The Vehicle Routing Problem with Time Windows (VRPTW), is an extension to the standard vehicle routing problem. VRPTW includes an additional constraint that restricts every customer to be served within a given time window. An approach for the VRPTW with the next three objectives is presented: 1)total distance (or time), 2)total waiting time, 3)number of vehicles. A data mining strategy, namely space partitioning, is adopted in this work. Optimal routes are extracted as features hidden in variable size regions where depots and customers are located. This chapter proposes the sector model for partitioning the space into regions. A new hybrid Particle Swarm Optimization algorithm (PSO), and combinatorial operators ad-hoc with space partitioning are described. A set of well-known benchmark functions in VRPTW are used to compare the effectiveness of the proposed method. The results show the importance of examining characteristics of a set of non-dominated solutions, that fairly consider the three dimensions, when a user should select only one solution according to problem conditions.

10.1 Introduction

In transportation management, there is a requirement to provide goods and/or services from a supply point to various geographically dispersed points with significant economic implications. The vehicle routing problem (VRP), first introduced by Dantzig and Ramser [1], is a well-known combinatorial optimization problem in the field of service operations management and logistics.

Vehicle Routing Problems (VRP) are all around us in the sense that many consumer products such as soft drinks, beer, bread, snack foods, gasoline, pharmaceutical products, etc., are delivered to retail outlets by a fleet of trucks whose operation fits the vehicle routing model. In practice, the VRP has been recognized as one of the great success stories of operations research and it has been studied widely since the late fifties.

The typical VRP can be stated as follows: design least-cost routes from a central depot to a set of geographically dispersed points (customers, stores, schools, cities, warehouses, etc.) with various demands. Each customer is to be serviced exactly once by only one vehicle, and each vehicle has a limited capacity. In practice, the problem is aimed at minimizing the total cost of the combined routes

C.A. Coello Coello et al. (Eds.): Swarm Intel. for Multi-objective Prob., SCI 242, pp. 233–257.
springerlink.com

for a fleet of vehicles. Since cost is closely associated with distance, in general, the alternative goal is to minimize the distance traveled by a fleet of vehicles with various constraints.

The Vehicle Routing Problem with Time Windows (VRPTW) is a generalization of the VRP involving the additional constraint that every customer should be served within a given time window. Examples of practical applications of the VRPTW include school bus, taxi scheduling, courier delivery/pickup, airline fleet scheduling, industrial refuse collection, etc. The objective of the VRPTW is to minimize the number of vehicles and total distance traveled to service the customers without violating the capacity and time window constraints. A vehicle may arrive early, but it must wait until start of service time is possible.

Many research works on VRPTW solve the problem by a weighted-sum approach in which the number of vehicles is given implicit priority, and consequently the scoring procedure must prioritize this dimension of the problem. The VRPTW is naturally multi-modal, and neither dimension is fundamentally more important than the other from a theoretical perspective and even from a practical aspect.

This chapter proposes a natural interpretation of the VRPTW as a multi-objective problem that prevents the introduction of solution bias towards either of the problem dimensions that commonly affect the weighted sum methods. The Multi-Objective Optimization (MOO) approach generates a set of equally valid VRPTW solutions. These solutions represent a range of possible answers, with different numbers of vehicles, distances and waiting times. The decision maker, who makes the final decision among the alternatives, can decide which kind of solution is preferable under different problem conditions.

Several VRP approaches use a Euclidean distance to form clusters of customers, and then approximate a route that visits all the customers in the cluster. In this work, the customers are clustered by applying spatial data mining principles. Spatial Data Mining (SDM) means to discover interesting relationships between space and the non-space data. A new space partitioning technique, called sectors model, is proposed for grouping the customers in neighborhoods (clusters) according to their spatial location.

A MOO algorithm based-on Particle Swarm Optimization (PSO) and Pareto dominance is introduced to solve the VRPTW. The rest of this chapter has the following structure: in Section 10.2 we give a formal definition of the VRPTW. Section 10.3 provides an overview of MOO research. The outline of our algorithm is presented in Section 10.4. Also Section 10.4 describes the sectors model technique. In Section 10.5 we present and analyze the experimental results performed. Finally, we draw some overall conclusions and suggest directions for future work in Section 10.6.

10.2 Vehicle Routing Problem

The Vehicle Routing Problem concerns the transportation of items between depots and customers by means of a fleet of vehicles. In general, solving a VRP

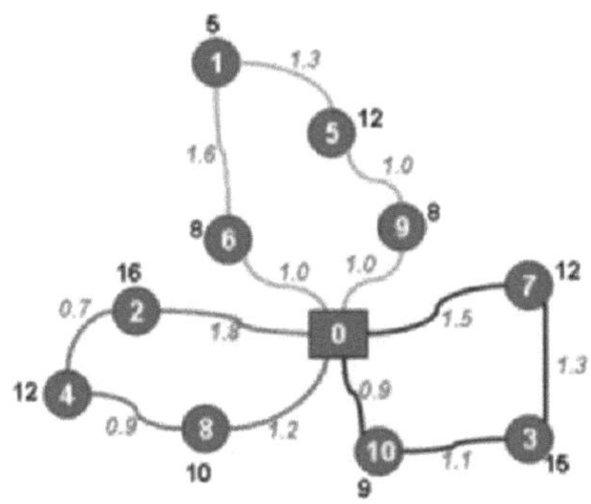

Fig. 10.1. VRP example

means to find the best route to service all customers using a fleet of vehicles (see Fig. 10.1). The solution must ensure that all customers are served, fulfilling the operational constraints, and minimizing the total transportation cost. Figure 10.1 shows a set of scattered points, where point 0 represents the depot and points 1, ..., 10 represent the customers. In Fig. 10.1 three routes supply the customers demand. For instance, a route attends customers 2, 4 and 8, whose demands are 16, 12 and 10 units, respectively. This route visits each customer exactly one time and then returns to the starting point (as a Hamiltonian cycle) accumulating 4.6 distance units.

In the VRP, the decisions to be made define the order of the sequence of visits to the customers; they are a set of routes. A route departs from the depot and it is an ordered sequence of visits to be made by a vehicle to the customers, fulfilling their orders. A solution must be verified to be feasible, checking that it does not violate any constraint, such as the one stating that the sum of the demands of the visited vertices shall not exceed the vehicle capacity.

10.2.1 Vehicle Routing Problem with Time Windows

The VRPTW is a variant of the VRP which considers the available time window in which either customer has to be supplied (see Fig. 10.2). The VRPTW is commonly found in real-world applications and is more realistic than the VRP that assumes the complete availability over time of the customers.

Each customer requests a given amount of goods, which must be delivered or collected at the customer location. Time intervals during which the customer can be served are specified. These time windows can be single or multiple. A vehicle cannot arrive later than a given time, but it can wait if arriving early. In such a case, the goal of the objective function is to minimize the distance traveled and the waiting time. The vehicle routing model can also include an estimation of the loading and unloading times at the customer location. These service times depend on the customer facilities, on the ordered quantity, and on bureaucratic

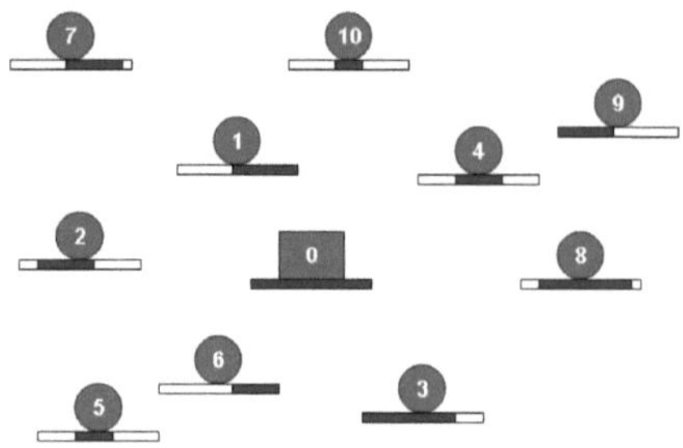

Fig. 10.2. VRPTW example

requirements. They are used to compute the delivery time, which is needed to compute the time at which the vehicle is ready to leave for the next customer in its tour.

The presence of time windows imposes a series of precedences on visits, which make the problem asymmetric, even if the distance and time matrices were originally symmetric. The objective is to minimize the vehicle fleet, the sum of travel time and waiting time needed to supply all customers in their required hours. In 1985, Salvelsberg [2] showed that even finding a feasible solution to the VRPTW, when the number of vehicles is fixed, is itself an NP-complete problem. An overview on the VRPTW formulation and related approaches can be found in Cordeau et al. [4].

10.2.2 VRPTW Formulation

The VRPTW is represented by a set of identical vehicles denoted by K, and a directed graph G, which consists of a set of customers and a depot. Node 0 represents the depot. The set of n vertices representing customers is denoted N. The arc set A denotes all possible connections between the nodes (including the node denoting depot). All routes start at node 0 and end at node 0. We associate a cost c_{ij} and a time t_{ij} with each arc $(i, j) \in A$ of the routing network. The travel time t_{ij} includes service time at customer i. Each vehicle has a capacity limit q_k and each customer i, a demand d_i, $i \in N$. Each customer i has a time window, $[a_i, b_i]$, where a_i and b_i are the respective opening time and closing times of i.

No vehicle may arrive past the closure of a given time window, b_i. Although a vehicle may arrive early, it must wait until the start of service time a_i is possible. It generates a waiting time w_k for the route. Vehicles must also leave the depot within the depot time window $[a_0, b_0]$ and must return before or, at time b_0. For eliminating any unnecessary waiting time [3], we assume that all routes start just-in-time $w_k = 0$, that is, we adjust the depot departure time of each vehicle $t_k = max(0, a_i - t_{0i})$, $i \in N$.

10.2.3 VRPTW Approaches

A brief review of the most important methodologies for solving the VRPTW is given. These methodologies can be divided into three main classes: exact methods, heuristic, and metaheuristic.

Exact Methods

The first optimization algorithm for the VRPTW can be attributed to Kolen et al. [5], who used dynamic programming and state space relaxation to compute lower bounds within a branch-and-bound algorithm. A branch-and-cut algorithm for the VRPTW was recently developed by Bard et al. [6], which incorporates four types of cuts: capacity constraints, comb inequalities, incompatible pair inequalities, and incompatible path inequalities. At each node of the search tree an upper bound is computed by means of a greedy randomized adaptive search procedure. Although these methods produce optimal solutions, they are computationally expensive, even for relatively small instances [7].

Obtaining exact optimal solutions in the VRPTW is computationally intractable due to its combinatorial explosion. Thus, heuristic and metaheuristic methods are the only feasible way to provide solutions for industrial scale problems.

Heuristics

Laporte and Semet [8] proposed the next classification of VRP heuristics:

- Route Construction Methods: they work by inserting customers one at a time into partial routes until a feasible solution is obtained. Routes can either be constructed sequentially or in parallel. Several sequential insertion heuristics for VPRTW were proposed by Solomon [9].
- Two-Phase Methods: customers are first grouped into distinct subsets based on some proximity measure; then, within each group, the customers are ordered to form a route (route-first cluster-second). In the cluster-first route-second method a giant route that visits all customers is first constructed; then, this giant route is partitioned into smaller feasible routes [12]. Kontoravdis and Bard [14] have described a two-phase greedy randomized adaptive search procedure (GRASP) for the VRPTW.
- Route Improvement Methods: iteratively improve an initial feasible solution by performing exchanges while maintaining feasibility. The first improvement heuristics for the VRPTW were in fact adaptations of the 2-opt, 3-opt and Or-opt edge exchange mechanisms originally introduced for the traveling salesperson problem (TSP) [10, 11].

As reported by Laporte and Semet [8], classical heuristics usually have low execution speed but often produce solutions having a large gap with respect to the best known.

Metaheuristics

The last 15 years witnessed increasing research efforts regarding the development of metaheuristic approaches. Unlike classical improvement methods, metaheuristics usually incorporate mechanisms to continue the exploration of the search space even after a local minimum is encountered. These methods typically perform a thorough exploration of the solution space, allowing deteriorating and even infeasible intermediate solutions. Gendreau et al. [13], have identified six families of metaheuristics for the VRP: simulated annealing, deterministic annealing, tabu search, genetic algorithms, ant systems, and neural networks.

Among the other types of metaheuristics that have been successful on different types of VRP, tabu search is certainly the most remarkable one. The first application of tabu search to the VRPTW can be attributed to Potvin et al. [15]. More recently, a tabu search heuristic was developed by Cordeau et al. [16] for the VRPTW and two of its generalizations: the periodic VRPTW and the multidepot VRPTW.

Bent and Van Hentenryck [17] have described a two-stage hybrid algorithm that first minimizes the number of routes by simulated annealing and then minimizes the total distance traveled by using a large neighborhood search.

Ant colony is rarely able to build acceptable VRPTW solutions, however, it is particularly apt for hybridization. Good candidate solutions found by the ant colony system can be further improved using various local search techniques. Gambardella et al. [18] have developed an ant colony optimization algorithm for the VRPTW which associates an attractiveness measure to arcs in the graph.

Blanton and Wainwright [19] were the first to apply a genetic algorithm to VRPTW. They hybridized a genetic algorithm with a greedy heuristic. Thangiah [20] describes a method called GIDEON that assigns customers to vehicles by partitioning the customers into sectors by genetic algorithm and customers within each formed sector are routed using an insertion method. In the next step the routes are improved using λ-exchanges introduced by Osman [21]. Bräysy [22] proposed several new crossover and mutation operators, testing different forms of genetic algorithms, selection schemes, scaling schemes and the significance of the initial solutions.

Homberger and Gehring [23] propose two evolutionary metaheuristics based on the class of evolutionary algorithms called Evolution Strategies and three well-known route improvement procedures Or-opt [24], λ-interchanges [21] and 2-opt* [25]. Metaheuristics have also been used to solve VRPTW with two or more objectives. This is described in the next section.

10.3 Multi-Objective Optimization

In Multi-Objective Optimization (MOO), two or more conflicting objectives contribute to the overall result. These objectives often affect one another in complex, nonlinear ways. The challenge is to find a set of values for them which yields an optimization of the overall problem at hand [26].

A MOO problem is defined as follows:

$$min_{\boldsymbol{x}} \quad \boldsymbol{f}(\boldsymbol{x}) = [f_1(\boldsymbol{x}), f_2(\boldsymbol{x}), \ldots, f_n(\boldsymbol{x})] \tag{10.1}$$

$$subject\ to \quad g_i(\boldsymbol{x}) \leq 0 \quad i = 1, 2, \ldots, m \tag{10.2}$$

$$h_j(\boldsymbol{x}) = 0 \quad j = 1, 2, \ldots, p \tag{10.3}$$

where $\boldsymbol{x} = [x_1, x_2, \ldots, x_d]$ is the vector of decision variables.

The notion of "optimum" in MOO problems differs from that of a single function in global optimization. Having several objective functions the aim is to find good compromises, rather than a single solution. In 1896, Vilfredo Pareto, an Italian economist, introduced the concept of Pareto efficiency [27].

10.3.1 Pareto Optimality

Pareto efficiency, or Pareto optimality, is an important concept in economics with broad applications in game theory, engineering and the social sciences. Given a set of alternative allocations of goods or income for a set of individuals, a movement from one allocation to another that can make at least one individual better off without making any other individual worse off is called a Pareto improvement. An allocation is Pareto efficient when no further Pareto improvements can be made. The Pareto efficient set in objective space is called Pareto set.

Definition 1. *Assuming minimization, a vector $\boldsymbol{x}^* \in F$ (F is the feasible region), is Pareto optimal if for every $\boldsymbol{x} \in F$ and $I = \{1, 2, \ldots, n\}$*

$$f_i(\boldsymbol{x}^*) \leq f_i(\boldsymbol{x}) \tag{10.4}$$

and, there is at least one $i \in I$ such that

$$f_i(\boldsymbol{x}^*) < f_i(\boldsymbol{x}) \tag{10.5}$$

A useful concept in MOO, related with Pareto efficiency, is the Pareto Front. A Pareto front is the projection of the Pareto set on function space (see Fig. 10.3). The Pareto front is particularly useful in engineering: by restricting attention to the set of choices that are Pareto-efficient, a designer can make tradeoffs within this set, rather than considering the full range of every parameter.

The Pareto efficiency is based on the Pareto dominance concept [28]. A vector $\boldsymbol{x}$ is preferred to (dominates) a vector $\boldsymbol{y}$ if each parameter of $\boldsymbol{x}$ is no greater than the corresponding parameter of $\boldsymbol{y}$ and at least one parameter is less (see Fig. 10.4).

Definition 2. *A vector $\boldsymbol{x}$ is said to dominate vector $\boldsymbol{y}$,*

$$\boldsymbol{x} \preceq \boldsymbol{y} \tag{10.6}$$

if and only if $\boldsymbol{x}$ is partially less than $\boldsymbol{y}$.

$$\forall i \in \{1, 2, \ldots, n\}, \boldsymbol{x}_i \leq \boldsymbol{y}_i \wedge \exists i \in \{1, 2, \ldots, n\} : \boldsymbol{x}_i < \boldsymbol{y}_i \tag{10.7}$$

The Pareto frontier is composed by a set of non-dominated vectors. Figure 10.5 shows a comparison between two individuals of the Pareto frontier.

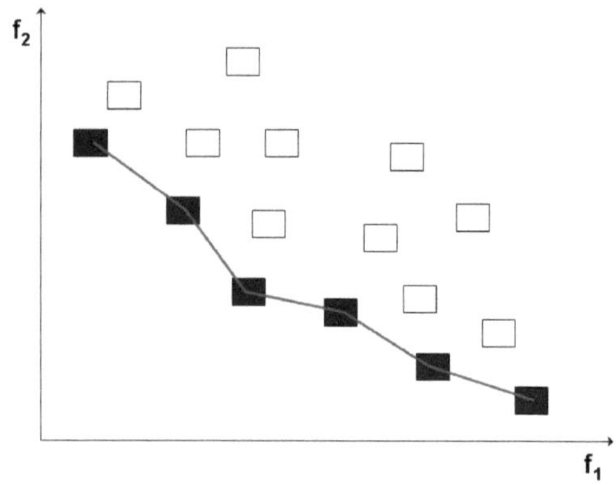

Fig. 10.3. Example of Pareto front

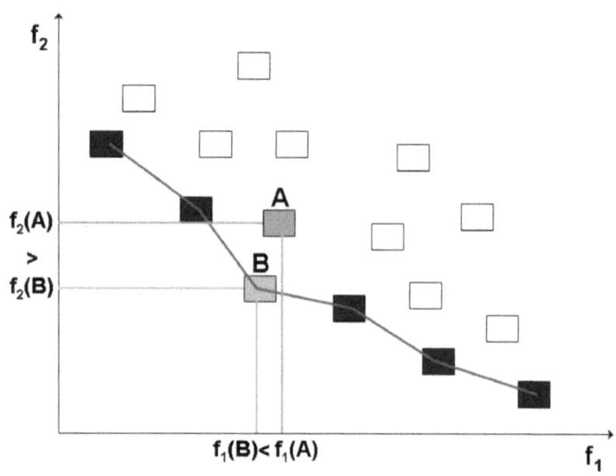

Fig. 10.4. Pareto dominance

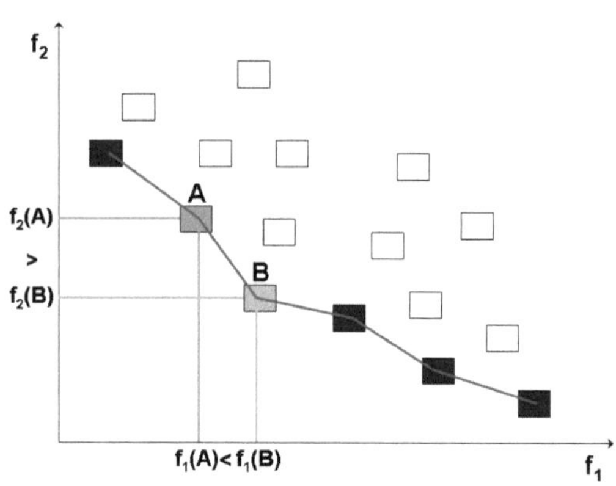

Fig. 10.5. Non-dominated vectors

10.3.2 MOO Approaches in VRPTW

In Section 10.2, the three objectives of the VRPTW were presented: total distance traveled, number of vehicles, and total waiting time. Many existing VRPTW techniques, however, are single objective-based heuristic methods that incorporate penalty functions, or combine the different criteria via a weighting function. Efficient solutions are difficult to obtain by integrating the objectives in a weighted sum. The difficulty with this is that this weighted sum necessitates the introduction of bias into both search performance and quality of solutions obtained. Finding an effective weighting for the multiple dimensions is hard, and often results in unsatisfactory performance and solutions.

The problem of VRPTW involves the optimization of routing multiple vehicles to meet all given constraints. It is required to minimize multiple conflicting cost functions concurrently, such as the travel distance and the number of vehicles. Therefore, the problem is best solved through a MOO approach.

In the last five years, some hybrid algorithms have been proposed to solve the VRPTW as a MOO problem. Tan et al. [29], and Ombuki et al. [30], employed the travel distance and the number of vehicles to be minimized. Chitty and Hernandez [31], tried to minimize the total mean transit time and the total variance in transit time. Murata and Itai [32] consider to minimize the number of vehicles and the maximum routing time among the vehicles in order to minimize the active duration of the central depot.

Other approaches report limited success at solving the VRPTW with multiple objectives. Saadah et al. [33], minimize the number of vehicles and the total distance traveled. One of the interesting results reported in this work is that the inclusion of a third objective, one that tries to maximize the customer waiting time, does produce a very significant improvement in the other two objectives. Recently, Moura [34] considers the number of vehicles, the total travel distance and volume utilization of the Vehicle Routing with Time Windows and Loading Problem (VRTWLP) as a MOO Problem.

This chapter proposes a hybrid particle swarm MOO algorithm that incorporates perturbation operators for keeping diversity in the evolutionary search. Pareto optimality is the principle adopted to facilitate the MOO of the VRPTW with 3 objectives: minimization of total travel distance, minimization of the number of vehicles, and minimization of total waiting time.

10.4 Particle Evolutionary Swarm for Multi-Objective Optimization (PESMOO)

Particle swarm optimization (PSO) algorithm is a population-based optimization technique inspired by the motion of a bird flock [35]. In the PSO model, every particle flies over a real valued n-dimensional space of decision variables $\boldsymbol{X}$. Each particle keeps track of its position $\boldsymbol{x}$, velocity $\boldsymbol{v}$, and remembers the best position ever visited, P_{Best}. The particle with the best P_{Best} value is called the leader, and its position is called global best, G_{Best}. The next particle's position is computed by adding a velocity term to its current position, as follows:

$$\boldsymbol{x}_{t+1} = \boldsymbol{x}_t + \boldsymbol{v}_{t+1} \tag{10.8}$$

The velocity term combines the local information of the particle with global information of the flock, in the following way.

$$\boldsymbol{v}_{t+1} = w * \boldsymbol{v}_t + \phi_1 * (P_{Best} - \boldsymbol{x}_t) + \phi_2 * (G_{Best} - \boldsymbol{x}_t) \tag{10.9}$$

The equation above reflects the socially exchanged information. It resumes PSO's three main features: distributed control, collective behavior, and local interaction with the environment [36]. The second term is called the cognitive component, while the last term is called the social component. w is the inertia weight, and ϕ_1 and ϕ_2 are called acceleration coefficients. The inertia weight indicates how much of the previous motion we want to include in the new one.

When the flock is split into several neighborhoods the particle's velocity is computed with respect to its neighbors. The best P_{Best} value in the neighborhood is called the local best, L_{Best}.

$$\boldsymbol{v}_{t+1} = w * \boldsymbol{v}_t + \phi_1 * (P_{Best} - \boldsymbol{x}_t) + \phi_2 * (L_{Best} - \boldsymbol{x}_t) \tag{10.10}$$

Neighborhoods can be interconnected in many different ways, some of the most popular are shown in Fig. 10.6. The star topology is, in fact, one big neighborhood where every particle is connected to each other, thus enabling the computation of a global best. The ring topology allows neighborhoods and it is therefore commonly used by local best PSO implementations.

PSO is a fast algorithm whose natural behavior is to converge to the best explored *local optima*. However, attainting flock's convergence to the *global optimum* with high probability implies certain adaptations. The approaches range from modifications to the main PSO equation, to the incorporation of reproduction operators.

For solving the VRPTW this chapter introduces the Particle Evolutionary Swarm Multi-Objective Optimization algorithm (PESMOO), a Particle Evolutionary Swarm Optimization () algorithm [38] with the following features:

- Perturbation operators to keep diversity. Although perturbations are not included in the original PSO model, they are quite common nowadays. The formal analysis of van den Bergh shows that the PSO algorithm will only converge to the best position visited, not the global optimum [37]. Therefore, the quest for high diversity is a sound approach to locate the global optimum since more diversity leads to increasing exploration. Figure 10.7 shows the P_{Best} position after applying the perturbation operators.
- Singly-linked ring topology. PESMOO creates several neighborhoods applying the ring topology used by PESO: the "Singly-Linked Ring" neighborhood (SLR) (see Fig. 10.8). Thus PESMOO promotes a flock with several local leaders to improve the exploration capacity [39].

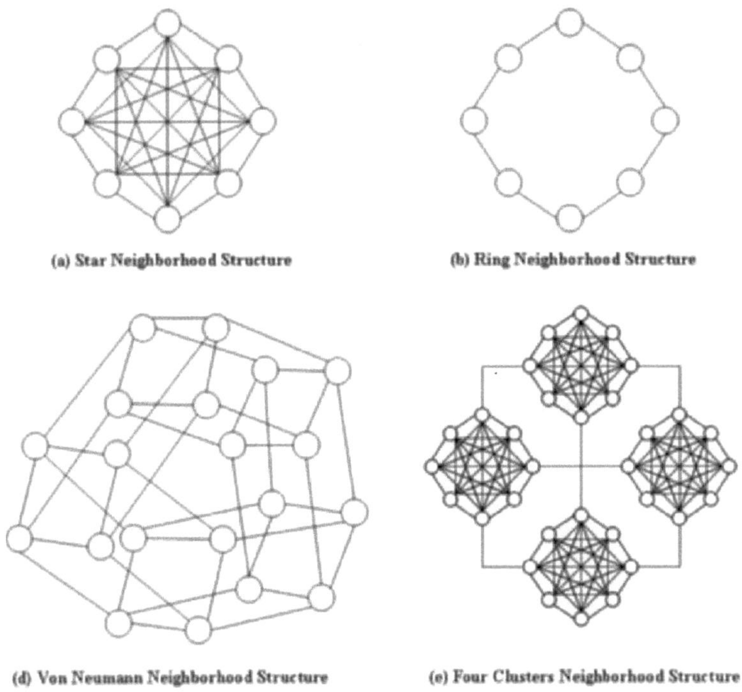

Fig. 10.6. Neighborhood structures for PSO

For each particle i, the members of a neighborhood of size k are selected as follows:

- $N_i = \emptyset$
- $Step = 1$
- $Switch = 1$
- **Repeat**
- $\quad N_i = N_i \cup P_{(i+Switch*Step)}$
- $\quad Step = Step + 1$
- $\quad Switch = -Switch$
- **Until** Size(N_i) = k

- Constraint handling. PSO lacks an explicit mechanism to bias the search towards the feasible region in constrained search spaces. For selecting a neighborhood leader, PESMOO picks the winner through a tournament that considers both feasibility and sum of constraint violations ("superiority of feasible solutions" [40]).
  1. Given two feasible particles, pick the one with better function value;
  2. From a pair of feasible and infeasible particles, pick the feasible one;
  3. If both particles are infeasible, pick the particle with the lowest sum of constraint violation.

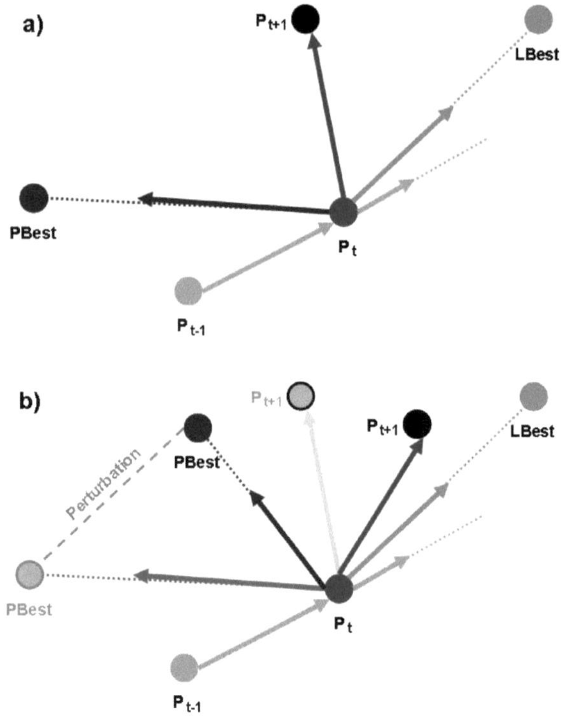

Fig. 10.7. Perturbation operators

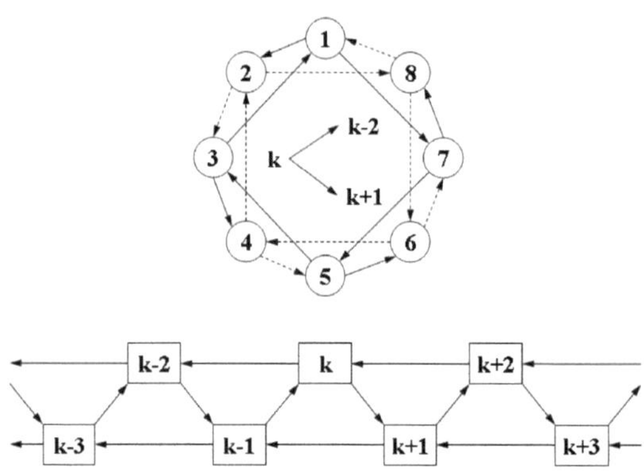

Fig. 10.8. Singly-linked ring neighborhood

The main PESMOO algorithm is shown in Fig. 10.9. A new approach based on *partition sector* is performed to solve the VRPTW. Details about this model, solution representation, perturbation operators, and adaptation of PESMOO for VRPTW are described along this section.

1. Generate P_0
2. Initialize $P_{Best} = P_0$
3. Find PF = non-dominated P_{Best}
4. P_{t+1} = Apply a PSO with **singly-linked ring** neighborhood to P_t
5. P_{Best} = *Feasibility tournament* P_{Best} vs P_{t+1}
6. PF = non-dominated $(P_{Best} \cup PF)$
7. P_C = Apply a **C-perturbation** to P_{Best}
8. P_{Best} = *Feasibility tournament* P_{Best} vs P_C
9. PF = non-dominated $(P_{Best} \cup PF)$
10. P_M = Apply a **M-perturbation** to P_{Best}
11. P_{Best} = *Feasibility tournament* P_{Best} vs P_M
12. PF = non-dominated $(P_{Best} \cup PF)$
13. $P_t = P_{t+1}$
14. Repeat steps 4 - 13 until stop criteria is met

Fig. 10.9. PESMOO algorithm

10.4.1 Spatial Data Mining: Sectors Model

The solution of a VRP instance is a set of least cost routes from one depot to a set of geographically scattered points. Every route is concerned with a customers subset located in the same neighborhood. A smart space partitioning of the customers set can save a lot of computational effort [41]. Spatial clustering can be applied to group similar spatial customers together; the implicit assumption is that patterns in space tend to be grouped rather than randomly located. A new spatial data mining technique for VRP, called **sectors model**, is proposed for clustering the customers according its spatial localization.

First, the customers are sorted according to its polar angle θ, geographical point in polar coordinates (r, θ) (see Fig. 10.10). Then, the customers and the depot can be inscribed by a circumference centered at D =depot, as shown in Fig. 10.11.

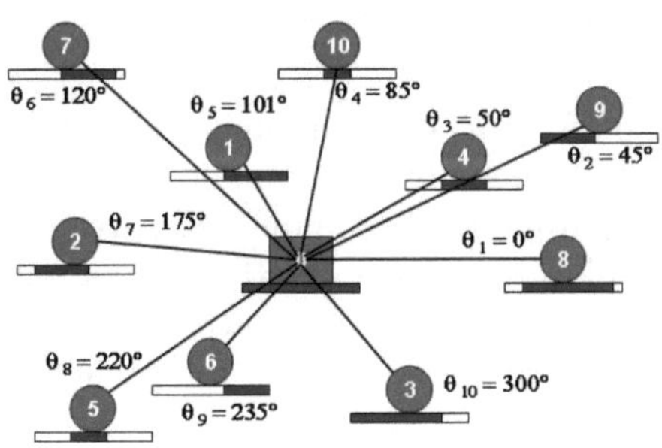

Fig. 10.10. Customer position in polar coordinates

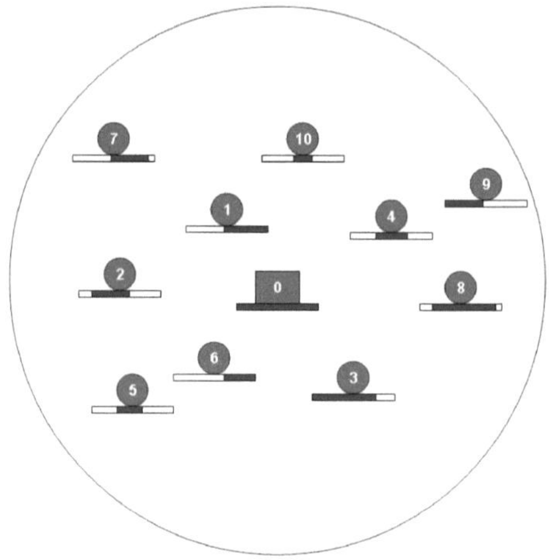

Fig. 10.11. VRP Representation

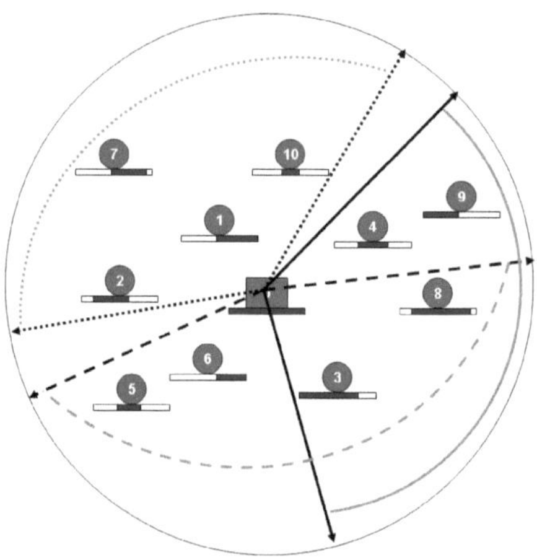

Fig. 10.12. Customers partitioned into sectors

Several approaches perform a cluster applying a Euclidean distance. We propose assign customers to vehicles by partitioning the customers into sectors according to their polar coordinates. The number of partitions (sectors) represents the number of vehicles allocated to find a VRPTW solution. The angles of sector limits (partition lines) are variables of the model. Figure 10.12 shows a partition of the customers into 3 sectors. The sectors can share a portion of the customer set as it is shown in Fig. 10.12.

Customers within each sector are routed minimizing distance and waiting time. Customers into sharing regions are assigned to only one sector with probability $p = 1/s_i$, where s_i is the number of sectors sharing customer i. Figure 10.13

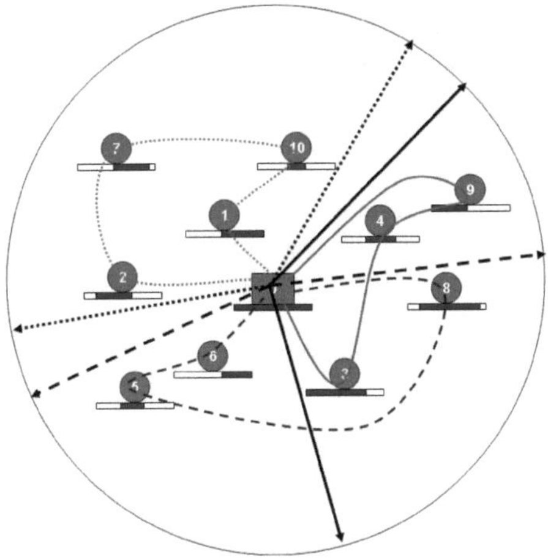

Fig. 10.13. Routes design

shows an instance of the posible customer assigment and routes obtained for the example shown in Fig. 10.12.

10.4.2 Representation

The sectors model (explained above) can be represented by a solution vector (particle) of dimension $d = 1 + 1.5 * \theta + n$

$$\boldsymbol{x} = \{x_0, x_1, \ldots, x_d\}$$

where x_0 represents the number of sectors (vehicles), components x_1 to x_θ represent the polar angle set ($\theta = 2 * x_0$), the components $x_{\theta+1}$ to $x_{\theta+n}$ are a permutation of the customer set $N = \{1, 2, \ldots, n\}$, and components $x_{\theta+n+1}$ to $x_{\theta+n+\theta/2}$ represent the route size set. The route size determines the end and the begining of each route in the customer set.

For instance, the next vector represents the solution shown in Fig. 10.13

$$\boldsymbol{x} = \{\underbrace{3}_{cars}, \underbrace{75, 190, 280, 60, 200, 15,}_{polar\ angles} \underbrace{\overbrace{2, 7, 10, 1,}^{route-1} \overbrace{3, 4, 9}^{route-2}, \overbrace{8, 5, 6}^{route-3},}_{customers} \underbrace{4, 3, 3}_{route\ sizes}\}$$

10.4.3 PESMOO for VRPTW

A PESMOO algorithm applying sectors model is performed to solve the VRPTW. Figure 10.14 presents the pseudocode of our proposal.

First, the number of partitions and the polar angle set are randomly generated. The number of partitions is limited to $maxcars = customers$,

1. Generate randomly $\boldsymbol{x}$ from x_0 to x_θ
2. $[\boldsymbol{y}, \boldsymbol{s}]$=Assign and count customers for sector set $\boldsymbol{x}$
3. $P_0 = \boldsymbol{x} \cup \boldsymbol{y} \cup \boldsymbol{s}$
4. Initialize $P_{Best} = P_0$
5. Find PF = non-dominated P_{Best}
6. $\boldsymbol{x}$ = Apply a PSO with SLR neighborhood to P_t from p_0 to p_θ
7. $[\boldsymbol{y}, \boldsymbol{s}]$=Reassign and count customers for sector set $\boldsymbol{x}$
8. $P_{t+1} = \boldsymbol{x} \cup \boldsymbol{y} \cup \boldsymbol{s}$
9. P_{Best} = *Feasibility tournament* P_{Best} vs P_{t+1}
10. PF = non-dominated $(P_{Best} \cup PF)$
11. P_C = Apply a **C-perturbation** to P_{Best}
12. P_{Best} = *Feasibility tournament* P_{Best} vs P_C
13. PF = non-dominated $(P_{Best} \cup PF)$
14. P_M = Apply a **M-perturbation** to P_{Best}
15. P_{Best} = *Feasibility tournament* P_{Best} vs P_M
16. PF = non-dominated $(P_{Best} \cup PF)$
17. $P_t = P_{t+1}$
18. Repeat steps 6 - 17 until stop criteria

Fig. 10.14. PESMOO algorithm for VRPTW

```
For k = 0 To swarmsize
  For j = 0 To θ
    r = U(0, 1)
    p1 = Random(n)
    p2 = Random(n)
    Temp[k, j] = P_Best[k, j] + r (P_Best[p1, j] - P_Best[p2, j])
  End For
  Temp[k,θ+1:d] = Assign and count costumers for sector set Temp[k,0 : θ]
  Apply fusion route on Temp[k,0:d]
End For
```

Fig. 10.15. Pseudocode of **C-Perturbation**

and $mincars = (totaldemand)/(vehiclecapacity)$. The polar angle limits are $0 \leq \theta \leq 360$. The customers are assigned to each route based on the sector partition generated.

After initialization, $P_{Best} = P_0$, and the current Pareto front PF is calculated, a PSO-SLR step is applied (10.9). The step is only performed in the first $\theta + 1$ variables of each particle. The customers are reassigned to each route based on the sector partition generated by the PSO-SLR step. These variables compose the P_{t+1} population. A feasibility tournament is performed between P_{Best} and P_{t+1} to update P_{Best} population. The current Pareto front PF is obtained from the non-dominated solutions of $PF \cup P_{Best}$.

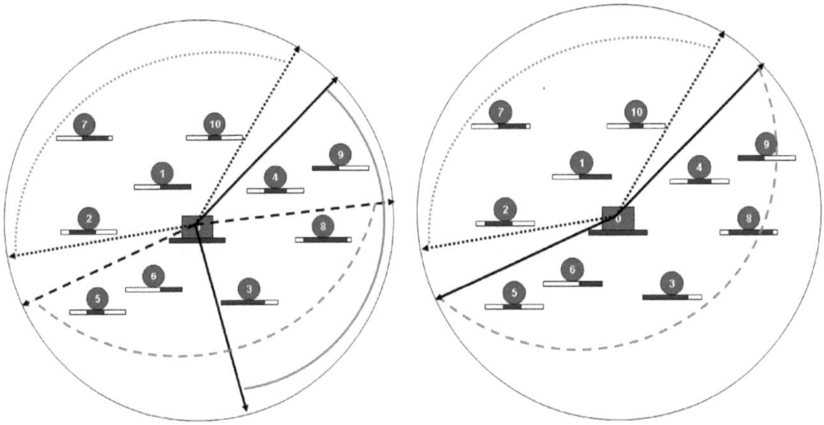

Fig. 10.16. Two sectors joined after applying the Fusion route step

For $k = 0$ **To** *swarmsize*
  For $j = 0$ **To** θ
    r = $U(0, 1)$
    If $r \leq 1/theta$ **Then**
      $Temp[k, j]$ = Rand(LL, UL)
    Else
      $Temp[k, j] = P_{Best}[k, j]$
  End For
  $Temp[k,\theta{+}1{:}d]$ = Assign and count costumers for sector set $Temp[k, 0 : \theta]$
  Apply *windows-time sorted* on $Temp$[k,0:d]
End For

Fig. 10.17. Pseudocode of **M-Perturbation**

Next, the first perturbation operator is performed on P_{Best}, called **C-perturbation**. The goal is to add a perturbation generated from the linear combination of three random vectors. This perturbation is preferred over other operators because it preserves the distribution of the population (also used for reproduction by the differential evolution algorithm [42]). Figure 10.15 shows the pseudocode of the **C-Perturbation** operator.

In the *fusion route* step (see Fig. 10.15), two sectors are joined giving place to one larger sector that reduces the number of partitions (vehicles). The customers of one route are inserted in the other route. Figure 10.16 presents an example of a fusion route process.

Finally, the second perturbation operator, **M-perturbation**, is performed on P_{Best}. Every vector is perturbed again so a particle could be deviated from its current direction as responding to external, maybe more promising, stimuli. This perturbation is implemented by adding small random numbers (from a uniform distribution) to the first $\theta + 1$ variables. Figure 10.17 shows the pseudocode of the **M-Perturbation** operator.

In the *windows-time sorted* step (see Fig. 10.17), the costumers into every sector are routed by its closing time b_i. It helps to accomplish time window contraints.

For every operator perturbation, a new population P_C and P_M is generated. After obtaining P_C and P_M, a feasibility tournament is performed to update P_{Best}. It perturbs the particle's memory and helps to avoid premature convergence.

10.5 Experiments

PESMOO uses a flock size of 100 members. PESMOO applies a PSO with parameters $c_1 = 1$ and $c_2 = 1$, and $w = U(0.5, 1)$. The total number of function evaluations is 1,000,000. A PC with Windows XP, C++ Builder Compiler, Pentium 4 processor running at 3.00GHz and 1GB of RAM was used for all our experiments. A well-know problem set proposed by Solomon [9] is used to test our model. In these problems, the travel times are equal to the corresponding Euclidean distances. Three problems were chosen for this experiment: problem C101 in dimensions 25, 50 and 100.

The Solomon's benchmark problems have been solved in the literature as a single objective problem. A few approaches have solved the Solomon's benchmark as a bi-objective problem, minimizing total distance and number of vehicles [29, 30]. Recently, Saadah et al. [33] solve this benchmark as a MOO problem with 3 objectives. Although the approach includes a third objective (maximize waiting time), it can not be considered a 3 objective problem. The third objective is not in conflict with the others, rather it is used as a tool to improve the results of the first objective (minimize total distance). The experiments presented below show a very interesting information about the optimal values reported with single objective approaches.

Figure 10.18 shows the Pareto front obtained by 10 runs of PESMOO for the problem C101 with 25 customers. The optimal route obtained by single objective approaches is marked with ***SO** in the figure. Figure 10.19 shows the projection of Pareto front on the planes: distance vs waiting time, distance vs cars, and waiting time vs cars.

The Pareto front obtained by PESMOO evidenced a huge waiting time accumulated by the optimal solution reported in the literature with single objective (total distance) approaches. The vector reports a total distance of 191.813620 units attained with only 3 cars, and a total waiting time of 2133.144826 units. As we mentioned above, in this problem, the travel times are equal to the corresponding Euclidean distances. Then, the total waiting time is at least 10 times greater than the total travel time performed for all routes. The total waiting time represents the 57.53% of the total available time (3708 units) of the 3 vehicles, if the depot's window time $[0, 1236]$ is taken as reference.

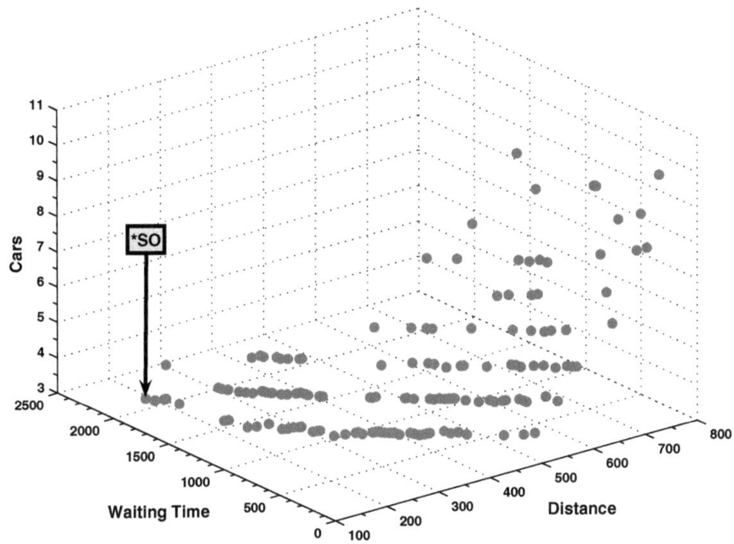

Fig. 10.18. Pareto front for problem C101 with 25 customers

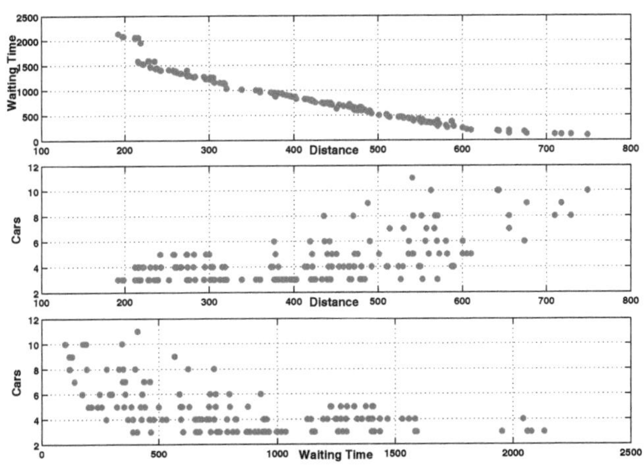

Fig. 10.19. Projected Pareto front for problem C101 with 25 customers

In Section 10.2 we explained that for eliminating any unnecessary waiting time, we assume that all routes start just-in-time $w_k = 0$, that is, we adjust the depot departure time of each vehicle $t_k = max(0, a_i - t_{0i})$, $i \in N$. Therefore, there is no waiting times at the beginning of the route.

Table 10.1 presents some non-dominated solutions of the Pareto set found by PESMOO for the problem C101 with 25 costumers. For example, the solution vector $(215.703725, 1584.387467, 4)$ presents a significant reduction in the

Table 10.1. Points in the Pareto front of problem C101 with 25 customers

Distance	Waiting Time	Vehicles	Distance	Waiting Time	Vehicles
191.813620	2133.144826	3	197.256586	2077.701861	3
215.703725	1584.387467	4	216.940011	1557.237840	4
259.554945	1386.403501	4	276.183581	1293.765800	3
305.926091	1158.898774	3	320.373466	1035.318654	3
359.858288	963.122881	3	377.999710	875.322687	5
402.591073	829.344048	3	421.982950	787.173144	4
443.570466	744.983131	3	450.638312	629.020781	5
480.144686	656.534658	3	511.430115	478.360000	4
526.659555	466.430155	3	550.021285	370.890864	4
570.268942	391.435229	3	581.731078	256.516096	5
589.610396	276.000750	4	609.975206	199.124629	5
655.237860	139.064867	7	673.520691	172.003649	6
717.592067	118.155401	9	748.901755	100.252129	10

total waiting time, but, the total distance and number of vehicles increased respect the solution vector (191.813620, 2133.144826, 3). The solution vector (359.858288, 963.122881, 3) has a waiting time of only 25.9% of the total available time (3708 units) of the three vehicles. Finally, the last solution vector (748.901755, 100.252129, 10) presents a reduction of 20 times the waiting time of the solution vector (191.813620, 2133.144826, 3), and it represents the 2.7% of the total available time. Nevertheless, this solution vector increases at least 3 times the total distance and the number of vehicles of the solution vector (191.813620, 2133.144826, 3).

Figure 10.20 shows the Pareto front obtained by 10 runs of PESMOO for the problem C101 with 50 customers. For the optimal solution reported in the literature with single objective approaches (363.246800, 3587.127672, 5), the total waiting time is 10 times greater than the total travel time performed for all routes and it represents the 58.04% of the total available time (6180 units) of the 5 vehicles, if the depot's window time $[0, 1236]$ is taken as reference.

Figure 10.21 shows the Pareto front obtained by 10 runs of PESMOO for the problem C101 with 100 customers. For the optimal solution reported in the literature with single objective approaches (***SO**) (828.936867, 6826.815787, 10), the total waiting time represents the 55.23% of the total available time (12360 units) of the 10 vehicles based on depot's window time $[0, 1236]$.

For the 3 dimensions of the problem C101, the total waiting time is over 50% of the available time of all vehicles occupied in each solution. In real-world problems, these levels of waiting times are not conforming to standard usage. Therefore, it is important to find a set of trade-off solutions to provide alternative solutions for satisfying the decision-maker's requirements.

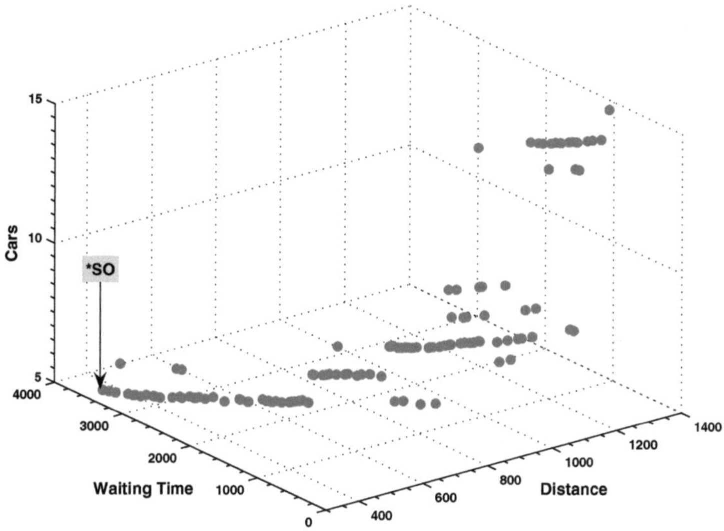

Fig. 10.20. Pareto front for problem C101 with 50 customers

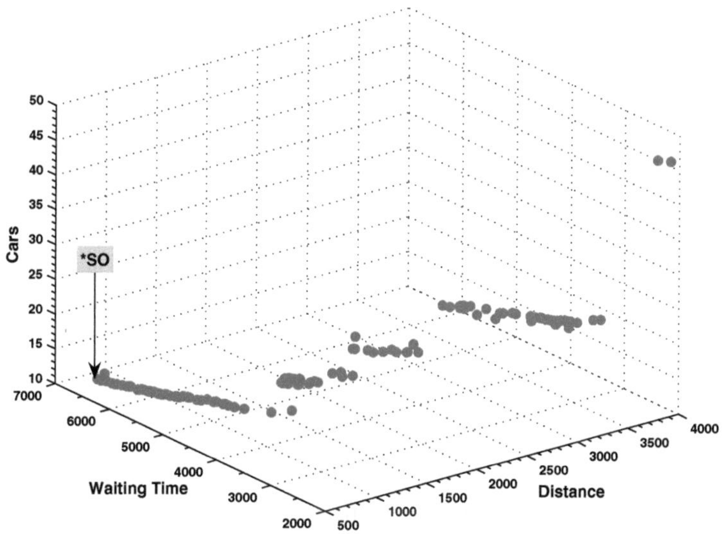

Fig. 10.21. Pareto front for problem C101 with 100 customers

10.6 Final Remarks

This chapter studies the VRPTW as a multi-objective optimization problem. Specifically, the three dimensions of the problem to be optimized are considered to be separate dimensions of a multi-objective search space: total distance traveled, total acumulated waiting time and number of vehicles engaged. Although

MOO approaches and evolutionary algorithms have both been applied to the VRPTW before, they only used two objectives. Solving the VRPTW as a MOO with 3 objectives involved the design of new operators that distribute customers into sectors. Our experiments indicate that waiting time must be included in the problem if more realistic solutions are pursued.

There are a number of advantages in using this literal MOO model of the VRPTW. First, by treating the number of vehicles, total distance, and total waiting time as separate entities, search bias is not introduced. Second, there is a strong philosophical case to be made for treating the VRPTW as a MOO problem. It is not necessary to numerically reconcile these problem characteristics with each another. In other words, we do not specify that either the number of vehicles or the total distance traveled take priority.

Most researchers clearly place priority on minimizing the number of vehicles. Minimizing the number of vehicles affects vehicle costs and fuel resources, while minimizing distance affects waiting times and vehicle availability. Therefore, the VRPTW is intrinsically a MOO in nature, and our MOO formulation recognizes these alternative solutions.

The Pareto front obtained by PESMOO represents a range of possible answers, with different numbers of vehicles, distances and waiting times. The choice of the preferable solution must be made by the decision-maker.

A simple algorithm to solve VRPTW problems is proposed in this chapter. PESMOO applies a spatial data mining technique for spatial clustering, called sectors model. The sectors model assigns customers to routes according to partitions performed in a polar coordinate system. This model is similar to a divide-and-conquer strategy. Combination of data mining and evolutionary approaches have shown a very good performance in other works [43].

The results of PESMOO are competitive with respect to other vehicle-biased results in the literature. However, the PESMOO algorithm is not producing detailed Pareto fronts in high dimensional problems. Our further work will be to look at how to improve the performance and the results of PESMOO for VRPTW through looking at different kinds of route elimination and customer ejection techniques.

References

[1] Dantzig, G., Ramser, R.: The truck dispatching problem. Management Science 6, 80–91 (1959)

[2] Salvelsberg, M.: Local search in routing problems with time windows. Annals of Operations Research 4, 285–305 (1985)

[3] Solomon, M.: Algorithms for the vehicle routing problem with time windows. Transportation Science 29(2), 156–166 (1995)

[4] Cordeau, J., Desaulniers, G., Desrosiers, J., Solomon, M., Soumis, F.: The VRP with time windows. In: Toth, P., Vigo, D. (eds.) The Vehicle Routing Problem. SIAM, Monographs on Discrete Mathematics and Applications, Philadelphia USA, pp. 157–193 (2002)

[5] Kolen, A., Rinnooy, K., Trienekens, H.: Vehicle routing with time windows. Operations Research 35, 256–273 (1987)
[6] Bard, J., Kontoravdis, G., Yu, G.: A branch-and-cut procedure for the vehicle routing problem with time windows. Transportation Science 36, 250–269 (2002)
[7] Laporte, G.: The vehicle routing problem: an overview of exact and approximate algorithms. European Journal of Operational Research 59, 345–358 (1992)
[8] Laporte, G., Semet, F.: Classical heuristics for the capacitated VRP. In: Toth, P., Vigo, D. (eds.) The Vehicle Routing Problem. SIAM, Monographs on Discrete Mathematics and Applications, Philadelphia USA, pp. 109–128 (2002)
[9] Solomon, M.: Algorithms for the vehicle routing and scheduling problems with time window constraints. Operations Research 35, 254–265 (1987)
[10] Russell, R.: An effective heuristic for the M-tour traveling salesman problem with some side conditions. Operations Research 25, 517–524 (1977)
[11] Baker, E., Schaffer, J.: Computational experience with branch exchange heuristics for vehicle routing problems with time window constraints. American Journal of Mathematical and Management Sciences 6, 261–300 (1986)
[12] Bräysy, O., Gendreau, M.: Vehicle routing problem with time windows, Part I: route construction and local search algorithms. Transportation Science 39(1), 104–118 (2005)
[13] Gendreau, M., Laporte, G., Potvin, J.: Metaheuristics for the VRP. In: Toth, P., Vigo, D. (eds.) The Vehicle Routing Problem. SIAM, Monographs on Discrete Mathematics and Applications, Philadelphia USA, pp. 129–154 (2002)
[14] Kontoravdis, G., Bard, J.: A GRASP for the vehicle routing problem with time windows. ORSA Journal on Computing 7, 10–23 (1995)
[15] Potvin, J., Kervahut, T., Garcia, B., Rousseau, J.: The vehicle routing problem with time windows - Part I: tabu search. INFORMS Journal on Computing 8, 158–164 (1996)
[16] Cordeau, J., Laporte, G., Mercier, A.: A unified tabu search heuristic for vehicle routing problems with time windows. Journal of the Operational Research Society 52, 928–936 (2001)
[17] Bent, R., Van Hentenryck, P.: A two-stage hybrid local search for the vehicle routing problem with time windows. Technical Report CS-01-06, Computer Science Department, Brown University (2001)
[18] Gambardella, L., Taillard, E., Agazzi, G.: MACS-VRPTW: a multiple ant colony system for vehicle routing problems with time windows. In: Corne, D., Dorigo, M., Glover, F. (eds.) New Ideas in Optimization, pp. 63–76. McGraw-Hill, New York (1999)
[19] Blanton, J., Wainwright, R.: Multiple vehicle routing with time and capacity constraints using genetic algorithms. In: Forrest, S. (ed.) Proceedings of the 5th International Conference on Genetic Algorithms, pp. 452–459. Morgan Kaufmann, San Francisco (1993)
[20] Thangiah, S.: Vehicle routing with time windows using genetic algorithms. In: Chambers, L. (ed.) Application Handbook of Genetic Algorithms: New Frontiers, vol. II, pp. 253–277. CRC Press, Boca Raton (1995)
[21] Osman, I.: Metastrategy simulated annealing and tabu search algorithms for the vehicle routing problems. Annals of Operations Research 41, 421–452 (1993)

[22] Bräysy, O.: A new algorithm for the vehicle routing problem with time windows based on the hybridization of a genetic algorithm and route construction heuristics. In: Proceedings of the University of Vaasa, Research papers, Vaasa, Finland (1999)
[23] Homberger, J., Gehring, H.: Two evolutionary metaheuristics for the vehicle routing problem with time windows. INFOR 37, 297–318 (1999)
[24] Or, I.: Traveling salesman-type combinatorial problems and their relation to the logistics of regional blood banking. Ph.D. Thesis, Northwestern University, Evanston, USA (1976)
[25] Potvin, J., Rousseau, J.: An exchange heuristic for routeing problems with time windows. Journal of the Operational Research Society 46, 1433–1446 (1995)
[26] Ghosh, A., Dehuri, S.: Evolutionary algorithms for multi-criterion optimization: a survey. International Journal of Computing & Information Sciences 2(1), 38–57 (2004)
[27] Pareto, V.: Cours d'Economie Politique, vol. I - II (1896)
[28] Coello, C.: A short tutorial on evolutionary multiobjective optimization. In: Proceedings of the First International Conference on Evolutionary Multi-Criterion Optimization, pp. 21–40 (1993)
[29] Tan, K., Chew, Y., Lee, L.: A hybrid multiobjective evolutionary algorithm for solving vehicle routing problem with time windows. Computational Optimization and Applications 34(1), 115–151 (2006)
[30] Ombuki, B., Ross, B., Hanshar, F.: Multi-objective genetic algorithms for vehicle routing problem with time windows. Applied Intelligence 24(1), 17–30 (2006)
[31] Chitty, D.M., Hernandez, M.L.: A hybrid ant colony optimisation technique for dynamic vehicle routing. In: Deb, K., et al. (eds.) GECCO 2004. LNCS, vol. 3102, pp. 48–59. Springer, Heidelberg (2004)
[32] Murata, T., Itai, R.: Multi-objective vehicle routing problems using two-fold EMO algorithms to enhance solution similarity on non-dominated solutions. In: Coello Coello, C.A., Hernández Aguirre, A., Zitzler, E. (eds.) EMO 2005. LNCS, vol. 3410, pp. 885–896. Springer, Heidelberg (2005)
[33] Sa'adah, S., Ross, P., Paechter, B.: Improving vehicle routing using a customer waiting time colony. In: Gottlieb, J., Raidl, G.R. (eds.) EvoCOP 2004. LNCS, vol. 3004, pp. 188–198. Springer, Heidelberg (2004)
[34] Moura, A.: A multi-objective genetic algorithm for the vehicle routing with time windows and loading problem. In: Bortfeldt, A., Homberger, J., Kopfer, H., Pankratz, G., Strangmeier, R. (eds.) Intelligent Decision Support, Current Challenges and Approaches, pp. 187–201. Gabler-Verlag, Wiesbaden (2008)
[35] Kennedy, J., Eberhart, R.: Particle swarm optimization. In: Proceedings of IEEE International Conference On Neural Networks, pp. 1942–1948 (1995)
[36] Eberhart, R., Dobbins, R., Simpson, P.: Computational Intelligence PC Tools. Academic Press Professional (1996)
[37] Van den Bergh, F.: An Analysis of Particle Swarm Optimizers. University of Pretoria, South Africa (2002)
[38] Muñoz, A., Hernández, A., Villa, E.: Robust PSO-based constrained optimization by perturbing the particle's memory. In: Chan, F., Tiwari, M. (eds.) Swarm Intelligence, Focus on Ant and Particle Swarm Optimization, pp. 57–76. I-Tech Education and Publishing, Vienna (2007)
[39] Kennedy, J., Eberhart, R.: The Particle Swarm: Social Adaptation in Information-Processing Systems. McGraw-Hill, London (1999)
[40] Deb, K.: An efficient constraint handling method for genetic algorithms. Computer Methods in Appplied Mechanics and Engineering 186(2-4), 311–338 (2000)

[41] Ester, M., Kriegel, H., Sander, J.: Algorithms and applications for spatial data mining. In: Miller, H., Han, J. (eds.) Geographic Data Mining and Knowledge Discovery, pp. 160–187. Taylor & Francis, London (2001)
[42] Price, K., Storn, R., Lampinen, J.: Differential Evolution: A Practical Approach to Global Optimization. Springer, Berlin (2005)
[43] Dehuri, S., Ghosh, A., Mall, R.: Genetic algorithms for multi-criterion classification and clustering in data mining. International Journal of Computing & Information Sciences 4(3), 143–154 (2006)

11

Combining Correlated Data from Multiple Classifiers

Lisa Osadciw, Nisha Srinivas, and Kalyan Veeramachaneni

Department of Electrical Engineering and Computer Science
Syracuse University
Syracuse, NY 13244, USA
{laosadci,nsriniva,kveerama}@syr.edu

Summary. Real time data collected from sensors and then having been processed by multiple classifiers are correlated due to common, inherent noise resulting from the sensor. The data is collected at different ranges and may have different statistical distributions and characteristics. Measurements from different classifiers are fused together to obtain more information about the phenomenon or environment under observation. Since the classifier fusion is attempting to improve the data through processing only, this chapter focuses on object refinement or fusion level 1 within the Joint Directors of Laboratories (JDL) data fusion process model and taxonomy of algorithms. Before fusion, the correlated data is normalized using a variety of normalization techniques such as z-score and min-max normalization. Further, by dynamically weighting and combining the measurements from these classifiers, performance is improved. Weights are found using a Particle Swarm Optimization search algorithm and are a function of required accuracy and degree of correlation. Results are presented for (a) synthetic data generated using conditional multivariate normal distribution with different covariance matrices and (b) NIST BSSR dataset. For correlated classifiers, the performance of the particle swarm optimization technique outperforms the traditional data level fusion i.e., z-normalization and min-max.

11.1 Introduction

The recent proliferation of sensor networks in civilian applications attempt to detect a signal in noise remotely. The signal of interest and type of noise depends on the application. For example, in radars, the signal is the transmitted pulse reflected by the target object, most often an aircraft or at times rain for weather detection. There are sensor networks responsible for detecting cracks in pipelines, weaknesses in buildings, bridges, etc. People can not be at these sites where these problems occur nor make the accurate measurements needed. Traditionally, the design of detection systems is the empirical selection of a single classifier following an experimental evaluation of a large variety of classifiers. The parameters of the selected classifiers are optimized to meet specified performance requirements. Single classifier systems have limited performance capabilities for applications operating in a wide variety of situations and environments. Thus, a different approach is needed altogether even after all classifier parameters and/or architectures are optimized. Classifier ensembles, one of the most

C.A. Coello Coello et al. (Eds.): Swarm Intel. for Multi-objective Prob., SCI 242, pp. 259–281.
springerlink.com

significant advances in pattern classification in recent years, prove to be more effective and efficient than a single classifier [13, 15, 17]. A classifier ensemble is a set of individually trained classifiers whose individual decisions/measurements are combined in some way to form the final decision. By taking advantage of complementary information provided by the constituent classifiers, we improve system performance beyond the best individual classifier. This involves combining and processing the data extracted from multiple classifiers, i.e., data fusion. Some of the primary advantages of data fusion are [2]:

1. More information is obtained about the object of interest, which may include aircraft, birds, bugs, rain, people, cars, etc.
2. Better spatio-temporal coverage.
3. Reducing ambiguity.
4. Increasing the dimensionality of the data space.

For example, let us assume that each classifier gives a measurement, x, which is a distance measure from the class of interest. For example, neural networks (NN) give a measurement x in support of class H_h. The range of the x given by a NN can vary significantly based on the parameters and the transition functions. In Fig. 11.1, two classifier measurements in support of two classes, class 1, absence of a phenomenon, class 2, presence of phenomena, for 100,000 observations are given. The *x-axis* corresponds to a measurement given by the first classifier, and the *y-axis* corresponds to measurement given by the second classifier. The gray 'x' are measurements by the classifiers when class 1 is present, and the black 'o' are measurements by the classifiers when class 2 is present. The measurements corresponding to both class l and class 2 overlap, making it impossible to separate the two. This is the primary reason for errors.

Data fusion can be divided into two categories, Centralized data fusion and Decentralized data fusion, often determined by constraints on the individual node's processing power, network's communication bandwidth, fused data accuracy, and time delay. Fusion of multiple classifiers can be done at different locations in the network as well as different processing impacting time delay and accuracy as shown in Fig. 11.2.

In the first fusion level illustrated at the bottom of the Fig. 11.2, the raw measurements, x, given by the classifiers, are fused first. This may be done using any fusion algorithm such as the averaged sum rule or, leading to greater complexity, another classifier based on support vector machines, linear discriminant analysis or decision trees.

In the second level shown in the middle of Fig. 11.2, the soft decisions from the individual classifiers are fused. Some classifiers output soft decisions, which in the form of posterior probabilities. Posterior probabilities can be derived from the likelihood probability density functions (normalized and conditioned using the measurements) under both hypotheses for the simple binary classification or detection.

In hard decision fusion as shown at the top of Fig. 11.2, a threshold is applied to each classifier's measurement, resulting in a binary decision as to which of 2 classes it belongs to. Then, the fusion processor fuses the decisions from multiple

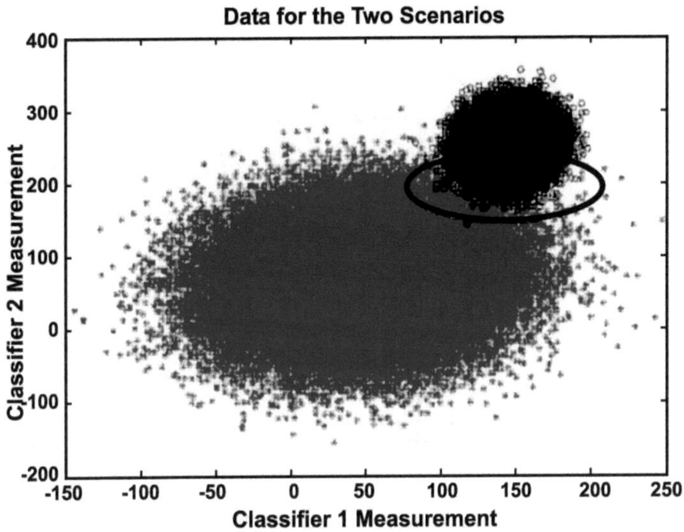

Fig. 11.1. Output data from the two classifiers

classifiers. The fusion processor uses one of the many fusion rules to fuse the multiple decisions into a global decision. The majority voting rule (MVR) or Chair-Varshney [16] (CVR) optimal fusion rule are used producing a global decision. The MVR selects one class from the two that has been chosen by the majority of classifiers. In contrast, the Chair-Varshney rule approximates the optimum fusion rule by estimating the constituent classifier's expected random error and forms a robust and more accurate rule.

Recent approaches in classifier fusion have shown performance advantages of data level fusion ([6, 7]) and, more specifically, averaged sum rule under the assumption of statistical independence. Statistical independence implies that the observations from multiple classifiers do not correlate to each other even if they are observing the same class. This is not a realistic assumption, since classifiers are designed using the same image or data. In this chapter, we focus on correlated classifiers and design simple weighting functions that improve the performance of the system. For a binary classification problem (two classes), there are two kinds of errors resulting from the two classes: class 1 declared when class 2 is present and vice-versa. This is a two hypotheses problem. As the number of classes increases, the number of hypotheses increases. This increases the number of possible errors and is equal to ${}^{N}C_2$, where N is the number of classes. We design a particle swarm optimization (PSO) algorithm to design the weights for the classifier measurements minimizing the two hypotheses errors for a 2-class problem.

The chapter is organized as follows. Section 11.2 presents the centralized fusion process. The fusion process is detailed, and the two errors are presented. The problems in the centralized fusion system, which are disparate and correlated measurements, are detailed in this section. Techniques to normalize the measurements are presented in this section. The weighting function to combine

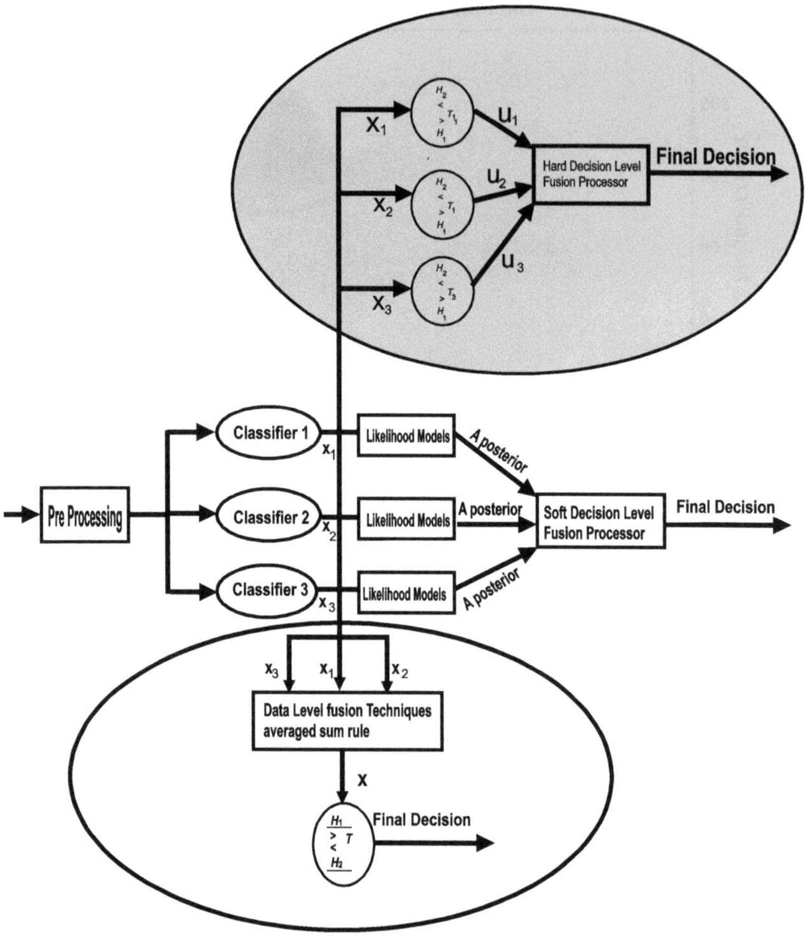

Fig. 11.2. Fusion levels for classifier measurements

the normalized measurements is presented in this section. Section 11.3 presents the multi-objective method adopted to solve the problem. Section 11.4 describes the design of the particle swarm optimization algorithm. Section 11.5 presents the results on the bi-variate and multivariate normally distributed dataset. A Biometrics application is presented in Section 11.6. The results achieved on the Biometrics dataset are presented in this section. Finally conclusions are presented in Section 11.7.

11.2 Centralized Fusion

The binary hypotheses problem of detecting a phenomenon is

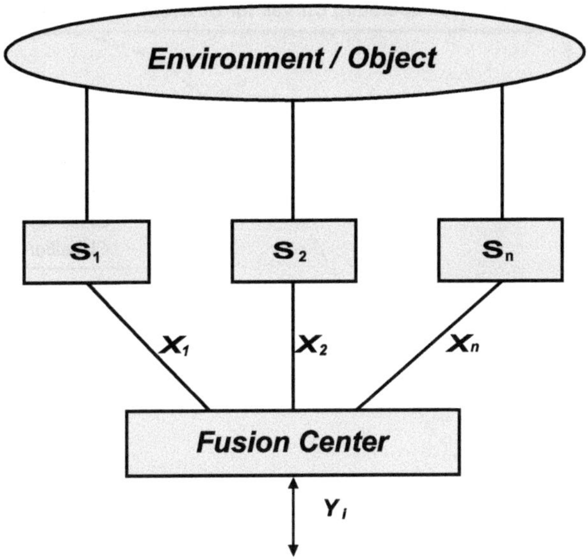

Fig. 11.3. Centralized fusion

- $\mathbf{H_0}$: absence of the phenomena;
- $\mathbf{H_1}$: presence of phenomena.

Given measurements from multiple classifiers, $[X] = (x_1, x_2, \ldots, x_n)$, a fusion function combining these scores is given by

$$z = f([X]) \tag{11.1}$$

This result is compared to a threshold determining the presence or absence of phenomena as in:

$$u_f = \begin{cases} 1 & z \geq \lambda \\ 0 & z < \lambda \end{cases} \tag{11.2}$$

where λ is the decision threshold value and u_f is 0 or 1, decision with respect to the presence or absence of a phenomena. A decision threshold on 'z' can be derived using a likelihood ratio test (LRT) as in

$$\log \frac{P(z|H_1)}{P(z|H_0)} \underset{H_1}{\overset{H_0}{\lessgtr}} \log \frac{P(H_1) \times C_{H_0|H_1}}{P(H_0) \times C_{H_1|H_0}} \tag{11.3}$$

where, H_1 is classification 1, H_0 is classification 0, and $C_{H_0|H_1}$ is the cost of making the error of choosing classification 0 when it should be classification 1. Each System can make two errors: the false alarm error, when presence of phenomena is falsely declared, and a miss, when absence of phenomena is declared

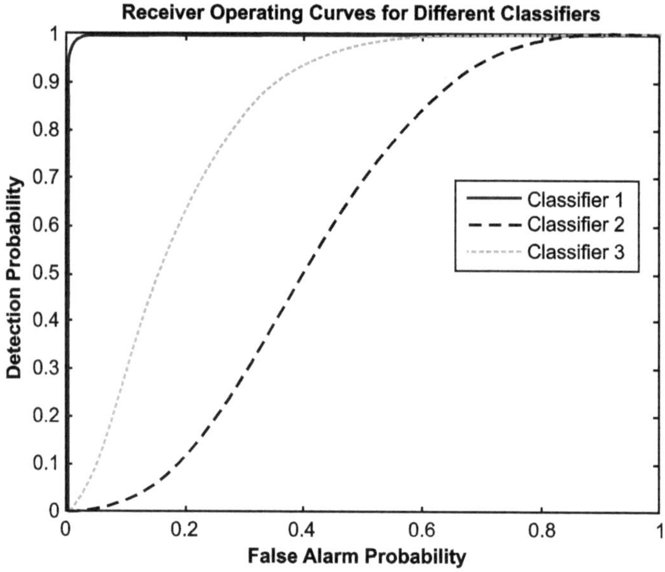

Fig. 11.4. Receiver operating curves for different classifiers

even though the phenomena is present. These are known as probability of false alarm and probability of miss and are given by

$$P_{FA} = P(u_f = 1|H_0), \tag{11.4}$$

and

$$P_M = P(u_f = 0|H_1). \tag{11.5}$$

A trade-off curve giving different sets of P_{FA} and $(1-P_M)$ is known as receiver operating characteristic curve (ROC). A simple average of the scores given by

$$z = \frac{1}{m}\sum_m x_m \tag{11.6}$$

where m is the number of classifiers, and x_m is the measurement from the mth classifier, has been found to be a beneficial fusion function for identically distributed classifiers [2]. However, as seen in Fig. 11.1, the two classifiers differ in measurement range and statistical distribution. An important metric for classifier performance is the area under the ROC curve. As an example, the ROCs of three different classifiers are shown in Fig. 11.4. The best performing classifier is classifier 1. Since different classifiers have varying levels of accuracy, the weights we use to combine the classifiers output impact and, hopefully, improve overall performance.

Data normalization is paramount for classifiers giving measurements from different ranges. Different methods are used to normalize the data such as *z-normalization*, *min-max normalization* [2], or *t-normalization*. A threshold is applied to the combined, normalized value, and then a decision is made regarding the presence or absence of threat/target.

Finally, because data from multiple classifiers is correlated, performance decreases. Even if all the classifiers are identical in performance independently, combining the least correlated classifier results in better performance. This is due to the addition of new information into the system. Correlation can be considered as a measure of how much new information is being introduced to the classification system. A correlation of 100% indicates identical inputs and identical processing so that the output is identical as well. Traditionally, a subset of least correlated classifiers is chosen from a classifier-suite to perform the fusion. In this chapter we make use of all the classifiers in the classifier suite. We adapt the weights for different correlation factors using PSO resulting in higher performance.

11.2.1 Data Normalization

Data normalization is a process of transforming the measurements from different, heterogeneous classifiers into a common range by altering them. Various types of normalization techniques exist such as *z-norm*, *min-max*, *tanh*, *decimal scaling*, *median*, *median absolute* (MAD), and the sigmoid function based normalization [2]. In this chapter, we use *z-norm* and *min-max* normalization. *Z-norm* is preferred over other normalization techniques because it is easily adapted to other standard problems and has the ability to minimize distortion. *Min-Max* is also used since it performs better for Gaussian distributions and is easy to implement since only the minimum and maximum values are required.

Z-Normalization

Z-norm normalizes the measurements using the mean and standard deviation computed from the training data. The normalized measurement for a classifier is obtained by

$$x_m^{'} = \frac{x_m - \mu_m}{\sigma_m} \tag{11.7}$$

where x_m is the measurement from the mth *classifier.* μ_m and σ_m are the mean and standard deviation estimated from the data. Once the measurements from the classifiers are transmitted to the fusion center, they are normalized according to the above transformation and combined into a single measurement using the average sum rule as in equation (11.6). *Z-norm* performs well if we have prior knowledge about the mean and standard deviation of the data associated with a classifier. If it is not available then the means and standard deviations of the data have to be estimated from the training data. It does not maintain the same distribution at the output if the data distribution is not Gaussian distorting it. This method is not robust since both mean and standard deviations are sensitive to outliers. This method is useful when the minimum and maximum values of the measurements are unknown [2].

Min-Max Normalization

Min-max, another type of normalization technique, performs linear transformation on the data so that it does not change the initial distribution type. It is used when the maximum and minimum values of the data produced by the classifiers are known. Each data point is normalized using the transformation,

$$x_m^{'} = \frac{x_m - \min}{\max - \min} \tag{11.8}$$

where *min* and *max* are the minimum and maximum values of the data. This method retains the statistical distribution and only scales and transforms the data into a common numerical range between [0, 1]. It performs best for Gaussian distributions. This method is not robust if the minimum and maximum values are estimated from the data.

Table 11.1. Parameters for the Marginal Normal Distributions under H_0 and H_1.

Parameter	Classifier 1	Classifier 2	Classifier 3
Mean, μ_0,: Imposter	47.3	67.7	50.41
Standard Deviation, σ_0: Imposter	43.8	52.6	26.2
Mean, μ_1: Genuine	144.5	251.2	167.4
Standard Deviation, σ_1: Genuine	12.843	23	10.18

11.2.2 Data Correlation

Correlation can be best explained by using data generated for different correlation factors demonstrating the affect of correlation. Fig. 11.5 shows an example measurement dataset generated using a bi-variate normal distribution assumed under both hypotheses. The bi-variate distribution under H_0 is given by

$$f_{H_0}(x_1, x_2) = \frac{1}{2\pi\sigma_{x_1}\sigma_{x_2}\sqrt{1-\rho_{H_0}^2}} \exp\left[\frac{-1}{2(1-\rho_{H_0}^2)}\left[\begin{array}{l}\left(\frac{x_1-\mu_x}{\sigma_{x1}}\right)^2 + \left(\frac{x_2-\mu_{x2}}{\sigma_{x2}}\right)^2 \\ -2\rho\left(\frac{x_1-\mu_{x1}}{\sigma_{x1}}\right)\left(\frac{x_2-\mu_{x2}}{\sigma_{x2}}\right)\end{array}\right]\right] \tag{11.9}$$

The parameters for the marginal distributions to be used for data generation are given in Table 11.1. Along the *x-axis* are the measurements by the first classifier and along the *y-axis* are the measurements from the second classifier. The black circles are the 100,000 samples pertaining to class 2 (H_1) and the gray 'x's are the 100,000 samples pertaining to class 1 (H_0). The overlap between these two classes of samples, i.e between H_1 and H_0, is the source of error. Figs. 11.5 and 11.6 show the effect of correlation. Fig. 11.5 shows samples for symmetric correlation, i.e., $\rho_{H_0} = \rho_{H_1}$. Fig. 11.6 shows samples generated with asymmetric correlation, i.e., $\rho_{H_0} \neq \rho_{H_1}$. When the data is symmetrically correlated, the overlap is lesser when compared to asymmetrically correlated data [16]. In both cases, as the value of correlation increases, the overlap in the 2-D region increases, resulting in a high error.

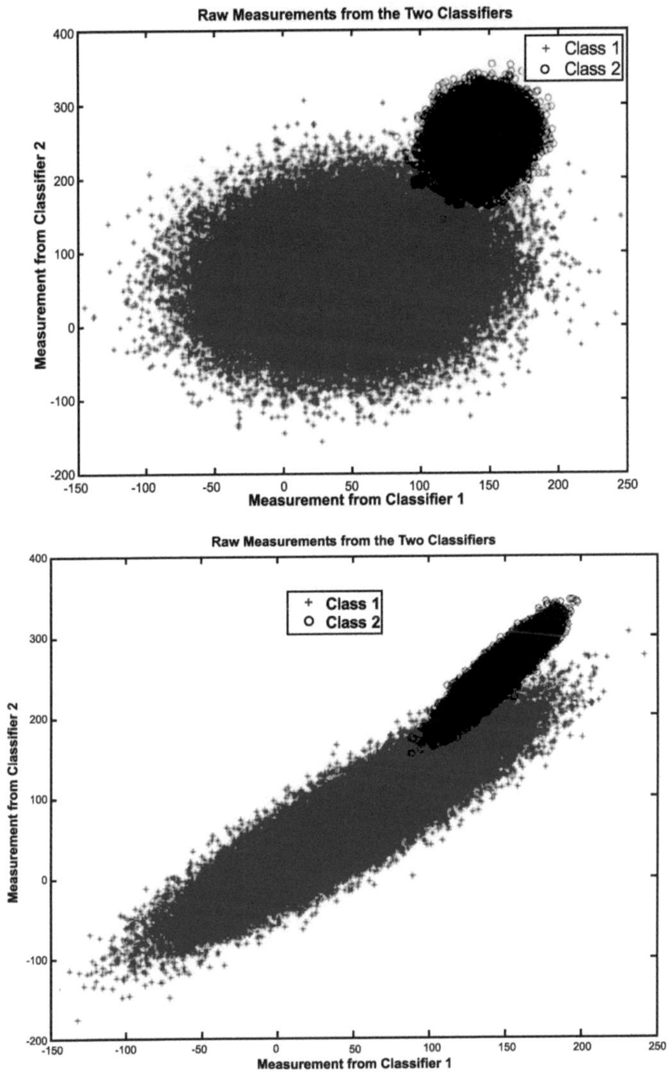

Fig. 11.5. Symmetrically correlated data under both hypotheses: (a) correlation of 0.1, and (b) Correlation of 0.9

11.2.3 Weighting to Enhance Performance

Weighting can enhance performance by mitigating the effects of correlation shown in Figs. 11.5 and 11.6. The weighting function provides a very simple alternative and transforms the original fusion function to

$$z = \frac{\sum_{m=1}^{n} w_m x'_m}{\sum_{m=1}^{n} w_m} \tag{11.10}$$

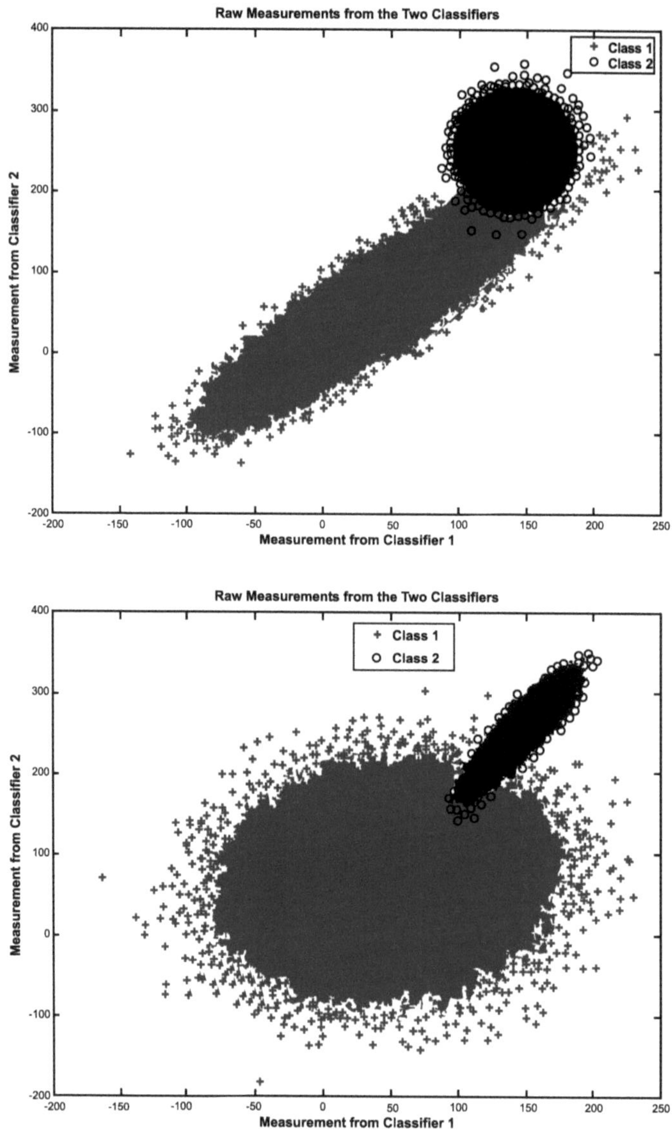

Fig. 11.6. Asymmetrically correlated data under both hypotheses

where x'_m is the normalized data point from the mth classifier. The denominator in equation (11.10) normalizes the weight set. In this chapter, we present a PSO based design of these weights for the multi-classifier fusion system which enhances the system performance. The two errors presented in the previous section conflict each other making a trade-off between the two necessary. In the next section we detail the two multi-objective methods that can be used for this problem.

11.3 Multi-objective Optimization Problem

Minimization of equations (11.4) and (11.5) form a classic multi-objective problem with two conflicting objectives. Traditionally, different methods have been employed to solve multi-objective optimization problems. The goal in a multi-objective optimization problem is to achieve a compromise solution for the different objectives. For our multi-objective optimization problem we call the solution vector as D defined by

$$D = \{w_1, w_2, \ldots, w_n\} \tag{11.11}$$

Let 'f_1' and 'f_2' be the functions which give the value for the 'PFA' and 'PM', respectively.

$$e_1 = f_1(D) \tag{11.12}$$

$$e_2 = f_2(D) \tag{11.13}$$

Most multi-objective optimization algorithms use the concept of domination. In the algorithms, two solutions are compared on the basis of whether one solution dominates other or not. The performance for D is used to determine the dominance of one solution over another. The definition for dominance is described in the *Non-dominance Definition.* An entire set of solutions may be non-dominant as described in the *"Non-dominant Solution Set Definition"*. The non-dominant solution set becomes the Pareto-optimal set when the solutions cover the entire design space as defined in *"Pareto-Optimal Set Definition"*. These definitions lay the foundation for Pareto optimality and global Pareto optimality, key concepts in multi-objective optimization.

Non-dominance Definition: A solution $D^{(1)}$ is said to dominate the other solution $D^{(2)}$ if and only if the two following conditions are true:

1. The solution $D^{(1)}$ is no worse than $D^{(2)}$ in all the objectives, or

$$f_j(D^{(1)}) \leq f_j(D^{(2)}) \ \ \forall j \tag{11.14}$$

2. The solution $D^{(1)}$ is strictly better than $D^{(2)}$ in at least one objective, or

$$\exists j f_j(D^{(1)}) < f_j(D^{(2)}) \tag{11.15}$$

Non-dominant Solution Set Definition: A solution set Q is called non-dominated set if no solutions in the set are dominated by any other solutions in the set.

Pareto-Optimal Set Definition: The non-dominated set of solutions of the entire feasible design space S is the global Pareto-Optimal Set.

The goal of multi-objective optimization algorithms is to support decision making through design compromises while still remaining on the Pareto optimal curve. Different methods are employed to achieve these solutions as detailed in the next section. Two common approaches are the method of weighted objectives and the e-constraint methods. In classical detection theory, these methods are employed and are known as minimum Bayesian error criterion and Neymen Pearson criterion, respectively.

11.3.1 Method of Weighted Objectives

One of the ways to solve this multi-objective optimization problem is by method of weighted objectives. The total error formulated is also known as Bayesian error and is also defined as

$$R = C_{FA} \times P(H_0) \times P_{FA} + C_{FR} \times P(H_1) \times P_{FR} \tag{11.16}$$

where, C_{FA} and C_{FR} are the cost of false acceptance and rejection; and $P(H_0)$ and $P(H_1)$ are the prior probabilities of class 1 and class 2. Different points on the Pareto curve can be obtained by varying the costs and evolving the weights for each cost structure. The weights for the fusion function are obtained using a Particle swarm optimization algorithm. PSO modifies the weights of the fusion function for various correlation factors, C_{FA}.

11.3.2 E-Constraint Method

Alternatively, the e-constraint method can be applied. In detection theory, this is known as Neymen Pearson criteria and is given by

$$\chi = P_{FA} + \beta \cdot (P_M - \alpha) \tag{11.17}$$

where β is the Lagrange multiplier and α is the requirement set for P_M. In this method, PSO can be used to achieve weights as a function of the requirement. Solutions achieved for different requirements result in different points on the Pareto curve.

11.3.3 Multi-objective Evolutionary Algorithms

Researchers have also designed algorithms that can simultaneously achieve different solutions on the Pareto front, known as evolutionary multi-objective optimization algorithms [10]. These algorithms evolve the particles in the swarm towards the Pareto front. These algorithms are highly beneficial when prior knowledge about the problem requirements is not available. For example, if the cost the system incurs due to the two errors is not available to apply the method of weighted objectives or if the requirement on one of the objectives is not available to apply the e-constraint method, one can resort to evolve the entire Pareto curve before making a decision.

In this data mining problem, the requirements are given by the application. For example, for a pipeline crack detection problem a requirement of 0.98 for probability of detection is specified. The costs and/or requirements are given by the mission manager. Also, the problem changes dynamically. For example, different sensors are used at different points in time changing the problem and the search space. This requires frequent redesign of the solution. Due to these factors we resort to the classical multi-objective methods such as method of weighted objectives.

11.4 Particle Swarm Optimization for Weighting Correlated Classifiers

The particle swarm optimization algorithm, originally introduced in terms of social and cognitive behavior by Kennedy and Eberhart in 1995 [8], has come to be widely used as a problem solving method in engineering and computer science. This technique is fairly simple and comprehensible as it derives its simulation from the social behavior of individuals. The individuals, called particles henceforth, are flown through the multidimensional search space, with each particle representing a possible solution. The movement of the particles is influenced by two factors. The first factor results in, particles storing in its memory the best position visited by it so far, called *pbest* and experiencing a pull towards this position as it traverses through the search space. As a result of the second

Algorithm PSO:
For t= 1 to the max. bound of the number on generations,
For i=1 to the population size,
For d=1 to the problem dimensionality
Apply the velocity update equation:

$$V_{id}^{t+1} = \omega \times V_{id}^{t} + \psi_1 \times (p_{id} - X_{id}) + \psi_2 \times (p_{gd} - X_{id})$$

where P_i is the best position visited so far by X_i,
and Pg is the best position visited so far by any particle;
Limit magnitude:

$$V_{id}^{(t+1)} = min(V_{max}, max(-(V_{max}, V_{id}^{(t+1)})))$$

Update Position:

$$X_{id}^{(t+1)} = min(Max_d, max(-(Min_d, X_{id}^{(t)} + V_{id}^{(t+1)})))$$

End-for-d;
Compute fitness of $X_i^{(t+1)}$;
If needed, update historical information regarding P_i and P_g;
End-for-i;
Terminate if Pg meets problem requirements;
End-for-t;
End algorithm.

Fig. 11.7. Pseudocode of the particle swarm optimization algorithm

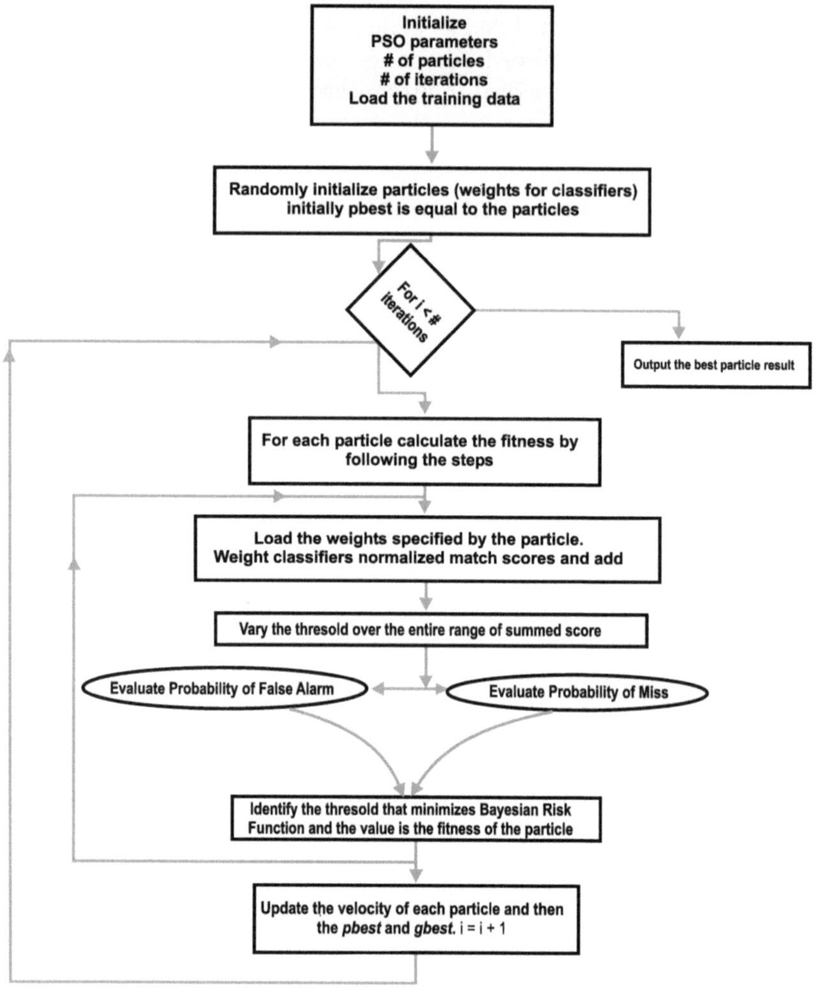

Fig. 11.8. Flowchart of the particle swarm optimization algorithm used to achieve weights for different classifiers

factor, the particle interacts with all the neighbors and stores in its memory the best position visited by any particle in the search space and experiences a pull towards this position, called *gbest*. The first and the second factors are called cognitive and social components respectively. After each iteration the *pbest* and *gbest* are updated if a more dominating solution (in terms of fitness) is found, by the particle and by the population respectively. Each particle, i, representing the weighting vector $w_i = (w_{i1}, w_{i2}, \ldots, w_{in})$, is randomly initialized. The *pbest* solutions are initially assigned the same values. The best position, *gbest*, is determined based on the fitness function (see equation (11.3)). A velocity term along each dimension is defined as $V_i = (V_{i1}, V_{i2}, \ldots, V_{in})$. In each iteration represented by index t, the velocity term is updated using,

$$V_{iq}^{(t+1)} = \omega \times V_{iq}^{(t)} + U[0,1] \times \psi_1 \times \left(w_{iq}^{p(t)} - w_{iq}^{(t)}\right) + U[0,1] \times \psi_2 \times \left(w_{gq}^{p(t)} - w_{iq}^{(t)}\right) \quad \forall q \tag{11.18}$$

where $U[0,1]$ is a sample from a uniform distribution. The weights (particle positions) are updated using

$$w_{iq}^{(t+1)} = w_{iq}^{(t)} + V_{iq}^{(t+1)} \tag{11.19}$$

The particle swarm optimization algorithm is applied to a training dataset. The final weights achieved by PSO are then tested on a test dataset. Particles are initialized randomly between 0 and 1. The weights represented by the particle are applied to the training data and errors are calculated by identifying both the false alarm rate and miss rate and then calculating the fitness of each particle by evaluating the Bayesian risk function. *Pbest* is initialized to be the same as the particles. The particle positions are updated using the velocity update and the position update functions. The particles at the new position are evaluated for the Bayesian error. Subsequently, the *pbest* is updated and *gbest* is updated. This is repeated for a preset number of iterations. The pseudo code of the PSO algorithm is given in Fig. 11.7 and the flowchart of the PSO algorithm specifically for identifying the weights in the fusion function is given in Fig. 11.8.

11.5 Results on Normally Distributed Datasets

In this section we present results achieved on Normally distributed datasets. We present the results achieved on the bi-variate dataset in the next subsection followed by the results achieved on multivariate dataset. The parameters for the marginal distributions are given in Table 11.1.

11.5.1 Bi-variate Normal Distribution

A bi-variate dataset represents a two classifier problem. Multiple "bi-variate normal datasets" containing 100,000 samples for H_0 and H_1 were generated by varying the Pearson correlation coefficient. Results are presented when the measurements from both the classifiers are symmetrically correlated, i.e., data under both H_0 and H_1 have the same correlation factor, as well as asymmetrically correlated, i.e., data under H_0 and H_1 have different correlation factors For symmetric correlation, the Pearson correlation coefficient is varied between 0.1 and 0.9 in steps of 0.2. For asymmetric correlation, the Pearson correlation factor is varied in such a way that if the correlation factor under H_0 is varied between 0.1 and 0.9 in steps of 0.2, then the correlation factor under H_1 is varied from 0.9

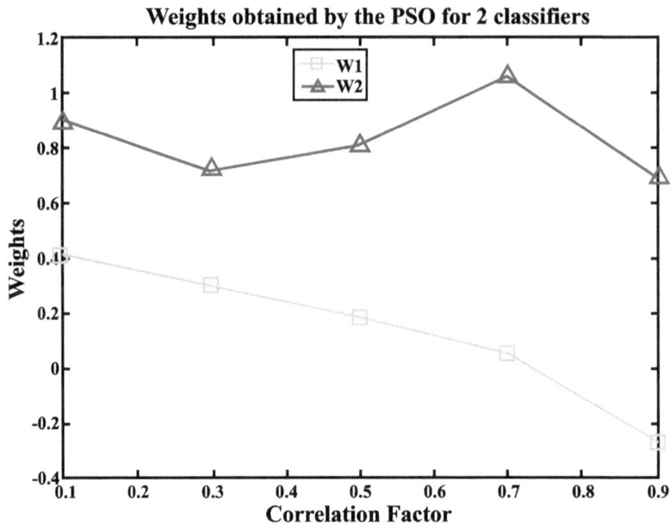

Fig. 11.9. W_1 and W_2 achieved for different correlation factors for the bi-variate dataset

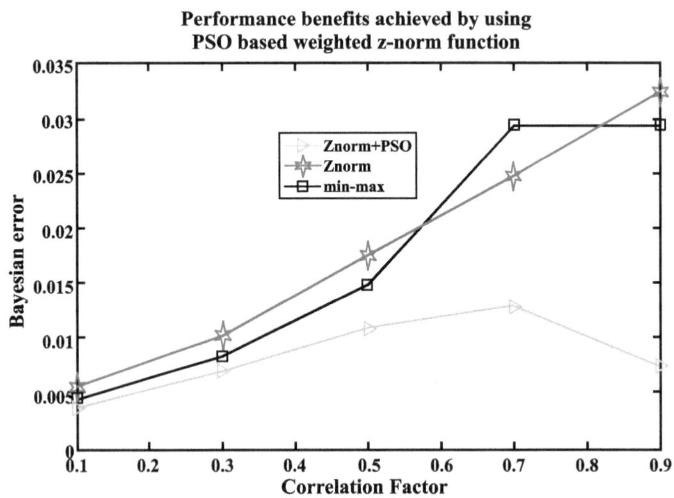

Fig. 11.10. Performance of the weighted fusion function on the symmetrically correlated bi-variate dataset

to 0.1 in steps of -0.2. This generates the datasets with correlation factor of (0.9, 0.1), (0.7, 0.3), (0.3, 0.7), (0.1, 0.9), where each set corresponds to (ρ_{H_0}, ρ_{H_1}).

The optimized weights achieved using PSO are shown in Fig. 11.9. Fig. 11.10 compares the performance of weighted *z-norm* with the classical *z-norm* method on these datasets for $C_{FA} = 1$ in equation (11.16). The performance of the PSO

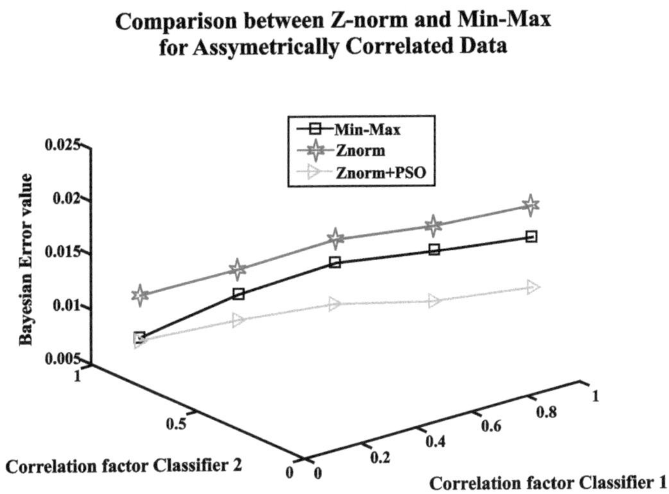

Fig. 11.11. Performance of the weighted fusion function on the asymmetrically correlated bi-variate dataset

based weighted *z-norm* is better than simple normalization techniques such as *'z-norm'* and *'min-max'*, since as the correlation increases, the performance benefits increase, too. Usually, performance of the fusion system degrades as the correlation increases; however, when the correlation factor is 0.9 we see that there is an improvement in performance. This is due to the fact that PSO inverts the addition operation in the fusion function by using negative weights reducing the overlapping region. The performance of the weighted *z-norm* on asymmetrically correlated dataset is presented in Fig. 11.11. For this problem, we evolve a single weight, w_1, and use $(1-w_1)$ for w_2. This reduces the complexity of the problem. Evolving both the weights and using equation (11.10) produces similar results. Performance benefits by using the weighted *z-norm* for the asymmetrically correlated dataset are inferior in comparison to that of a symmetrically correlated dataset. Asymmetrically correlated data set presents higher overlap between the two hypotheses making the detection process difficult. The increased difficulty of the classification process is presented in Figs. 11.5 and 11.6.

11.5.2 Multi-variate Normal Distribution

Multiple datasets using the multivariate normal distribution under both hypotheses are given by

$$f_{X_{H_h}}(x_1, x_2, \ldots, x_N) = \frac{1}{(2\pi)^{N/2} |\Sigma|^{1/2}} \exp\left(\frac{-1}{2}(x-\mu)^T \Sigma^{-1}(x-\mu)\right) \tag{11.20}$$

where $\sum$ is the covariance matrix and μ is the mean vector for the different marginals. The means and standard deviations for the marginal distributions are

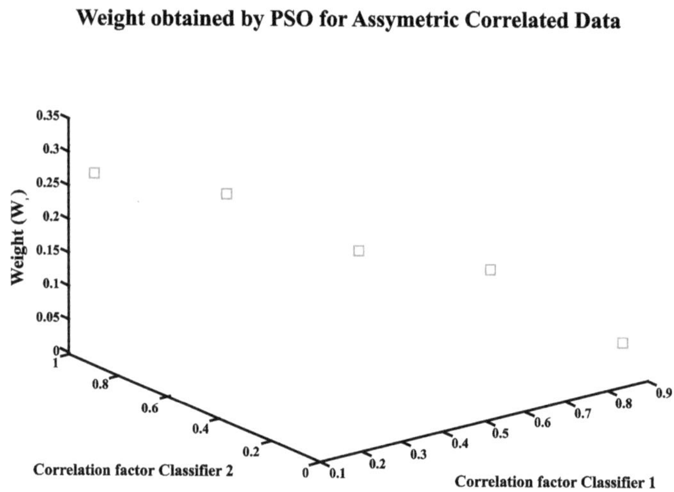

Fig. 11.12. Weights achieved by PSO for the asymmetrically correlated bi-variate dataset

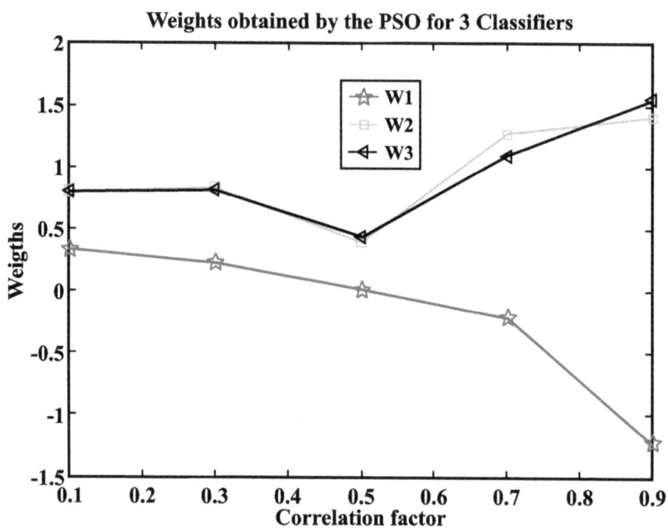

Fig. 11.13. Weights achieved by the PSO algorithm for symmetrically correlated 3-classifier dataset

given in Table 11.1. The weights achieved by PSO for the three classifiers are shown in Fig. 11.13. The performance of the three fusion techniques is shown in Fig. 11.14. The results are presented for symmetrically correlated dataset and $C_{FA} = 1$. The weight for classifier 1 reduces as the correlation increases. The weights for classifier 2 and classifier 3 first decrease and then increase.

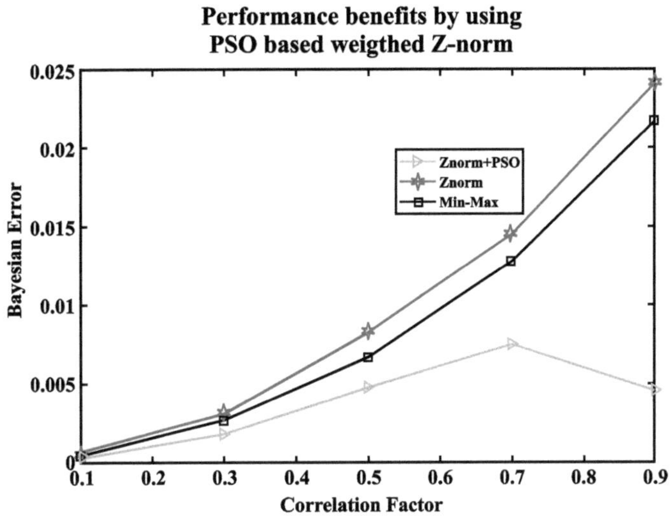

Fig. 11.14. Performance of the weighted fusion function on the symmetrically correlated 3-classifier dataset

The performance of the weighted *z-norm* (weights derived using PSO) is the best. The *min-max* technique performs better than the *z-norm*. Again, at higher correlation, the weighted fusion technique performs best. For a correlation factor of 0.9, PSO achieves a negative weight for classifier 1 and the performance gets better. Usually, as the correlation increases the performance becomes worse, since the information quality of multiple classifiers is low.

11.6 Application: Improving the Performance of a Multi-biometric Verification System

Biometrics identifies a person based on the unique physical and behavioral traits of a person. Some common biometrics include face, fingerprint, iris, hand geometry, earlobe geometry, voice and handwriting. As the ability to forge/hack passwords is becoming commonplace, Biometrics has become more suitable for identification and verification systems. Accuracy of a low cost Biometric device is a major concern for its large scale deployment. We analyze a multi-biometric verification system. We employ the fusion technique to fuse data from multiple biometric classifiers. This improves the performance of the system by mitigating problems that usually exist if one relies on a single biometric system, such as noisy data, undesirable error rates and/or non-universality and absence of distinctness. Similar to our discussion in previous sections, the biometric verification problem is posed as a binary hypothesis-testing problem with the matching score(s) being observations. The two hypotheses are H_0: Imposter; H_1: Genuine user. Given, scores from multiple classifiers, $[X] = (x_1, x_2, \ldots, x_n)$, a fusion function combining these scores is given by

$$z = f[X] \tag{11.21}$$

This result is compared to a threshold determining the presence or absence of a genuine individual as in

$$u_f = \begin{cases} 1, \, z \geq \lambda \\ 0, \, z < \lambda \end{cases} \tag{11.22}$$

Systems can make two errors: *false acceptance error*, when an imposter is identified as a genuine user, and *false rejection error*, when a genuine user is identified as an imposter. These are given by, $P_{FA} = P(u_f = 1|H_0)$, $P_{FR} = P(u_f = 0|H_1)$.

The total error known as Bayesian error is defined as

$$R = C_{FA} \times P(H_0) \times P_{FA} + C_{FR} \times P(H_1) \times P_{FR} \tag{11.23}$$

where, C_{FA}, C_{FR} are the cost of false acceptance and false rejection, and $P(H_0)$, $P(H_1)$ are the prior probabilities of imposter and genuine users. We use the weighted *z-norm* technique developed in the previous sections. We present the results for the face-recognition dataset provided by NIST.

11.6.1 NIST Biometrics Dataset

The results for the NIST BSSR face data are presented in this section. Only the face classifier of the dataset was used which consists of 3000 match scores

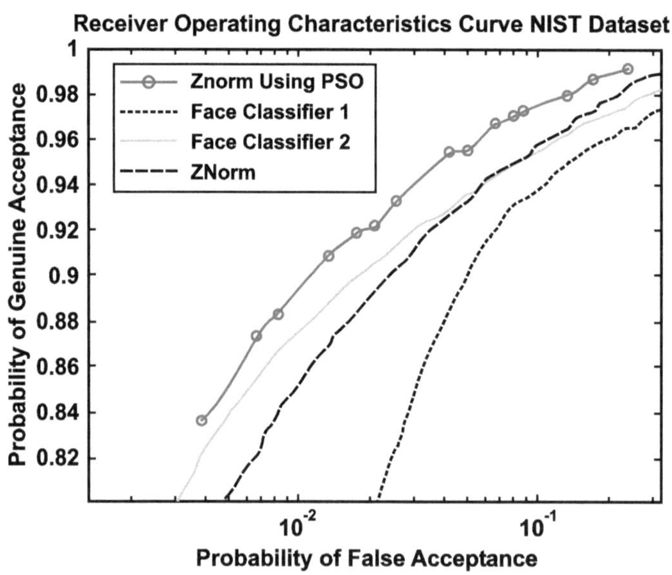

Fig. 11.15. Receiver operating characteristic curve for the NIST dataset

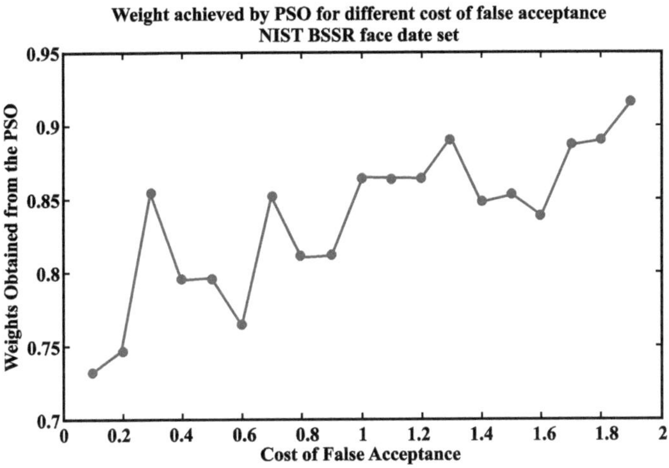

Fig. 11.16. Weight (W_1) achieved by PSO for different C_{FA} for the NIST dataset

corresponding to the number of individuals. The data is obtained from 2 different face classifiers. There are 6,000 genuine scores and almost 18 million imposter scores for each classifier. We use 50% of the genuine scores and 0.53% of the imposter scores from each classifier to compose the training set. The test set also contains the same number of scores. Two face recognition systems are used to generate genuine and imposter scores. The Pearson correlation under the two hypotheses is approximately 0.18 and 0.476, respectively. The performance of the system is expressed in terms of its receiver operating curves (ROC) which is a plot of $(P_{FA}, 1 - P_{FR})$. Multiple points on the ROC are achieved by varying C_{FA} in equation (11.10) and running PSO for each set of (C_{FA}, C_{FR}). Fig. 11.15 compares the performance of the system using *z-norm* and weighted *z-norm*. We again observe that the weighted *z-norm* performs better than the classical *z-norm* method. In Fig. 11.15, the better performing system has ROC towards the upper left corner. It is interesting to observe that the classical *z-norm* performs poorer in comparison to Classifier 2 for lower false acceptance probability. This implies that the information/score from Classifier 1 is not enhancing but degrading the system. Once the measurements are weighted, using the weights designed by PSO, higher performance is achieved.

Weights achieved by PSO for different C_{FA} in the multi-objective cost function are presented in Fig. 11.16. The weights achieved across different C_{FA} vary significantly and do not exhibit any specific pattern. This demonstrates the need of a PSO based technique for this problem.

11.7 Summary and Conclusions

In this chapter we presented the design of weights for fusing correlated measurements from multiple classifiers. The classifiers are usually derived from the same sensor data. The measurements from the classifiers are correlated requiring careful dynamic design of the weights for different correlation factors. A multi-objective optimization function is formulated for the binary classification problem. The classical multi-objective method, i.e., the "method of weighted objectives" is used. This method is suitable for this application since the costs associated with each error made by the system are readily available. Also, the problem changes dynamically, hence a point solution satisfying the requirements is needed.

The weights for the multiple classifier measurements are designed using particle swarm optimization, minimizing the objective function that ties in both objectives. The proposed algorithm is run for different datasets pertaining to various correlation levels (factors). First, the algorithm is used to generate weights for bi-variate (2 classifiers) normally distributed dataset. The weights are designed for both symmetrically and asymmetrically correlated datasets. The performance of the algorithm is shown to be better than the classically used *z-norm* and *min-max* fusion techniques. The algorithm is then extended to multivariate (more than 2 classifiers) normally distributed datasets. This increases the design variables. Again the performance of the proposed method is better than the classical fusion methods.

Finally, a multi-biometrics application is presented. The dataset for the multi-biometrics is obtained from the NIST database. The data corresponds to two face recognition classifiers. The performance of the swarm intelligence based fusion mechanism is compared to the classical *z-norm* fusion technique. The weighted z-norm (weights designed by PSO) attain an higher performance than the *z-norm* technique. The weights are derived for different C_{FA}. The resultant points (F_{AR}, F_{RR}) are plotted as the Receiver operating characteristic curve (ROC). The ROC shifts to the upper left corner. The performance of classical *z-norm* fusion function degrades for lower F_{AR} and falls below the performance of the individual classifier 2. However, the weighted *z-norm* performs better than both the classifiers for all F_{AR}.

References

[1] Veeramachaneni, K., Osadciw, L., Ross, A., Srinivas, N.: Decision-level fusion strategies for correlated biometric classifiers. In: 2008 IEEE Computer Society Conference on Computer Vision and Pattern Recognition, pp. 1–6. IEEE Computer Society Press, Los Alamitos (2008)

[2] Jain, A., Nandakumar, K., Ross, A.: Score normalization in multimodal biometrics system. Pattern Recognition 38(12), 2270–2285 (2005)

[3] Veeramachaneni, K.: An Evolutionary Algorithm Based Dynamic Thresholding for Multimodal Biometrics, Masters Thesis, School of Engineering and Computer Science, Syracuse University, USA (2005)

[4] Veeramachaneni, K., Osadciw, L., Varshney, P.K.: An adaptive multimodal biometric management algorithm. IEEE Transactions on Systems, Man, and Cybernetics Part C: Applications and Reviews 35(3), 344–356 (2005)

[5] Veeramachaneni, K., Osadciw, L.: Dynamic Sensor Management Using Multi Objective Particle Swarm Optimizer. In: Dasarathy, B.V. (ed.) Multisensor, Multisource Information Fusion: Architectures, Algorithms, and Applications 2004. Proceedings of the SPIE, vol. 5434, pp. 205–216 (2004)
[6] Ross, A., Jain, A.: Information Fusion in Biometrics. Pattern Recognition Letters 24, 2115–2125 (2003)
[7] Kuncheva, L.I.: A Theoretical Study on Six Classifier Fusion Strategies. IEEE Transactions on Pattern Analysis and Machine Intelligence 24(2), 281–286 (2002)
[8] Kennedy, J., Eberhart, R., Shi, Y.H.: Swarm Intelligence. Morgan Kaufmann Publishers, San Francisco (2001)
[9] Roli, F., Giacinto, G., Vernazza, G.: Methods for designing multiple classifier systems. In: Kittler, J., Roli, F. (eds.) MCS 2001. LNCS, vol. 2096, pp. 78–87. Springer, Heidelberg (2001)
[10] Deb, K.: Multi-Objective Optimization Using Evolutionary Algorithms. John Wiley and Sons, Chichester (2001)
[11] Willett, P., Swaszek, P.F., Blum, R.S.: The good, bad and ugly: distributed detection of a known signal in dependent Gaussian noise. IEEE Transactions on Signal Processing 48(12), 3266–3279 (2000)
[12] Kitler, J., Hanef, M., Duin, R.W., Matas, J.: On Combining Classifiers. IEEE Transactions on Pattern Analysis and Machine Intelligence 20(3), 226–239 (1998)
[13] Opitz, D.W., Maclin, R.F.: Popular ensemble methods: An empirical study. Journal of Artificial Intelligence Research 11, 169–198 (1999)
[14] Waltz, E.L., Llinas, J.: Multisensor Data Fusion. Artech House Radar, Norwood (1990)
[15] Guerra-Salcedo, C., Whitley, D.: Genetic approach to feature selection for ensemble creation. In: Banzhaf, W., et al. (eds.) Proceedings of the 1999 Genetic and Evolutionary Computation Conference (GECCO 1999), pp. 236–243. Morgan Kaufmann, San Francisco (1999)
[16] Opitz, D.W.: Feature selection for ensembles. In: Proceedings of the 16th International Conference on Artificial Intelligence, pp. 379–384. AAAI, Menlo Park (1999)
[17] Krogh, A., Vedelsby, J.: Neural network ensembles, cross validation, and active learning. In: Tesauro, G., Touretzky, D., Leen, T. (eds.) Advances in Neural Information Processing Systems, vol. 7, pp. 231–238. The MIT Press, Cambridge (1994)
[18] Varshney, P.K.: Distributed Detection and Data Fusion. Springer, New York (1996)
[19] Xu, L., Krzyzak, A., Suen, C.Y.: Methods of combining multiple classifiers and their applications to handwriting recognition. IEEE Transactions on Systems, Man and Cybernetics 22(3), 418–435 (1992)
[20] Peebles Jr., P.Z.: Probability, Random Variables and Random Signal Principles, 2nd edn., USA. McGraw -Hill Series in Electrical Engineering (1987)
[21] Chair, Z., Varshney, P.K.: Optimal Data Fusion in Multiple Sensor Detection Systems. IEEE Transactions on Aerospace and Electronic Systems AES-22(1), 98–101 (1986)
[22] Tenney, R.R., Sandell, N.R.: Detection with Distributed Sensors. IEEE Transactions on Aerospace and Electronic Systems AES-17(4), 501–510 (1981)

Index

MIX
Papier aus verantwortungsvollen Quellen
Paper from responsible sources
FSC® C105338

If you have any concerns about our products,
you can contact us on
ProductSafety@springernature.com

In case Publisher is established outside the EU,
the EU authorized representative is:
Springer Nature Customer Service Center GmbH
Europaplatz 3, 69115 Heidelberg, Germany

Printed by Libri Plureos GmbH
in Hamburg, Germany